MATERNAL EFFECTS AS ADAPTATIONS

MATERNAL
EFFECTS AS
ADAPTATIONS

Edited by Timothy A. Mousseau & Charles W. Fox

New York   Oxford

Oxford University Press

1998

Oxford University Press

Oxford   New York
Athens   Auckland   Bangkok   Bogota   Bombay
Buenos Aires   Calcutta   Cape Town   Dar es Salaam
Delhi   Florence   Hong Kong   Istanbul   Karachi
Kuala Lumpur   Madras   Madrid   Melbourne
Mexico City   Nairobi   Paris   Singapore
Taipei   Tokyo   Toronto   Warsaw

and associated companies in
Berlin   Ibadan

Copyright © 1998 by Oxford University Press, Inc.

Published by Oxford University Press, Inc.,
198 Madison Avenue, New York, New York 10016

Cover illustration: A female marbled salamander (*Ambystoma opacum*)
caring for a clutch of eggs. Drawing by Rosemary Bennetts, based on a
photograph by David Scott.

Library of Congress Cataloging-in-Publication Data
Maternal effects as adaptations / edited by Timothy A. Mousseau
   and Charles W. Fox.
      p.   cm.
   This volume was spawned by a symposium held in St. Louis in June 1996.
   Includes bibliographical references and index.
   ISBN 0-19-511163-X
   1. Adaptation (Biology)—Congresses.   2. Evolution (Biology)—Congresses.
I. Mousseau, Timothy A.   II. Fox, Charles W.
QH546.M27   1998
576.8'5—dc21   97-20875

9 8 7 6 5 4 3 2 1

Printed in the United States of America
on acid-free paper

Preface

Maternal Effects as Adaptations: A First Synthesis

A primary objective of evolutionary biology is to understand the mechanisms responsible for the expression of phenotypic variation on which Darwinian natural selection may act. We are accustomed to envisioning phenotypic variation as the sum of direct environmental and genetic effects (and their interactions), but rarely do we consider, or measure, the impact of environmental effects experienced in previous generations on contemporary phenotypic expression. Most often, such transgenerational interactions are caused by maternal effects. In the simplest terms, maternal effects occur when the phenotype of a mother (in the broadest sense), or the environment she experiences, causes phenotypic effects in her offspring. Such effects are independent of the genes contributed to offspring by the mother, or direct environmental effects on developing offspring, but may include genetic or environmental effects from the father if female mate choice is acting, or environmental effects mediated by the mother if, for example, parental care is provided.

The study of *adaptive* maternal effects is concerned with the mechanisms mothers employ to enhance their offspring's fitness. Maternal care, choice of mate, oviposition behavior, and provisioning to seeds or eggs often have significant effects on offspring fitness. To the extent that there exists heritable variation among mothers in their abilities to discern high-quality mates, pick an appropriate host to place seeds or eggs, or provide protection from predators, and so on, such traits are expected to evolve in much the same way as any other trait subject to the inevitable consequences of Darwinian natural selection.

Adaptations involving maternal effects are different from "normal" traits in that the genes underlying their expression are contained within the maternal generation while the phenotypic variation on which natural selection may act is expressed by the offspring. In theory, this time lag between cause and effect can lead to some very interesting and counterintuitive evolutionary dynamics. Maternal effects can enhance, deter, or even reverse evolutionary response to selection, and they can even lead to sustained oscillations in population density or phenotype such as those observed in many insects and small mammals.

The objective of this book, and the symposium that spawned it (Maternal Effects as Adaptations, a symposium organized by T.A. Mousseau and C.W. Fox for the annual meeting of the Society for the Study of Evolution, in St. Louis, June 1996), is to provide a first synthesis of the current state of knowledge concerning the evolution of adaptive maternal effects in nature. The contention of contributors to this book is that many maternal effects have been shaped by natural selection and often serve an adaptive role in an organism's life history; they may even play a large role in rapid population differentiation, host race formation, and speciation. In the past, most investigators have focused on reporting the range of expression and the mechanisms underlying the maternal effects observed in a wide variety of organisms without regard to their evolutionary history. In many cases, the primary interest was prediction of response to artificial selection in domestic or cultivated species where maternal effects often confound predictions based solely on classical quantitative genetic models. Usually, such maternal effects are treated as a source of environmental noise to be controlled or eliminated from the experimental design. As the renown quantitative geneticist D.S. Falconer put it: "Maternal effects are a frequent, and often troublesome source of environmental resemblance...." Evolutionary biologists now recognize maternal effects as a cause of phenotypic variation that may play an important role in organismal evolution.

Similarly, it is now widely acknowledged that many aspects of behavior, life history, and development that had been examined and modeled independently actually fall within the rubric of maternal effects (e.g., oviposition behavior, mate choice, environmental sex determination, dormancy in plants and animals, regulation of developmental events, etc.). Here, we hope to convince the reader by force of overwhelming (and ever-growing) evidence that in the wild, many maternal effects have evolved into adaptations for life in heterogeneous environments. A central thesis of this volume is that maternal effects often provide a mechanism by which mothers program future offspring development and consequent phenotype in response to contemporary, predictive cues experienced by the mother, resulting in the expression of transgenerational phenotypic plasticity (TPP).

This volume is loosely organized around four primary themes. Part I addresses recent theoretical developments in the maternal effects arena. Part II deals with the often challenging methods for the assessment and measurement of maternal effects. Part III presents several reviews of the scope and range of maternal effects in a variety of taxa, while part IV presents four particularly insightful case studies of maternal effects expression and their adaptive significance. As is the case for any compilation of this nature, a great deal of overlap exists among many of the chapters; several of the theoretically oriented chapters present excellent reviews of the maternal effects literature, while many of the more empirical chapters present exciting and novel models of general conceptual interest. In total, this volume represents a comprehensive review of the current theoretical and empirical knowledge concerning the evolution of maternal effects.

The editors gratefully thank the many contributors to this volume. We also thank and acknowledge the following individuals and organizations for their support of this venture: The Society for the Study of Evolution, Mark Courtney and the Division of Environmental Biology at the National Science Foundation, Wayne Getz for suggesting Oxford University Press as a publisher, Kirk Jensen, Susan Day, and Lisa Stallings at Oxford University Press, Udo Savalli for preparing the author index, Hugh Dingle for his enthusiastic and insightful support, and Ray Huey for collegial advice. The editors (and many of the contributors) are especially grateful to Bruce Riska, whose early expositions of the causes and consequences of maternal effects truly incited much of the modern work and ideas in this field. This volume is dedicated to our mothers, who gave us much more than genes.

Columbia, South Carolina   T.A.M.
December 1996   C.W.F.

Contents

Contributors

D. Max Blouw, Faculty of Natural Resources and Environmental Studies, University of Northern British Columbia, 3333 University Way, Prince George, British Columbia V2N 4Z9, Canada.

Edmund D. Brodie III, Department of Biology, Indiana University, Bloomington, Indiana 47405.

Diane L. Byers, Department of Ecology, Evolution and Behavior, University of Minnesota, 1987 Upper Buford Circle, St. Paul, Minnesota 55108.

Mertice M. Clark, Department of Psychology, McMaster University, Hamilton, Ontario L8S 4K1, Canada.

David L. Denlinger, Department of Entomology, Ohio State University, 1735 Neil Avenue, Columbus, Ohio 43210.

Kathleen Donohue, Department of Ecology and Evolutionary Biology, Box G-W, Brown University, Providence, Rhode Island 02912.

Charles W. Fox, The Louis Calder Center Biological Field Station of Fordham University, 53 Whippoorwill Road, Box K, Armonk, New York 10504.

Bennett G. Galef, Jr., Department of Psychology, McMaster University, Hamilton, Ontario L8S 4K1, Canada.

Lev R. Ginzburg, Department of Ecology and Evolution, State University of New York, Stony Brook, New York 11794.

Daniel D. Heath, Faculty of Natural Resources and Environmental Studies, University of Northern British Columbia, 3333 University Way, Prince George, British Columbia V2N 4Z9, Canada.

Robert H. Kaplan, Department of Biology, Reed College, Portland, Oregon 97202.

Elizabeth P. Lacey, Department of Biology, P.O. Box 26174 University of North Carolina, Greensboro, North Carolina 27412.

Susan J. Mazer, Department of Ecology, Evolution and Marine Biology, University of California, Santa Barbara, California 93106.

Frank J. Messina, Department of Biology, Utah State University, Logan, Utah 84322-5305.

Allen J. Moore, Department of Entomology, S-225 Agricultural Sciences Center North, University of Kentucky, Lexington, Kentucky 40546-0091.

Timothy A. Mousseau, Department of Biological Sciences, University of South Carolina, Columbia, South Carolina 29208.

Peter Niewiarowski, Department of Biology, Buchtel College of Arts and Science, University of Akron, Akron, Ohio 44325.

Trevor Price, Biology Department 0116, University of California at San Diego, La Jolla, California 92093.

Derek A. Roff, Department of Biology, McGill University, 1205 Dr. Penfield Avenue, Montreal, Quebec H3A 1B1, Canada.

Bernard D. Roitberg, Behavioral Ecology Research Group and Center for Pest Management, Department of Biosciences, Simon Fraser University, Burnaby, BC V5A 1S6, Canada.

Willem M. Roosenburg, Department of Biological Sciences, Ohio University, Athens, Ohio 45701.

MaryCarol Rossiter, Institute of Ecology, University of Georgia, Athens, Georgia 30602.

Johanna Schmitt, Department of Ecology and Evolutionary Biology, Box G-W, Brown University, Providence, Rhode Island 02912.

Ruth G. Shaw, Department of Ecology, Evolution and Behavior, University of Minnesota, 1987 Upper Buford Circle, St. Paul, Minnesota 55108.

Barry Sinervo, Department of Biology, University of California, Santa Cruz, California 95064.

Michael J. Wade, Department of Ecology and Evolution, University of Chicago, Chicago, Illinois 60637.

Jason B. Wolf, Department of Biology, Indiana University, Bloomington, Indiana 47405.

Lorne M. Wolfe, Department of Biology, Georgia Southern University, Statesboro, Georgia 30460.

MATERNAL EFFECTS AS ADAPTATIONS

Part I

CONCEPTUAL ISSUES

Although models for the estimation and measurement of maternal effects are not new (e.g., models by R.L. Willham, J.J. Rutledge, E.J. Eisen, D.S. Falconer; for citation information, see individual chapters), contemporary interest in their importance to evolutionary process and pattern stems primarily from the recent works of B. Riska, M. Kirkpatrick, R. Lande, J. Cheverud, W. Atchley, and D. Cowley. In this opening part, the authors build on these early approaches in original and exciting ways to offer completely novel insights to the unique attributes of maternal effects and their impact on the evolution of adaptive organismal responses to heterogeneous environments.

In chapter 1, Michael Wade, starting from first principles, develops a new population genetic model for the evolution of maternal effects. Wade's model generates two exciting insights. First, significantly more genetic variation can be maintained at equilibrium via mutation (usually twice as much) for maternally affected traits than for "ordinary" traits. Second, Wade finds that maternal effects can lead to a special kind of genotype x genotype epistasis between maternal and offspring genotypes. These two unique attributes of maternal effects may dramatically influence the rate of local adaptation and may be particularly important for host race formation and speciation. In chapter 2, Allen Moore et al. build on earlier models by J. Cheverud and M. Lynch to develop two quantitative genetic models to explore the importance of social and sexual selection for the evolution of maternal effects on offspring fitness. In essence, Moore et al. extend the boundaries of simple maternal effects to include the indirect genetic influences of fathers, kin, and other associates in the population. Of particular interest is their synthesis of maternal effects and sexual selection theory. This makes good sense if one considers the consequences of maternal behaviors such as mate choice to offspring fitness: a mother with the ability to select a mate on the basis of a high-quality territory, good parenting abilities, or perhaps even his "good genes" would likely have higher quality offspring than mothers not exerting such choices. Similarly, the ability to choose an advantageous social milieu for one's offspring is a trait likely to have a fitness benefit. Moore et al.'s models suggest that such behaviors could evolve very quickly.

Chapter 3 addresses one of the great mysteries of ecology: the causes of population density cycles. Using inertial growth models, Lev Ginzburg elegantly demystifies the likely culprit (maternal effects!) underlying the dramatic population cycles observed in many rodents and insects. In chapter 4, Elizabeth Lacey undertakes the challenging task of conceptual and semantical clarification of the complex and sometimes incomprehensible language contained within the maternal effects literature. Of particular concern is the need to correctly define the sources of maternal effects, which is critical for rigorous tests of adaptation. The final chapter of this section (chapter 5 by Bernie Roitberg) addresses questions relating to the evolution of oviposition behaviors in insects, with a special interest in conservation issues. When, where, and how mothers lay their eggs is often the single greatest determinant of offspring fitness, at least for many herbivorous insects. Maternal effects mediated via oviposition behavior (especially host choice) are likely to play a major role in local adaptation, host race formation, and speciation. Roitberg makes a compelling case for the inclusion of maternal effects in population models dealing with conservation issues.

1

The Evolutionary Genetics
of Maternal Effects

MICHAEL J. WADE

Maternal effects occur when the phenotype of an individual is determined not only by its own genotype and the environmental conditions it experiences during development, but also by the phenotype or environment of its mother (Kempthorne 1969; Wright 1969; Falconer 1989). Indeed, in many species, the mother's phenotype is *the* most important environmental condition experienced by an individual during development. The mother's phenotype determines when and where an offspring is produced as well as the offspring's size and quality, and in some instances the offspring's sex (Trivers and Willard 1973; Janzen 1993). By clustering or dispersing offspring, the maternal phenotype has effects on offspring density, the postnatal competitive environment, and the vulnerability of the offspring to predators and disease. Maternal behaviors determine in many species whether and how density regulation occurs at the level of the family and the intensity of postnatal competition among the offspring for limited maternal resources. Offspring mortality is often clustered at the level of maternal families owing to transovariolar transmission of pathogens. These maternal effects on the distribution of offspring mortality influence N_e, the effective size of the population, and hence the balance between natural selection and random genetic drift for all loci (Wade 1996). The maternal phenotypes that have effects on offspring phenotypes and offspring fitnesses can themselves be partitioned into genetic and environmental components, like any other phenotype. Genetic maternal effects are particularly interesting because they represent a special mechanism for the coevolution of the maternal environment and the offspring phenotype, as I will discuss.

Among both plants and animals, maternal effects are common as a fundamental consequence of anisogamy (Westneat and Sargent 1996). Because maternal effects provide the context for offspring development and the expression of the offspring genotype, they add an important aspect of continuity of the maternal soma to the standard dogma of germ-line continuity. In many instances, the ecology of the maternal environment is as important to offspring development and fitness as the offspring genotype. Because the mother's gamete is larger than that of the father (anisogamy), there tends to be a sex difference in the magnitude of parental environmental effects that can manifest themselves at three different stages of offspring development:

1. *prezygotic maternal effects*, wherein the maternal phenotype, especially maternal nutritional condition, affects gametic size and quality;
2. *postzygotic-prenatal maternal effects* that arise from the more or less intimate prenatal or preripening nutritional relationship between mothers and developing embryos; and
3. *postzygotic-postnatal* or *postgerminational maternal effects*, wherein maternal care and maternal competition affect offspring phenotype.

Roach and Wulff (1987) discuss the many opportunities for maternal genetic contributions to the developing embryo in plants, including direct cytoplasmic inheritance of organelles, direct Mendelian inheritance of nuclear genes, and indirect genetic contributions to the nutritive endosperm, to the tissues surrounding the seed, and to the seed coat. In some animal species, such as the matrilineal social primates, the period of maternal influence can extend throughout the lifetime of the individual because a female's dominance rank is inherited through her mother (McNamara and Houston 1996). Altmann et al. (1988) have shown in baboons not only that dominance rank is maternally inherited, but also that mother's rank influences the sex ratio of her offspring, probably mediated proximately by social effects on levels of maternal hormones and ultimately by sexual selection (Alberts and Altmann 1995). Indeed, most learned behaviors exhibit a significant maternal contribution such that maternal effects are *inseparable* from issues in cultural evolution. Male paternal effects can manifest themselves at similar stages of embryo development, but the opportunities for and examples of paternal care from nature are rarer overall (Darwin 1871, p. 200), except in the case of cultural evolution (Feldman and Laland 1996). Maternal effects are expressed in many dimensions of the phenotype and include the kinds of nurturing behaviors typically associated with parenting as well as behaviors such as the blood-sucking, disease-vectoring behavior of many mosquitos (Gillett 1971) and sexual cannibalism (Elgar 1992).

1.1 The Ubiquity of Maternal Effects in Evolutionary Biology

Yamanaka (1928) was the first to demonstrate that maternal social status determined offspring sex in the social hymenopteran *Polistes fadwigae*. It is common in the social hymenoptera for maternal caste to determine offspring sex: unfertilized eggs of worker females develop into sons while the fertilized dominant queen produces both diploid daughters and haploid sons. Park (1935), using experimental populations of flour beetles, showed that changing the quality of the maternal environment causes a threefold change in primary fecundity and thus affects the ensuing competitive milieu of the offspring. Since that time, research in several areas of ecology and evolution explicitly or implicitly has involved maternal effects (table 1.1). Wulff (1986) showed that population regulation by density-dependent effects and community structure exhibits a significant component owing to maternal effects. Maternal effects can influence offspring numbers and offspring competitive ability (Wulff 1986) as well as offspring vulnerability to predators or pathogens (Augspurger 1984; Burdon 1987). Studies of life-history evolution have found that the dormancy, germination, and dispersal strategies of many plants are affected as much or more by the maternal phenotype as by the individual's own genotype (Schaal 1986; Roach and Wulff 1987). State-dependent life

Table 1.1 Areas of current research in ecology and evolution that explicitly or implicitly involve maternal effects.

Topic	Maternal effect	Author
Population regulation	Maternal density effects on offspring numbers	Park (1935)
Community structure	Competitive ability	Wulff (1982)
Life history	Germination, dispersal	Schaal (1986), Roach and Wulff (1987)
Eusociality	Maternal care	Wilson (1971, 1975), Alexander (1974)
Kin selection	Parental care	Cheverud (1984), Cheverud and Moore (1994)
Sex ratio	Parental investment, decision to fertilize	Charnov (1982)
	Environmental sex determination	Janzen and Paukstis (1991), Janzen (1993)
Parent-offspring conflict	Parental investment	Westneat and Sargent (1996), Alexander (1974)
Sexual selection	Sexual conflict	Westneat and Sargent (1996)
	Sexual cannibalism	Elgar (1992)
Development	Timing of gene action	Haldane (1932)
Cultural evolution	Parental inheritance	Feldman and Cavalli-Sforza (1976), Feldman and Laland (1996)
Selfish genes	Maternal-embryonic lethal genes	Wade and Beeman (1994)
Hybrid dysgenesis	Maternal cytoplasmic effects	Engels (1983)
Population dynamics	Herbivore outbreaks, time lags	Rossiter (1994, 1995)
Self-incompatibility, cytoplasmic sterility, speciation	Exceptions to Haldane's Rule, maternal oviposition behavior, host-race formation	Sawamura (1996)

history theory (in the sense of McNamara and Houston 1996) proposes a fundamental trade-off between maternal effects on offspring (derived from maternal investment and provisioning of the young) and a mother's future reproductive success. There are elegant experimental demonstrations of maternal effects on insect life histories (Mousseau and Dingle 1991a,b) and on many aspects of lizard life histories (Sinervo 1990; Sinervo et al. 1992). In theories of the evolution of social behaviors, maternal care is believed by some to be one of the essential precursors of eusociality (Wilson 1971, 1975; Alexander 1974). That is, the eusociality manifested by the bees, ants, wasps, and termites is an extremely exaggerated maternal phenotype. More recently, the mathematical relationship between kin selection theory and the evolution of genetic maternal effects has been developed and made explicit by the work of Cheverud, Moore, and collaborators (Cheverud 1984; Cheverud and Moore 1994; Moore et al, this volume).

Learned behaviors in the context of cultural evolution (e.g., Feldman and Cavalli-Sforza 1976; Boyd and Richerson 1985) can be modeled explicitly as parental effects on offspring phenotype. These clearly can exhibit a strong maternal component in addition to components from father and other (particularly older) siblings, as well as unrelated conspecifics. Several of the theoretical results unique to the evolution of maternal effects (e.g., Cheverud 1984;

Kirkpatrick and Lande 1989; Riska 1991; Cheverud and Moore 1994) were first derived or noticed in models of cultural evolution (e.g., Feldman and Cavalli-Sforza 1976; Boyd and Richerson 1985; Feldman and Laland 1996). Thus, there is a body of preexisting theory that obliquely addresses the evolution of maternal effects that might be used to inform a more synthetic framework.

Optimization strategies in game theoretic models of behavioral evolution generally view offspring quality as a phenotype determined or heavily influenced by maternal decisions regarding present and future investments (see review by Westneat and Sargent 1996). Evolutionary genetic theories of parent-offspring conflict (Alexander 1974) are especially explicit involving the coevolution of maternal and offspring traits and trade-offs between current and future maternal investment in offspring. Sex ratio in the social and asocial parasitic hymenoptera is believed to be determined by maternal behaviors that influence the rate of fertilization of ova, and natural selection acts on parental ability to affect offspring sex ratio (Trivers and Willard 1973). Maternal oviposition behavior determines offspring sex ratio in those reptiles with environmental sex determination, which includes many turtles and all crocodilians (Bull 1983; Janzen and Paukstis 1991; Janzen 1995; Janzen et al. 1995). Roosenburg (1996) has proposed an explicit model of the evolution of ESD as a maternal effect. Theories of the evolution of sex invoke sex differences in parental, generally maternal, investment as the fundamental biology behind the evolutionary dynamic (Charnov 1982; Bull 1983). Genomic incompatibilities, such as hybrid dysgenesis (Engels 1983), cytoplasmic incompatibility, sex ratio distorters (Hurst 1993), and maternal-effect selfish genes (Beeman et al. 1992; Wade and Beeman 1994), have been reported in a variety of taxa. Host race formation and speciation can be the long-term genetic consequences of the evolution of maternal oviposition preferences and habitat choice, both important maternal effects. Indeed, natural selection can establish a positive genetic covariance between genes for maternal oviposition preference and genes for habitat-specific larval survival, which opens the opportunity for a runaway process of host race formation (see section 1.5).

1.2 Lack of Attention to Maternal Effects

Research in all areas of evolutionary genetics is pervaded with maternal effects. However, despite their ubiquity and potential impact, maternal effects tend to be overlooked in current textbooks (e.g., they are not mentioned in Ridley 1993 or Price 1996). When they are mentioned in textbooks, it is usually in passing and most often in the contexts of cytoplasmic inheritance (e.g., Futuyma 1986), problems in the estimation of heritability (e.g., Cockburn 1991), or fecundity selection (Hartl and Clark 1989). This diminished emphasis may date back to Darwin (1859), who emphasized that "seedlings from the same fruit, and the young of the same litter, sometimes differ considerably from each other, though both the young and the parents . . . have been exposed to exactly the same conditions of life; this shows how unimportant the direct effects of the conditions of life are in comparison with the laws of reproduction, and of growth, and of inheritance" (p. 10). Darwin was arguing here about the importance of heredity relative to direct environmental effects and habits in shaping the phenotype. Nevertheless, he is diminishing the importance of *direct effects of the conditions of life*, including maternal effects. His emphasis on the phenotypic differences among members of the same family diminishes the role of maternal effects, which in many cases cause members of the same family to be more similar. (A notable exception is dormancy that may be achieved by increased variance in the timing of germination within families.)

Because the existence of maternal effects has several important consequences for our understanding of the evolutionary process (see section 1.3), it is important to redress the current lack of emphasis in textbooks. First, when maternal effects exist, the approach most commonly used for measuring selection in natural populations is inadequate because selection within one generation cannot be separated from the genetic response to selection from the previous generation(s). Second, maternal effects bring the problem of the units of selection to the forefront of evolutionary discussion because they constitute a significant component of among-family selection and response to selection. Third, the genetic variance for maternal effects maintained at equilibrium by the balance between mutation and family selection can be two to three times as large as that classically expected for autosomal genes. Fourth, the components of epistatic genetic variance derived from maternal-offspring interactions for fitness can contribute to the response to selection for maternal effects but are not available to the response to direct individual selection alone. Thus, epistatic gene interactions play a more important role in the evolution of maternal effects than in the evolution of ordinary traits. Fifth, maternal effect genes that influence offspring fitness will often exhibit genetic correlations with other genes for offspring survival, which can accelerate or retard the rate of evolution. These genetic correlations result from linkage disequilibrium: the genes causing maternal effects, by definition, cannot be randomly distributed across offspring genotypes in nature.

Environmental correlations and phenotypic effects stemming from the special circumstances of the mother are sometimes viewed as the bane of heritability estimates. They can give rise to nongenetic and somewhat uncontrollable correlations between the phenotypes of maternal full siblings and half-siblings (e.g., Falconer 1989, p. 137). When we estimate genetic influences on the phenotype by means of resemblances between genetic relatives, heritable and nonheritable maternal effects contribute to this resemblance in plants and animals. They are transmissible from parent to offspring and result in phenotypic correlations between those genetic relatives sharing the same maternal environment, but they do not belong to the suite of segregating genes. This is the reason that paternal full- and half-sibling families are preferred for estimating heritabilities from controlled breeding studies. This bias, originating from experimental quantitative genetics applied to animal and plant breeding and reinforcing that of Darwin (1859), is another reason that maternal effects have been overlooked as a significant evolutionary factor. Nevertheless, like other phenotypes, maternal effects on offspring phenotypes can themselves be partitioned into separate genotypic and environmental components.

The origins of the bias against maternal effects should not prevent us from learning about maternal effects from animal breeding studies, however. In particular, studies investigating artificial selection on milk yield in cattle have shown not only that heritable maternal effects exist but also that there can be negative genetic correlations among maternal effects. For example, selection for increased milk yield often leads to decreased "mothering ability" owing to negative pleiotropic effects of the genes for increased milk yield on mother-calf behavioral interactions! More recently, studies from animal breeding emphasize the use of embryo transplants to estimate the magnitude of maternal genetic influences on economically important offspring phenotypes and to maximize the positive maternal effects on the economically important phenotypes.

Mother's Genotype	Offspring Genotype			Family Values
	AA	Aa	aa	
AA	M(AA)	M(AA)	M(AA)	M(AA)
Aa	M(Aa)	M(Aa)	M(Aa)	M(Aa)
aa	M(aa)	M(aa)	M(aa)	M(aa)
Offspring Average	M	M'	M"	

Figure 1.1 Genetic maternal effects only: no offspring effects.

1.3 Models of Simple Genetic Maternal Effects

The genetic model of maternal effects could be one of asexual, uniparental inheritance, like mitochondria or chloroplasts, or normal, diploid sexual inheritance. I use the simplest case of a maternal effect caused by alternative alleles at a single autosomal locus as an illustrative example (figure 1.1) of some of the special features involved in the evolutionary genetics of maternal effects. In figure 1.1, the maternal phenotypes are indicated by $M(X)$, where X indicates the maternal genotypes AA, Aa, or aa. To avoid the infinite regress into past maternal environments, imagine that the two alleles, A and a, cause the maternal phenotype, be it oviposition behavior, size, or duration of care. There are thus three genotypically different kinds of mothers, though not necessarily three maternal phenotypes. If A were dominant to a, then $M(AA)$ would equal $M(Aa)$. The genotypic configuration of each mother's offspring depends on the mating system. However, regardless of the mating system, the shaded cells in figure 1.1 cannot occur naturally because an AA mother cannot have an offspring of genotype aa, nor can an offspring of AA genotype have a mother of genotype aa. These shaded cells could only be produced by experimental cross-fostering to obtain a factorial design, crossing maternal and filial genotypes. The *row means* illustrate the important point that the effect of a given maternal genotype, $M(X)$, is common to all offspring genotypes reared by that maternal genotype. Similarly, the *column means* show the average across families of the direct maternal effect experienced by each offspring genotype.

Average fitness, W_{AA}, of a randomly chosen offspring with genotype AA depends on this genotype's average experience of maternal effects. This is obtained from the sum over the maternal genotypes X,

$$W_{AA} = \sum_{X} W_{AA}(X)\Pr(X/AA), \tag{1.1}$$

where $W_{AA}(X)$ is the maternal effect on the fitness of an offspring of genotype AA born to a mother of genotype X. $\Pr(X|AA)$ is the conditional probability that an AA offspring has a mother of genotype X and so experiences the fitness effect of that maternal genotype. Thus, the maternal effects on offspring fitness are necessarily frequency dependent as pointed out by Kirkpatrick and Lande (1989). Offspring of the same genotype but in different families will experience different maternal effects (see figure 1.1), and the sum over X of $\sum \Pr(X|AA)$ must equal 1.0 because every AA offspring has a mother.

Several of the special dynamical properties of the evolution of maternal effects mentioned earlier can be illustrated with this most simple model. The first property is that *the evolutionary process cannot be partitioned into the separate components of selection within a generation and response to selection across generations* (Lande and Arnold 1983; Wade and Kalisz 1990). Consider the conditional probabilities, $\Pr(X|AA)$, necessary for calculating offspring fitnesses, and let X be AA. Then, we have

$$Pr(AA/AA) = \frac{Pr(AA\text{mother}) * Pr(AA\text{offspring}/AA\text{mother})}{Pr\ (AA\text{offspring})}. \tag{1.2}$$

The $\Pr(AA$ mother$)$ is the distribution of maternal genotypes after selection in the maternal generation, which is represented as $G'(X)$ (see table 1.2). In other words, selection on mothers in generation $t-1$ appears explicitly in our expression (eq. [1.1]) for selection on offspring in generation t in the form of a conditional probability. This is a Mendelian example of the result obtained by Kirkpatrick and Lande (1989). Note also that the $\Pr(AA$ offspring$/AA$ mother$)$ is affected by the mating system. Positive assortative mating by genotype or inbreeding will increase this term, and negative assortative mating or outbreeding will decrease it.

With the standard models of natural selection, the assumptions of random mating and simple selection coefficients guarantee that the effects of natural selection on the distribution of gene frequencies can be described by a simple recursion. In these standard models, parents (i.e., mothers) only contribute genes to the next generation. Hence, we know from the Hardy-Weinberg equilibrium the genotypic distribution of the offspring and, from the selection coefficients, the offspring genotypic fitnesses. In contrast, *with maternal effects, parents contribute more than genes to the next generation.* They also contribute the special environmental effects unique to each maternal genotype. Differently put, there is no true separation of maternal soma and germ line with respect to the expression of the offspring genotype. For this reason, we need to know not only the mating system but also the *distribution of maternal genotypes after selection* in order to calculate the offspring genotypic fitnesses. As a result, we lose the simple form of the gene frequency recursion so familiar in textbooks. We also lose the ability to partition the evolutionary process into the two components of (1) selection within generations and (2) transmission across generations. These same considerations apply to quantitative genetic models that assume weak selection, additive across many loci (e.g., Lande and Arnold 1983; Cheverud and Moore 1994).

Equally important is the second property of the evolution of maternal effects, namely, that *the evolution of maternal effects involves selection among families.* Each row of figure 1.1 has the same entry for $M(X)$, showing that all of the offspring genotypes within the same family experience a common maternal effect (aside from maternal-offspring genotype interactions, discussed in 1.5). Differently put, there is no within-family variation in offspring fitness owing to maternal effects (table 1.2). All selection arises from the fitness differences among families;

Table 1.2 Family types, family (maternal genotype) frequencies, allele (A) frequency per family, and mean family fitness under the model of simple maternal effects shown in figure 1.1.

Family		Freq. of A	Offspring genotype			Family
Type	Frequency	in family	AA	Aa	aa	fitness
AA	$G'(AA)$	$(1+p)/2$	$W(AA)$	$W(AA)$	—	$W(AA)$
Aa	$G'(Aa)$	$(1+2p)/4$	$W(Aa)$	$W(Aa)$	$W(Aa)$	$W(Aa)$
Aa	$G'(aa)$	$p/2$	—	$W(aa)$	$W(aa)$	$W(aa)$

that is, all selection is among-family selection. Gene frequencies change due to the covariance between-family mean fitness, resulting from the maternal effect, and within-family gene frequency, resulting from Mendelian transmission of maternal genes (Wade 1980). From table 1.2, it is clear that the family frequencies are the maternal genotype frequencies after selection, $G'(X)$. The covariance across families between gene frequency and fitness is given by

$$\mathrm{Cov}[F(A), W(AA)] = \sum_x G'(X)* F(X)*W(X) \\ -[\sum_x G'(X)* F(X)]*[\sum_x G'(X)*W(X)], \tag{1.3}$$

where $F(X)$ is the frequency of the A allele within family type X as defined by the maternal genotypes. [The values of $F(X)$ are given in table 1.2, column 3.] Note that the within-family frequencies of allele A, $F(X)$, all share the additive constant, $p/2$ (see table 1.2), which is the frequency of the A allele from the paternal genetic contribution to the offspring in a randomly mating population. Because, by definition, it does not vary among families, this constant contributes nothing to the covariance across families between the within-family gene frequencies and the maternal effects on offspring fitness. It is the *maternal*, Mendelian genetic contribution to the offspring that varies among families, taking values 1/2, 1/4, and 0, for families of the AA, Aa, and aa mothers, respectively (see table 1.2, column 3). The covariance between gene and fitness defined in equation (1.3) arises because genetic maternal effects cannot be randomly distributed across families without the experimental manipulation of cross-fostering. This covariance is responsible for the mathematical congruence between models for the evolution of maternal effects and models for kin selection as first pointed out by Cheverud (1984; see also Cheverud and Moore 1994; Moore et al., this volume).

Evolution by both natural selection and mutation results in a higher equilibrium gene frequency and heterozygosity for maternal effects than is typically associated with mutation selection balance. Deleterious alleles persist in a population as a result of the balance between natural selection against the alleles and mutation reintroducing them into the population. The among-family selection involved in the maternal effects evolution, as depicted in figure 1.1 and table 1.2, screens only those genes found in mothers and not those in fathers. Selection is sex limited because the male parents harbor genes for maternal effects but do not express them. If we let the a allele have a maternal fitness effect s in aa mothers relative to the effect in AA mothers, then 1, $1 - (s/2)$, and $1 - s$ are the maternal genotypic or family mean fitnesses in table 1.2 for AA, Aa, and aa mothers, respectively. When the frequency, q, of the a allele is low, equilibrium gene frequency at the mutation-family selection balance is, to a good approximation,

$$\hat{q} \approx \frac{4\mu}{s}, \tag{1.4}$$

which is *twice* the frequency expected for an autosomal gene expressed in both sexes (Whitlock and Wade 1995). As a result, the equilibrium additive genetic variance for maternal effects on fitness is $2ms$, *twice* as large as that for a "regular" autosomal gene expressed in both sexes. This results from selection being only half as effective for autosomal genes with sex-limited expression, as is the case with the maternal effects genes of figure 1.1 and table 1.2.

Not only is the equilibrium genetic variance for fitness higher, but there can also be a sex difference in the equilibrium variance for X-linked maternal effect genes. Whether the female is the homogametic or the heterogametic sex, the equilibrium gene frequency at mutation-family selection balance is

$$\hat{q} \approx \frac{3\mu}{s}, \tag{1.5}$$

However, the equilibrium additive variance for offspring fitness due to maternal effects differs for the two X-linked cases, depending on whether mothers are homogametic or heterogametic. When the female is the heterogametic sex, the additive genetic variance for fitness at equilibrium is $3ms$, but when she is the homogametic sex, the equilibrium variance for fitness is $(3/2)ms$ (Whitlock and Wade 1995). As a consequence of the higher equilibrium gene frequency and variance for fitness, isolated or allopatric populations can be expected to diverge more rapidly for maternal effects than for other traits by the combined action of random genetic drift and local selection. If we consider speciation to be that evolutionary process that converts genetic variation within populations to more or less permanent variation among them, then the evolutionary dynamics suggest that genetic maternal effects contribute uniquely to speciation. Mousseau and Dingle (1991b) found significant variation among populations in maternal effects in a large number of insect species and argued that much of this variation could be interpreted as adaptive: "The observed geographic variation in many species strongly supports the hypothesis that natural selection has acted to produce local adaptation and genetic divergence among populations" (p. 756). Mousseau (1991) demonstrated adaptive variation in maternal age effects on embryonic diapause among 10 populations of the striped ground cricket, *Allonemobius fasciatus*. We would predict that reproductive isolation between species should arise relatively frequently owing to genetic divergence for maternal effects. This is not the runaway process alluded to earlier. As shown in 1.5, when further complications are added to the model, there are additional reasons to expect the rapid divergence of isolated populations regarding maternal effects.

1.4 Models of Simple Genetic Maternal Effects with Direct Offspring Effects

Yet more interesting evolutionary properties emerge when direct offspring genotypic effects are combined with the simple genetic maternal effects model discussed in section 1.3. Figure 1.2 illustrates the combination of direct offspring genotypic effects with genotypic maternal effects. Again, notice that certain mother-offspring genotypic combinations are not possible (shaded cells in figure 1.2), so heritable maternal effects will not be randomly dis-

Mother's Genotype	Offspring Genotype			Family Values
	AA	Aa	aa	
AA	M(AA) + O(AA) + MxO	M(AA) +O(Aa) + MxO	M(AA) + O(aa) + MxO	M(AA) + O + MxO
Aa	M(Aa) + O(AA) + MxO	M(Aa) + O(Aa) + MxO	M(Aa) + O(aa) + MxO	M(Aa) + O' + MxO'
aa	M(aa) + O(AA) + MxO	M(aa) + O(Aa) + MxO	M(aa) + O(aa) + MxO	M(aa) + O'' + MxO''
Offspring Average	M + O(AA) + MxO	M' + O(Aa) + M'xO	M'' + O(aa) + M''xO	

Figure 1.2 Genetic maternal effects and direct offspring effects.

tributed across offspring genotypes. The genetic correlation between maternal and offspring mean effects will be 1/4, just like the paternal half-sibling mean correlation in typical quantitative genetic studies. When the genetic effects on fitness of mothers and offspring are congruent in sign, the selective response will be accelerated because the among-family component of selection arising from maternal effects enhances the force of direct selection on offspring genotypes. That is, selection among families and selection among individual offspring genotypes can operate in concert with one another. If we use the paradigm of ascribing all adaptation to the lowest level of selection (e.g., Williams 1966; Thornhill and Alcock 1983), then we would overlook this important contribution of maternal effects to evolution and overestimate the strength of selection among individuals by about 50% and possibly more, depending on the relative magnitudes of the maternal and direct effects.

Clearly, direct genotypic and maternal genotypic effects need not operate in concert with one another. When they do not, there is conflict between the levels of selection within and between families. Although selection between individuals within families has been suggested on certain grounds (e.g., Williams 1966) to always override selection for maternal effects between families, that view is incorrect. Selection within families is only half as strong as selection among individuals across an entire population because the within-family component of genetic variance is only half of the total genetic variance. The magnitude of this effect on within-family selection is comparable to the effect of sex-limited selection on maternal effects alone (see section1.3). As a result, both levels of selection are equally matched a priori as far as the available additive genetic variance for an evolutionary response is concerned.

A surprising form of epistasis for fitness can arise from the interaction between direct offspring genotypic effects and genotypic maternal effects (see figure 1.2, MxO terms): epistasis between genotypes at the same locus. This epistatic interaction can be understood by analogy with genotype × environment interactions. Consider the maternal genotypes to specify a series of environments and consider the offspring genotypes at the same locus to determine how well each survives in a given environment. Each offspring genotype can be viewed as having a "norm of reaction" specified by rearing it across the series of maternal genotypes. Changes in scale or relative ranking of the fitnesses of offspring genotypes across the maternal (genotypic) environments represent a form of epistasis. This component of epistasis, arising from maternal-offspring genotype × genotype interactions, is known to be an important component of competitive fitness or so-called "intermixing ability" (Griffing 1977, 1989; McCauley and Wade 1980; Goodnight 1991; Craig and Muir 1996). In the context of maternal effects and in keeping with previous usage in quantitative genetic studies, we might refer to this component of genetic maternal effects as the "maternal-offspring intermixing ability." By analogy with other studies of competition, this kind of genetic epistasis might play a particularly important role in the evolution of parent-offspring conflict, the maternal mediation of within-family competition between offspring, and the evolution of maternal-offspring interactions in general. It is important to note that previous theoretical and experimental works on traits affected by this component of genetic variance have demonstrated (1) that it does not contribute to the response to selection between individuals, and (2) that it contributes directly to the genetic response to group selection (Griffing 1977, 1989; Goodnight 1991). Because selection on genetic maternal effects involves interfamily or group selection, *this epistatic component of the genetic variance is available for the evolution of maternal effects but not for the evolutionary response of "ordinary" traits.*

Recent empirical findings indicate that negative epistatic interactions between mothers and offspring can evolve. Empirical evidence for the widespread existence of maternal-effect selfish genes has been reported in studies of the flour beetle *Tribolium castaneum* by Beeman et al. (1992), and their evolution has been modeled by Wade and Beeman (1994). These genes are "selfish" in the sense that they self-select by maternal-effect lethality to all offspring genotypes not inheriting at least one copy of the gene. The gene in the maternal genotype imposes direct viability selection against those offspring genotypes with the alternative allele. As might be expected, these genes are selected against by the among-family component of selection that attends the evolution of maternal effects because they reduce family size (see Wade and Beeman 1994, their eq. [5]). However, the strength of the effect on survivorship of the opposite homozygote is sufficiently strong (lethality) that they enjoy a net selective advantage. In addition, whenever there is strong within-family density regulation, such genes may be favored because they replace an "ecological death in the family" owing to competition with an earlier "genetic death," thereby conserving family resources. Wade and Beeman (1994) showed how this type of soft selection accelerates the evolution of this class of maternal-effect selfish genes.

Overall, the reasonable addition of direct genotypic effects to the simple one-locus model of maternal effects introduces two new features. First, it opens the possibility of evolutionary synergism and conflict between the levels selection within and between families. Second, the interaction of direct and maternal effects introduces a surprising component of epistatic genetic variance into the evolution of maternal effects. Unlike ordinary individual selection, the among-family component of selection involved in the evolution of maternal effects permits this new component of genetic variation to contribute to the evolutionary response. When artificial selection studies have used intergroup selection to focus on this particular component of genetic variation, they have uniformly produced very large responses to selection in short periods of time (see McCauley and Wade 1980; Goodnight 1991; Craig and Muir 1996).

Last, when the timing of gene expression is delayed such that a gene formerly expressed by the offspring genotype early in development is delayed in its expression to the time of reproduction by the maternal genotype, the level of selection affecting the gene changes from individual to family selection. Thus, in the evolution of early development, there is the interesting question of whether a gene product should be produced by the maternal genotype or directly by the genotype of the developing embryo. The answer to this question affects the level of selection experienced by the gene.

1.5 Genetic Correlations between Loci Owing to Direct and Maternal Effects

Epistasis between genotypes at different loci is a possibility when some genes are responsible for the maternal effects and others are responsible for the direct offspring effects. Let us complicate the simple model one step further by introducing an additional locus so that alternative alleles at locus A determine maternal effects while alternative alleles at locus B determine the direct offspring genotypic effect. Figure 1.3 illustrates this more complicated model for the nine possible two-locus maternal genotypes and the nine possible offspring genotypes. Note again the shaded cells, which are combinations of maternal-offspring genotypes not permitted by Mendelian genetics. In figure 1.3, 31 of the 81 (9×9) possible maternal-offspring genotype combinations do not and cannot occur, but all nine of the interlocus genotype combinations can be found. It is these interlocus genotypes that give rise to linkage disequilibrium between the loci *if* they become nonrandomly associated. A genetic correlation, owing to this disequilibrium, can cause a runaway process to occur in the evolution of maternal effects. That is, there can be an accelerated coevolution of the maternal and direct offspring effects. When will this genetic correlation develop between loci in the evolution of maternal effects?

Disequilibrium will arise whenever selection causes the two-locus genotype, *AABB*, to occur more frequently than expected under the assumption of independent or random association of the *AA* and *BB* genotypes. Disequilibrium can occur with or without epistasis. Examination of figure 1.3 suggests that whenever the family combination of fitness effects is additive, $M(AA) + O(BB)$, and has a higher fitness than any other, a positive genetic correlation between the A and B alleles across the two loci develops as a result of the linkage disequilibrium caused by selection. Mothers of genotype *AA*, with the greatest maternal effect on offspring fitness, leave more offspring than mothers with weaker maternal effects. Furthermore, within the families of *AA* mothers, *BB* offspring are best suited to survive and enjoy the additional advantage of the favored maternal effect. Because of the nonrandom distribution of maternal and offspring genotypes *within loci* (discussed in 1.3), these same *BB* offspring harbor an excess of copies of the most favorable maternal effect gene, *A*, which they will express when they become mothers at maturity. In this way, a genetic correlation can become established even in the absence of the epistatic interactions between the maternal and offspring genotypes discussed in section 1.3.

Epistasis for fitness between the *AA* maternal effect genotype and the *BB* offspring genotype occurs whenever the *BB* genotype is better suited than other offspring genotypes to develop in the *AA* maternal environment. That is, within-family variation among offspring

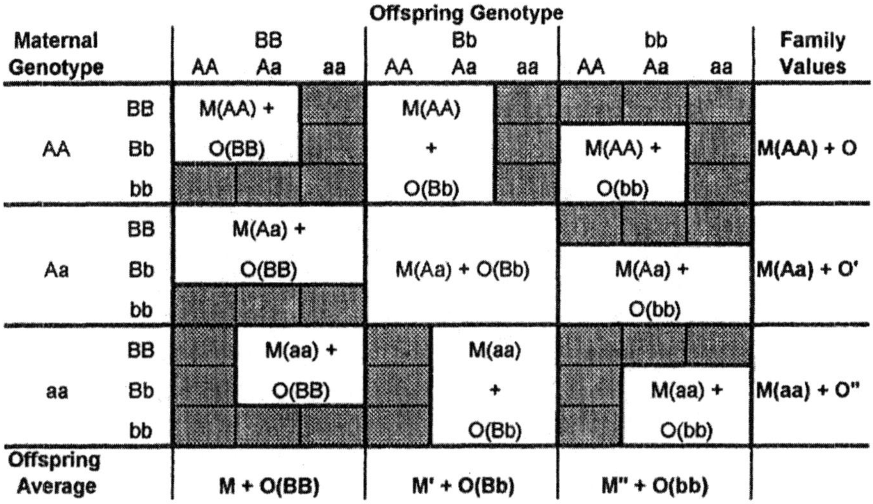

Maternal Genotype		BB AA	Aa	aa	Bb AA	Aa	aa	bb AA	Aa	aa	Family Values
AA	BB	M(AA) +			M(AA)						
	Bb	O(BB)			+			M(AA) +			M(AA) + O
	bb				O(Bb)			O(bb)			
Aa	BB	M(Aa) +									
	Bb	O(BB)			M(Aa) + O(Bb)			M(Aa) +			M(Aa) + O'
	bb							O(bb)			
aa	BB		M(aa) +			M(aa)					
	Bb		O(BB)			+			M(aa) +		M(aa) + O''
	bb						O(Bb)			O(bb)	
Offspring Average		M + O(BB)			M' + O(Bb)			M'' + O(bb)			

Figure 1.3 Genetic correlation between a maternal effect and offspring survival.

genotypes in the response to the maternal environment is genotype × genotype epistasis. I emphasize again that this kind of genotypic epistasis is different from the standard epistasis that arises from interactions between genes within the same individual, that is, when phenotypic differences between individuals are more or less than additive (Wade 1992). This will also precipitate linkage disequilibrium between the two loci.

By creating disequilibrium between the A and B alleles by either means, selection initiates a runaway process between the genes for maternal effects and the genes affecting offspring fitness. The relative fitness advantage of AA mothers over aa mothers is enhanced by the disproportionate number of BB offspring produced and surviving from their families. This accelerates the rate of increase of frequency of the A maternal effect allele. The greater the frequency of AA mothers, the greater the relative fitness advantage of BB offspring because they tend to occur disproportionately in "best environments," namely, those associated with AA mothers.

Epistatic interactions between maternal and offspring genotypes may be viewed as the simultaneous evolution of the environment and the offspring adaptations to that environment. In this context, maternal oviposition behavior is one trait that is very likely to experience a maternal effects runaway process. In this instance, the maternal effects derive not only from the internal resources of the mother but also from the external environmental conditions in which she places her offspring. Plants, too, have a type of "oviposition behavior" because the maternal genotype determines not only the size of the seed but also its pericarp and other maternal architectural features of the seed that affect dispersal. A positive genetic correlation between the environment of oviposition and the offspring genotypes most suited to the environment will result in the accelerated coevolution of maternal oviposition behavior and subsequent offspring survival.

Returning to the basic mathematical formulation introduced in section 1.3, we find that whenever offspring survival is conditional on maternal genotype as well as on offspring genotype, the expression for offspring fitness is given by

$$W_{BB} = \sum_x W_{BB}[M(X)]\Pr[M(X)/O(BB)]. \tag{1.6}$$

Notice that the fitness of offspring genotype *BB* depends on the distribution of maternal genotypes. The conditional probability that an *AA* mother has a *BB* offspring, $\Pr(M(AA)/O(BB))$, can be written as

$$\Pr[M(AA/O(BB)] = \frac{\Pr(AA\text{mother}) * \Pr(BB\text{offspring}/AA\text{mother})}{\Pr(BB\text{offspring})}. \tag{1.7}$$

Not only is the $\Pr[M(AA)]$ affected by selection on *BB* in the previous generation, $t - 1$, but also $\Pr[O(BB)/M(AA)]$ is affected by prior selection, linkage disequilibrium, recombination, and the mating system! Whenever offspring tend to mate in the local environment in which they developed, assortative mating will also assist in establishing and maintaining the linkage disequilibrium on which the runaway process of maternal effects evolution is based. Note, too, that expression (1.7) means that the fitnesses given by equation (1.6) will be doubly frequency dependent, that is, frequency dependent regarding both the *A* and *B* loci.

Runaway coevolution for maternal oviposition behaviors should play an important role in speciation because it can lead to rapid host race formation. This kind of coupling of habitat choice by maternal effects with the genes best suited to survive in the chosen habitat has been suggested by many authors without explicit mention of the special dynamical properties of maternal effects evolution. In the earlier sections, I discussed how the increased genetic variance at mutation-family selection equilibrium would contribute to the diversification of isolated populations by random drift and selection for genes involved in maternal effects. In this section, using somewhat more complicated two-locus models, I illustrated how these same maternal effect genes can interact with other genes affecting offspring survival by specialization to a maternal environment to further accelerate the genetic divergence of local populations. In these two ways, the evolution of maternal effects should contribute to the process of speciation.

Summary and Conclusion

Existing textbooks do not adequately discuss maternal effects or their pervasive effects on the evolutionary process for a wide variety of traits. This omission can be traced back to Darwin (1859), who emphasized the heritable variations among family members experiencing a common maternal environment at the expense of the differences among families owing to maternal effects. In the absence of genetic knowledge and without a firm grasp of experimental design and statistical analysis, it is difficult to see how anyone could clearly partition heritable from nonheritable maternal effects. This omission was reinforced by the definite bias against maternal effects that emerged from the practice of quantitative genetic methods that viewed maternal effects as a complicating factor in the estimation of heritabilities. Closer examination of maternal effects not only shows them to be widespread among both plants and animals but also indicates that their evolutionary dynamics are unusual relative to standard theory in several respects. The evolution of maternal effects involves two levels of selection, within and between families, and bears a compelling mathematical similarity to the much more widely studied models of kin selection. As a result of selection among families, there can be much more genetic variance maintained at equilibrium with mutation than for ordinary genes. The increased heterozygosity, in turn, permits random genetic drift to differentiate local populations and open opportunities for local adaptation via individual selection (see Mousseau and Dingle 1991a, b). In addition, a special kind of genotype × genotype epistasis can emerge even from the most basic one-locus model of interaction between maternal and offspring genotypes. This is a novel component of genetic variation available for the evolution of maternal effects and not for many other traits. The maternal-offspring intermixing ability represented by these transgenerational genotypic interactions can initiate a runaway process of maternal effects evolution. Indeed, a runaway process is possible even in the absence of this component of epistasis. Accelerated local divergence of isolated populations for maternal effects, especially when it involves maternal oviposition behavior, will contribute to speciation.

Acknowledgments   I acknowledge John Kelly and Jim Cheverud for helpful discussions, and Allen Moore, Kathleen Donohue, and Fred Janzen for comments. I am particularly indebted to the symposium organizers for the invitation to participate, which opened my eyes to the fascinating world of genetic maternal effects. This work was supported in part by NIH Grant GM22523.

References

Alberts, S., and J. Altmann. 1995. Preparation and activation: Determinants of age at reproductive maturity in male baboons. Behav. Ecol. Sociobiol. 36:397–406.

Alexander, R. D. 1974. The evolution of social behavior. Annu. Rev. Ecol. Syst. 5:325–383.

Altmann, J., S. Altmann, and G. Hausfater. 1988. Determinants of reproductive successs in savannah baboons (*Papio cynocephalus*). Pp. 403–418 T. H. Clutton-Brock (ed.), Reproductive Success. University of Chicago Press, Chicago.

Augspurger, C. 1984. Seedling survival of tropical tree species: Interactions of dispersal distance, light gaps, and pathogens. Ecology 65:1705–1712.

Beeman, R. W., K. S. Friesen, and R. E. Denell. 1992. Maternal-effect selfish genes in flour beetles. Science 256:89–92.

Boyd, R., and P. J. Richerson. 1985. Culture and the Evolutionary Process. University of Chicago Press, Chicago.

Bull, J. J. 1983. Evolution of Sex Determining Mechanisms. Benjamin/Cummings, Menlo Park, Calif.

Burdon, J. J. 1987. Diseases and Plant Population Biology. Cambridge University Press, Cambridge.

Charnov, E. 1982. The Theory of Sex Allocation. Princeton University Press, Princeton, N.J.

Cheverud, J. M. 1984. Evolution by kin selection: A quantitative genetic model illustrated by maternal performance in mice. Evolution 38:766–777.

Cheverud, J. M., and A. J. Moore. 1994. Quantitative genetics and the role of the environment provided by relatives in behavioral evolution. Pp. 67–100 C. R. B. Boake (ed.), Quantitative Genetic Studies of Behavioral Evolution. University of Chicago Press, Chicago.

Cockburn, A. 1991. An Introduction to Evolutionary Ecology. Blackwell Scientific, London.

Craig, J. V., and W. M. Muir. 1996. Group selection for adaptation to multiple-hen cages: Beak related mortality, feathering, and body weight response. Poult. Sci. 75:294–302.

Darwin, C. 1859. On the Origin of Species. Harvard University Press, Cambridge, Mass. (reprinted 1964).

Darwin, C. 1871. The Descent of Man and Selection in Relation to Sex. Vol. 2. Princeton University Press, Princeton, N.J. (reprinted 1981).

Elgar, M. A. 1992. Sexual cannibalism in spiders and other invertebrates. Pp. 128–155 in

M. A. Elgar and B. J. Crespi (eds.), Cannibalism Ecology and Evolution among Diverse Taxa. Oxford University Press, New York.

Engels, W. R. 1983. The P family of transposable elements in *Drosophila*. Annu. Rev. Genet. 17:315–344.

Falconer, D. S. 1989. Introduction to Quantitative Genetics. Wiley, New York.

Feldman, M. W., and L. L. Cavalli-Sforza. 1976. Culture and biological evolution: Selection for a trait under complex transmission. J. Theor. Biol. 9:238–259.

Feldman, M. W., and K. N. Laland. 1996. Gene-culture coevolutionary theory. Trends Ecol. Evol. 11:453–457.

Futuyma, D. J. 1986. Evolutionary Biology. Sinauer, Sunderland, Mass.

Gillett, J. D. 1971. Mosquitos. Weidenfeld and Nicolson, London.

Goodnight, C. 1991. Intermixing ability in two-species communities of flour beetles. Am.Nat. 138:342–354.

Griffing, B. 1977. Selection for populations of interacting genotypes. Pp. 413–434 in E. Pollack, O. Kempthorne, and T. B. Bailey (eds.), Proceedings of the International Congress on Quantitative Genetics, August 16–21, 1976. Iowa State University Press, Ames.

Griffing, B. 1989. Genetic analysis of plant mixtures. Genetics 122:943–957.

Haldane, J. B. S. 1932. The time of action of genes, and its bearing on some evolutionary problems. Am. Nat. 66:5–24.

Hartl, D. L., and A. G. Clark. 1989. Principles of Population Genetics. Sinauer, Sunderland, Mass.

Hurst, L. D. 1993. The incidences, mechanisms and evolution of cytoplasmic sex ratio distorters in animals. Biol. Rev. 68:121–193.

Janzen, F. J. 1993.The influence of incubation temperature and family on eggs, embryos, and hatchlings of the smooth softshell turtle (*Apalone mutica*). Physiol. Zool. 66:349–373.

Janzen, F. J. 1995.Vegetational cover predicts the sex ratio of hatchling turtles in natural nests. Ecology 75:1593–1599.

Janzen, F. J., and G. L. Paukstis. 1991. Environmental sex determination in reptiles: Ecology, evolution, and experimental design. Q. Rev. Biol. 66:149–179.

Janzen, F. J., J. C. Ast, and G. L. Paukstis. 1995. Influence of the hydric environment and clutch on eggs and embryos of two sympatric map turtles. Funct. Ecol. 9:913–922.

Kempthorne, O. 1969. An Introduction to Genetics Statistics. Iowa State University Press, Ames.

Kirkpatrick, M., and R. Lande. 1989. The evolution of maternal characters. Evolution 43:485–503.

Lande, R., and S. J. Arnold. 1983. The measurement of selection on correlated characters. Evolution 36:1210–1226.

McCauley, D. E., and M. J. Wade. 1980. Group selection: The genetic and demographic basis for the phenotypic differentiation of small populations of Tribolium castaneum. Evolution 34:813–821.

McNamara, J. M., and A. I. Houston. 1996. State-dependent life histories. Science 380:215–221.

Mousseau, T. A. 1991. Geographic variation in maternal-age effects on diapause in a cricket. Evolution 45:1053–1059.

Mousseau, T. A., and H. Dingle. 1991a. Maternal effects in insect life histories. Annu. Rev. Ecol. Syst. 36:511–534.

Mousseau, T. A., and H. Dingle. 1991b. Maternal effects in insects: Examples, constraints, and geographic variation. Pp. 745–761 in E. C. Dudley (ed.), The Unity of Evolutionary Biology. Dioscorides Press, Portland, Ore.

Park, T. 1935. Studies in population physiology. IV. Some physiological effects of conditioned flour upon *Triboliu confusum* Duval and its populations. Physiol. Zool. 8:91–115.

Price, P. W. 1996. Biological Evolution. Saunders College Publications, New York.

Ridley, M. 1993. Evolution. Blackwell Scientific. Cambridge, Mass.

Riska, B. 1991. Introduction to the symposium, maternal effects in evolutionary biology. Pp. 719–724 in E. C. Dudley (ed.), The Unity of Evolutionary Biology. Dioscorides, Portland, Ore.

Roach, D. A., and R. D. Wulff. 1987. Maternal effects in plants. Q. Rev. Biol. 18:209–235.

Roosenburg, W. M. 1996. Maternal condition and nest site choice: An alternative for the maintenance of environmental sex determination. Am. Zool. 36:157–168.

Rossiter, M. C. 1994. Maternal effect hypothesis of herbivore outbreak. Bioscience 44:752–763.

Rossiter, M. C. 1995. Impact of life-history evolution on population dynamics: Predicting the presence of maternal effects. Pp. 251–275 in N. Cappucino and P. W. Price (eds.), Population Dynamics. Academic Press, San Diego.

Sawamura, K. 1996. Maternal effect as a cause of exceptions to Haldane's Rule. Genetics 143:609–611.

Schaal, B. 1986. Life history variation, natural selection, and maternal effects in plant populations. Pp. 188-206 in R. Dirzo and J. Sarukhan (eds.), Perspectives in Plant Population Ecology. Sinauer, Sunderland, Mass.

Sinervo, B. 1990. The evolution of maternal investment in lizards: An experimental and comparative analysis of egg size and its effects on offspring fitness. Evolution 44:279–294.

Sinervo, B., P. Doughty, R. B. Huey, and K. Zamudio. 1992. Allometric engineering: A causal analysis of natural selection on offspring size. Science 258:1927–1930.

Thornhill, R., and J. Alcock. 1983. The Evolution of Insect Mating Systems. Harvard University Press, Cambridge, Mass.

Trivers, R. L., and D. E. Willard. 1973. Natural selection of parental ability to vary sex ratio of offspring. Science 179:90–92.

Wade, M. J. 1980. Kin selection: Its components. Science 210:665–667.

Wade, M. J. 1992. Sewall Wright: Gene interaction and the Shifting Balance Theory. Pp. 35–62 in J. Antonovics and D. Futuyama (eds.), Oxford Series on Evolutionary Biology.Vol.8. Oxford University Press, Oxford.

Wade, M. J. 1996. Adaptation in subdivided populations: Kin selection and interdemic selection. Pp. 1–25 in M. Rose and G. Lauder (eds.), Adaptation. Academic Press, New York.

Wade, M. J., and R. W. Beeman. 1994. The population dynamics of maternal-effect selfish genes. Genetics 138:1309–1314.

Wade, M. J., and S. Kalisz. 1990. The causes of natural selection. Evolution 44:1947–1955.

Westneat, D. F., and R. C. Sargent. 1996. Sex and parenting: The effects of sexual conflict and parentage on parental strategies. Trends Ecol. Evol. 11:87–91.

Whitlock, M. C., and M. J. Wade. 1995. Speciation: Founder events and their effects on X-linked and autosomal genes. Am. Nat. 145:676–685.

Williams, G. C. 1966. Adaptation and Natural Selection. Princeton University Press, Princeton, N.J.

Wilson, E. O. 1971. The Insect Societies. Harvard University Press, Cambridge, Mass.

Wilson, E. O. 1975. Sociobiology: The New Synthesis. Harvard University Press, Cambridge, Mass.

Wright, S. 1969. Evolution and the Genetics of Populations. Vol. 2. University of Chicago Press, Chicago.

Wulff, R. 1986. Seed size variation in *Desmodium paniculatum*. III. Effects on reproductive yield and competitive ability. J. Ecol. 74:99–114.

Yamanaka, M. 1928. On the male of a paper wasp, *Polistes fadwigae* Dalla Torre. Sci. Rep. Tohoka Imperial Univ., (ser. 4, Biol.) 3:265–269.

2

The Influence of Direct and Indirect Genetic Effects on the Evolution of Behavior

Social and Sexual Selection Meet Maternal Effects

ALLEN J. MOORE, JASON B. WOLF, & EDMUND D. BRODIE III

Intraspecific behavioral interactions are ubiquitous. Parents caring for their offspring, dominant individuals maintaining their status over subordinates, and males using odors or sounds to signal to potential mates are just a few examples of the many types of social interchanges that require interactions among at least two individuals. In this chapter we describe two different models that address how genetically based variation among individuals might influence such interactions and therefore the evolution of interactions. To do this we adopt a quantitative genetic approach and build quantitative genetic models that are similar in perspective to maternal effects models. Our chapter is devoted to describing two different models that characterize interactions expected to result in two related but different types of selection. Our models provide insights into the potential role of genetics in the evolution of interactions. In section 2.1 we describe how genetic variation might influence the evolution of social interactions among unrelated individuals. In section 2.2 we outline the role of genetics in the evolution of a sexually selected male signal of parental quality.

To analyze the role of inheritance in responses to social and sexual selection, our models consider indirect genetic effects that can arise as a result of interactions. Indirect genetic effects are genetically based environmental influences that are generated whenever the phenotype of one individual acts as an environment for another (Riska et al. 1985; Moore et al. 1997). The most commonly modeled and studied indirect genetic effects are those that result from interactions between mothers and their offspring, that is, maternal effects (for a recent review, see Cheverud and Moore 1994). Mothers make a direct genetic contribution to their offspring by providing one-half of the additive genetic effects influencing an offspring's phenotype (i.e., Mendelian inheritance). Mothers also contribute indirect genetic effects because they provide environmental influences—shelter, food, and so on—to their young. These environmental contributions are genetic when differences among mothers reflect additive genetic variation underlying parental care. The genes that determine the environment provided by the mother indirectly determine phenotypes of their offspring in the next generation—an indirect genetic effect. Unlike direct (Mendelian) genetic effects, genes with indirect effects are expressed in a different individual (in this example, the mother) than the individual whose phenotype and fitness is affected (in this example, the offspring).

Because the environments provided by others complicate predictions and genetic estimates, quantitative geneticists, especially breeders, have treated indirect genetic effects arising from maternal effects as non-Mendelian influences that need to be statistically controlled (Eisen 1967; Willham 1972). However, indirect genetic effects have important influences on both fitness and responses to selection and have accordingly begun to interest evolutionary biologists. Maternal effects models allow environmental effects to pass across generations and therefore allow persistent environments to play a role in response to selection. Evolution of interacting phenotypes in response to kin selection (Cheverud 1984; Lynch 1987) and natural selection (Riska et al. 1985; Kirkpatrick and Lande 1989; Lande and Kirkpatrick 1990) have been modeled from a quantitative perspective that has included maternal (or kin, or parental; Arnold 1994) effects. These models have shown that nonintuitive evolutionary trajectories can result because of the indirect genetic effects of interactions between relatives.

Although maternal genetic effects are the most commonly considered form of indirect genetic effect, the use of the adjective "maternal" has had the unfortunate consequence of focusing research almost exclusively on the interactions between mothers and offspring. This is not because maternal effects are expected to be the only cause of indirect genetic effects in nature (Arnold 1994). Indeed, noting that "maternal effect" is misleading but well entrenched in the literature (Cheverud and Moore 1994; Bernardo 1996), reviews on the evolutionary importance of maternal effects often begin with an apology for the term itself (e.g., see Cheverud and Moore 1994, p. 72; Bernardo 1996, p. 85). Interactions between mothers and their offspring are fascinating but represent a special case. There has been considerable theoretical work indicating that all relatives should be considered important sources of indirect genetic effects, and additional terms such as "parental effects," "kin effects," or "associate effects" have been suggested as alternatives (Griffing 1967, 1977; Cheverud 1984; Lynch 1987; Cheverud and Moore 1994). We prefer to use the general term "indirect genetic effect," which provides a contrast with the commonly used "direct genetic effect," that describes simple Mendelian inheritance. Kin effects, parental effects, associate effects, and maternal effects then are simply special cases of indirect genetic effects that can arise as a result of interactions (Moore et al. 1997; Wolf et al. 1998).

Typically, there have been two different approaches to studying social interactions. First, the interaction itself can be examined. This topic is, and continues to be, well explored by behavioral ecologists (Huntingford and Turner 1987; Krebs and Davies 1991). Alternatively, the consequences of the interaction can be evaluated. Our chapter focuses on the separate consequences of interactions by providing an analysis of the genetic (direct and indirect) and special environmental (kin, maternal, paternal, parental, or interacting) effects themselves. Focusing on the indirect genetic effects of interactions is important because interactions are known to be ubiquitous, but the evolution of these interactions is less well understood (Wolf et al. 1998). Considering how pervasive interactions are in nature, the role of indirect genetic effects in evolution may be severely underestimated because genetic studies are uncommon. We hope to stimulate genetic studies of characters that influence, or are influenced by, interactions.

To underscore the generality of our approach, we have specifically chosen to review two of our models where mothers are not even considered (Moore et al. 1997; Wolf et al. 1997). In fact, our first model assumes that the interacting individuals are not related. Our approach is to treat each phenotype as a character whose expression is determined, at least in part, by interactions. Interacting phenotypes therefore have an environmental component that results from the actions of the social partners. Within interactions, the phenotype of one individual (e.g., a dominant male) is expressed in the presence of a specific environment, which itself is influenced by another individual's phenotype (e.g., a subordinate male). Within this perspective, each individual or trait can be both a phenotype and an environment.

We have three specific goals in this chapter. First, we hope to illustrate how a quantitative genetic approach to modeling behavioral interactions can provide additional insights into many issues of interest to those who study interactions other than mother-offspring interactions. We believe that adopting the outlook and experimental designs required to examine indirect genetic effects will stimulate new ideas and data. Because indirect genetic effects are not limited strictly to mothers (or even relatives) but are simply genetically influenced environmental effects of one individual on another, any study that addresses persistent environmental effects on behavior is appropriate for a quantitative genetic approach. Second, we hope to provide a "road map" to quantitative genetic terminology and symbols. In each of our models we use commonly encountered symbols and terminology. Although our models are complementary in that they both address the role of indirect genetic effects, the first analyzes evolution of interactions influenced by social selection while the second considers a specific case of evolution by sexual selection. Thus, the same terms and symbols are presented in two different contexts. Finally, we hope our examples illustrate that indirect genetic effects are often mediated by behavior and therefore should be of fundamental interest to behavioral ecologists (see also Stamps 1991; Cheverud and Moore 1994). We expect that by modeling widely studied behaviors, the utility of our approach will be apparent.

2.1 Indirect Genetic Effects of Interacting Phenotypes: Evolution by Social Selection

West-Eberhard (1979, 1983, 1984) argued that social competition—interactions among conspecifics within a social group directed toward gaining access to limited resources—can be a potent evolutionary force. Such "competition by social interaction" (West-Eberhard 1979, p. 222) incorporates social competition for resources that influence survival and includes competition for access to mates (as traditionally considered under sexual selection). Such competition should be strong, resulting in rapid divergence of behaviors that influence social interactions. Social selection is expected to influence the evolution of competitive rituals and displays (social signals), specialization of tasks, and even speciation (West-Eberhard 1979, 1983, 1984). Further, social selection is predicted to result in a runaway process of evolution, such that most characters involved in mediating social behavior should be extravagant relative to traits that do not influence social interactions (West-Eberhard 1979, 1983, 1984; Tanaka 1991, 1996).

Although there have been a number of empirical studies that support the arguments made by West-Eberhard, she did not present a formal mathematical model of her ideas. Tanaka (1991, 1996) has developed mathematical models of the selection component acting on interacting phenotypes. Griffing (1967, 1977, 1981) has offered simple population genetic models of interacting phenotypes but has not considered how these might relate to the evolution of social interactions. We have tried to formalize the genetics and evolution of phenotypes exa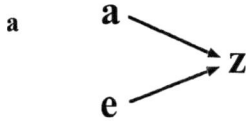

b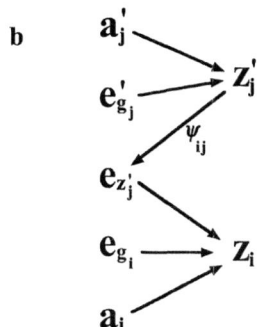

Figure 2.1 Illustration of the sources contributing to the expression of an interacting phenotype. Panel a presents the simplest case where the phenotype reflects only additive genetic (a) and environmental (e) influences. Panel b presents the interacting phenotypes model, where one phenotype (z_j') acts as a environmental factor influencing the expression of the phenotype in a different individual (z_i). Genetic and environmental factors also influence both phenotypes. The interaction effect coefficient (ψ) reflects the strength of the effect of the interaction on the phenotype (z_i).

pressed in social interactions in a quantitative genetic model that illustrates the special nature of interacting phenotypes (Moore et al. 1997). Our perspective is that of behavior as a phenotype, influenced by both genes and environments. Specifically, we hope to identify the genetic conditions under which rapid evolution of traits that influence social interactions might occur.

The central tenet of social selection is that interactions among social partners affect each other's fitness. We show that such interactions also generate indirect genetic effects because in interactions, individuals affect the phenotype of their social partner. Any trait (phenotype) has an indirect genetic effect if it is heritable and contributes an environmental effect to the phenotype of a second individual. Note that this definition does not specify any genetic relatedness among interacting individuals. Thus, such effects could just as easily (though perhaps more awkwardly) be referred to as "effects of interacting phenotypes."

The Evolution of Interacting Phenotypes:
A Quantitative Genetic Model

Phenotypes of individuals reflect both the genes they inherit from their parents and the environment that these genes experience whenever they are expressed. Mathematically, this can be expressed very simply by partitioning the contributions to the phenotype (see Falconer and Mackay 1996):

$$z = a + e, \tag{2.1}$$

where z is the phenotype, a reflects additive genetic effects, and e reflects environmental effects (figure 2.1a). Equation (2.1) describes simple Mendelian inheritance of direct genetic effects (Fisher 1918) and forms the foundation for all of quantitative genetics. Other parti-

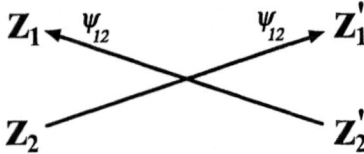

Figure 2.2 Diagram illustrating a social interaction with nonreciprocal effects. In this example the expression of one trait (e.g., aggression, z_1) is influenced by a second trait in another individual (e.g., body size, z_2'). In this example the second trait is unaffected by the interaction. The strength of the effect of the interaction on the first trait is quantified by the interaction effect coefficient (ψ_{12}).

tionings are possible; in particular, specific environmental subdivisions as well as other genetic effects such as dominance and epistasis are often examined (Falconer and Mackay 1996). It is one of these specific environmental subdivisions, the environmental effect provided by a social partner, that we are interested in here.

Consider two individuals and two phenotypes where one individual provides an environment that contributes to the expression of a trait in the other individual. For example, the level of aggression expressed by one individual is often modified by the body size of the other individual (Huntingford and Turner 1987; see figure 2.2). We can express this in symbols with different phenotypes noted by subscripts and different individuals indicated by the presence or absence of a prime (i.e., two traits z_i and z_j in the focal individual and z_i' and z_j' in the social partner). Following a quantitative genetic approach, we can model the contribution of the general environmental effect (e_{gi}) and the environment influences contributed by phenotype z_j of an interacting individual, that is, the environment ($e_{z'_j}$), as well as a direct additive genetic effect (a_i), on the behavioral (interacting) phenotype in the first individual (z_i):

$$z_i = a_i + e_{gi} + e_{z'_j}, \tag{2.2}$$

Because the environmental trait e_{zj} is determined by the phenotype of another individual, we can substitute the definition for the phenotype z_j of the second individual:

$$z_i = a_i + e_{gi} + \psi_{ij} z_j', \tag{2.3}$$

where $\psi_{ij} z_j'$ is the effect of phenotype z_j' in another individual on the expression of z_i in the first individual. The interaction effect coefficient, ψ_{ij}, describes the extent that z_i changes as a result of an interaction with z_j' (figure 2.1b).

The preceding assumed that a z_i phenotype is only expressed in the context of an interaction with a z_j' phenotype (though the different phenotypes can take on any value). However, we do not need to assume anything about the importance of the interaction. The interaction effect coefficient ψ_{ij} describes the strength of the effect that the phenotype of another (z_j') has on determining the phenotype of interest (z_i). In practice, ψ_{ij} describes the amount z_i is changed from its expected expression in the absence of an interaction. The interaction effect coefficient ψ_{ij} can take on any value from -1 to 1. Thus, ψ_{ij} is analogous to m, the maternal effect coefficient (Kirkpatrick and Lande 1989), and would be equivalent if the interaction occurred between a mother and her offspring. Finally, we also assume here that ψ is a population parameter that doesn't vary among individuals and doesn't evolve. Clearly, this is a simplifying assumption and is unlikely to be true for all interactions. The implications of this assumption are being explored.

In equation (2.3) we have defined a phenotype as a combination of the contributions of genetics, environments, and the influences of the social environment provided by the phenotype of another. However, the phenotype of the other individual also reflects genetic and environmental components. We can therefore further substitute into equation (2.3):

$$z_i = a_i + e_{g_i} + \psi_{ij}(a'_j + e'_{gj}),   \tag{2.4}$$

Note that in this example, we have assumed that the environmental trait (z_j) is not changed as a result of the interaction. This assumption allows us to illustrate the role of indirect genetic effects in the simplest example. We can easily relax this assumption, and Moore et al. (1997) analyze interactions with reciprocal effects (i.e., where both traits act as a behavior and an environment).

Given our definition of an interacting phenotype (eq. [2.4]), we can now derive the mean values for the characters. Taking the expectation, the mean of the behavioral trait is

$$\bar{z}_i = \bar{a}_i + \psi_{ij}\bar{a}_j,   \tag{2.5}$$

and for the environmental trait, which is unaffected by the interaction,

$$\bar{z}_j = \bar{a}_j.   \tag{2.5b}$$

From these equations we can define variance components so that we can evaluate the potential for evolutionary change in a population. Following standard quantitative genetic theory, general environmental effects are assumed to be independent and normally distributed with a mean of zero and variance E. Further, assuming that environmental effects are independent of the additive genetic effect, the overall phenotypic variance, P, is simply the sum of the additive genetic variance (G) and environmental variance (E). However, for the interacting phenotype, indirect genetic effects now contribute to P (Moore et al. 1997). In the population, then, variation among individuals expressing the behavioral trait reflects not only the genetic variation contributing to the trait itself, but also the indirect genetic influences of the environmental trait:

$$P_{ii} = G_{ii} + E_{ii} + \psi_{ij}(G_{ij} + E_{jj})   \tag{2.6}$$

For all further analyses we drop the environmental variation reflecting undescribed (random) influences (E_{ii} and E_{jj}), as these do not contribute to evolutionary responses to selection.

The evolution of quantitative characters is usually described as selection acting on the covariance between the breeding value of a trait and the phenotypic value for that trait (Falconer and Mackay 1996). For interacting phenotypes, this term is made more complicated because of the contributions of indirect genetic effects (for an example of how indirect genetic effects influence this covariance when relatives are involved, see Cheverud and Moore 1994). The response to direct selection on z_i is

$$\Delta\bar{z}_i = (G_{ii} + \psi_{ij}G_{jj})\beta_i,   \tag{2.7}$$

where G_{ii} is the genetic variance for z_i and G_{ij} is the genetic covariance between the traits. Selection is the gradient defined by $\beta_i = s_i/P_{ii}$, where P_{ii} is the phenotypic variance, s_i is the selection differential, and β_i is the regression of fitness on the phenotypic value (Lande and Arnold 1983; Brodie et al. 1995). Note, however, that the translation of selection during interactions into a simple selection gradient (β) is not always straightforward because interactions change the phenotypic variance-covariance structure and may in fact generate covariances among individuals participating in interactions (Wolf et al. in preparation). In addition, because interactions themselves can generate selection, the selection coefficients may contain components of social selection that are generated by the fitness effects of behaviors expressed in social partners during interactions. For example, aggression expressed by an opponent may affect the fitness of an individual. An analogous situation occurs when there is maternal selection, where a maternal trait directly affects the fitness of the offspring (e.g., nest defense). A result similar to equation (2.7), without an interaction coefficient, is presented by Griffing (1967, 1977, 1981).

Of course, equation (2.7) only considers selection acting directly on z_i. Even in traits without indirect genetic effects, changes in z_j cause changes in z_i whenever there is a genetic covariance between the two traits (Lande 1979). Therefore, to understand all possible evolutionary responses, we have to consider correlated responses to selection (Lande 1979; Falconer and Mackay 1996), that is, direct and indirect selection. A full equation, considering all forces of selection and complete inheritance, describes the response to both direct and indirect selection (direct and correlated responses to selection):

$$\Delta \bar{z}_i = [(G_{ii} + \psi_{ij}G_{ij})\beta_i + (G_{ij} + \psi_{ij}G_{jj})\beta_j] \tag{2.8}$$

For our model, correlated responses also include changes in the mean environmental value described by the indirect genetic influences.

The Evolution of Interacting Phenotypes: Interpreting the Model

Interacting phenotypes differ from other traits because they are determined in part by an environment that can evolve, that is, traits of other individuals. The evolution of the environmental component is transmitted through the indirect genetic effects into phenotypic change in the focal trait. These indirect genetic effects result in evolutionary consequences that would not be obvious from any other perspective.

We can highlight these consequences by considering how direct and indirect genetic effects contribute separately to an evolutionary response to selection. To illustrate, we rearrange equation (2.8):

$$\Delta \bar{z}_i = (G_{ii}\beta_i + G_{ij}\beta_j) + \psi_{ij}(G_{ij}\beta_i + G_{jj}\beta_j) \tag{2.9}$$

The first part of this equation now reflects the contribution of direct genetic effects and is the standard quantitative genetic equation for phenotypic change (Lande 1979; Lande and Arnold 1983). The second part reflects the role of indirect genetic effects. It is now clear that whenever there is an effect of the interaction (i.e., ψ_{ij} is nonzero) the evolutionary response to selection will be altered because of indirect genetic effects.

The first major difference between evolution of traits influenced by indirect genetic effects and the evolution of typical traits is that the relative rate of phenotypic change is affected by the interaction. This can occur because in addition to causing phenotypic change by acting through the genetic variance, direct selection on z_i (β_i) is now also filtered through the genetic covariance between the two traits. In this example, selection can have up to twice the effect that might otherwise be seen. Whenever there are reciprocal interactions between traits, the increased response to selection can be even more dramatic (Moore et al. 1997).

There is some justification for expecting a genetic covariance between the interacting behavior and the environmental trait at the population level. If there are combinations of genes that are more fit, then linkage is expected (see Wade, this volume). In genetically structured populations, selection will create a positive genetic covariance between the genes for the behavioral trait and the genes for the environmental trait, assuming that groups with the highest frequency of the environmental trait will have the highest fitness as a result of the behavior. In a nonstructured population, pleiotropy would be required for a positive genetic covariance. Whenever the covariance is positive, evolutionary changes in the behavioral trait will occur much more rapidly. The potential exists for social selection to be one mechanism by which interactions among individuals cause the expression of "adaptive" responses in a social partner. Indirect genetic effects can hasten adaptation.

The second consequence of indirect genetic effects is even more unexpected. Equation (2.9) shows that z_i can respond to selection even with little or no genetic variation underlying that behavior. Selection on z_j, through the genetic variance underlying the environmental trait G_{jj}, results in a phenotypic change in (Δz_i) whenever the interaction effect coefficient is nonzero. Even when the first half of equation (2.9) is zero (because of no genetic variance underlying z_i), phenotypic change can still occur because of the role of z_j as an environmental effect. An analogous result can result from maternal effects. A genetically variable maternal trait, such as maternal performance, can create variation in an offspring trait such as body size. Thus, even if the offspring trait has no additive genetic variance, body size evolves whenever maternal performance evolves.

Our model suggests that indirect genetic effects are likely to be important whenever social interactions affect the expression of a trait. Rates of evolutionary change are much different for interacting phenotypes except when $\psi_{ij} = 0$. This result provides a genetic context for West-Eberhard's insights stimulated by a consideration of social selection. West-Eberhard suggested that social selection occurring in social interactions should result in greater evolutionary changes in behavior than might be expected for other, nonbehavioral, traits. We find that evolution can also be much greater as a result of the unique effects of inheritance on social behavior. Of course, depending on the sign of the genetic covariance, evolution can also be slowed. However, as we have argued above, there is some theoretical justification for expecting positive covariances between interacting phenotypes.

The Evolution of Interacting Phenotypes: Empirical Support

Empirical evidence, based on artificial selection studies, supports our prediction that behaviors involved in interactions can respond quickly to selection. A selected list of artificial selection experiments conducted on *Drosophila melanogaster* shows that artificial directional selection has resulted in strong responses in aggression and response (Hemmat and Eggleston 1988), mating speed (Manning 1961, 1963), courtship behavior (Welbergen and van Dijken 1992), and territoriality (Hoffmann 1994). More recently, an elegant study on *D. melanogaster* by Rice (1996) has shown that male and female phenotypes related to sperm competition are likely to coevolve rapidly because members of each sex act as both a phenotype and an environment. Rapid changes in behavior involved in social interactions, such as aggression and social dominance, in response to artificial selection have been found in domestic chickens (*Gallus domesticus*), paradise fish (*Macropodus opercularis*), sticklebacks (*Gasterosteus aculeatus*), and cockroaches (*Nauphoeta cinerea*; Guhl et al. 1960; Craig et al. 1965; Francis 1984; Bakker 1986).

The model we have presented in this section is purposefully simplistic to illustrate the value of considering indirect genetic effects. Nonetheless, our model is a general one and easily modified to accommodate different or special cases. Although we incorporate only directional selection here, Tanaka (1996) presents models of more complicated social selection that are easily incorporated into our model (Wolf et al. in preparation). More complicated models of selection and more focused analyses of the number of traits and level of the effect from each are considered elsewhere (Moore et al. 1997). While we assume that the interacting individuals are unrelated, Cheverud (1984) and Lynch (1987) present models of interactions among relatives in a maternal effects framework. Finally, although the model presented here considers only a single phenotype, we can expand our model to a multivariate version (Moore et al. 1997). Accounting for sex-limited expression of a behavior is also easily accomplished (Cheverud and Moore 1994; Moore et al. 1997).

2.2 Indirect Genetic Effects and Parental Investment: The Evolution of Indicator Traits by Sexual Selection

One of the areas where indirect genetic effects have been most extensively studied is the effect of parental care (typically mother's) on offspring (Cheverud and Moore 1994). Parental care has also been extensively studied from the perspective of fitness effects and the type of selection involved, particularly sexual selection (Clutton-Brock 1991; Andersson 1994). Surprisingly, however, there has been no unified consideration of sexual selection, parental care, and indirect genetic effects. Here we present our model of sexual selection where females discriminate among males based on the parental investment that they provide to their offspring. Again, we consider only male parental care and ignore the nongenetic contributions of the mother to her offspring. This is done to highlight the need to consider all potential sources of indirect genetic effects. The contributions of indirect genetic effects of mothers are incorporated elsewhere and generate similar conclusions (Wolf et al. 1997).

Signals of male parental quality are generally considered under the rubric of indicator traits. The evolution of indicator traits has been modeled from game theoretic (e.g., Michod and Hasson 1990; Grafen 1991; Grafen and Johnstone 1993; Krakauer and Johnstone 1995), population genetic (Hoelzer 1989), and quantitative genetic (Price et al. 1993) frameworks. However, previous models either ignored genetics or assumed that either the parental investment or the indicator trait itself was not influenced by genetic variation. We adopt a quantitative genetic approach and use many of the equations developed by Kirkpatrick (1985) and Price et al. (1993) but allow all of the traits, rather than a subset, to be under genetic influences. Thus, in our model, indirect genetic effects can occur and all traits (rather than just a subset) can evolve—a critical facet when the environment provided by parents is one of the traits of interest.

Mate Choice for Indicator Traits and Parental Investment: The Model

In this model we consider three phenotypes (figure 2.3a): the potential indicator trait of the male (z_O), the trait of parental investment provided by the male (z_F), and the mate choice trait expressed by the female (z_y). We specifically follow the evolution of the indicator trait in the

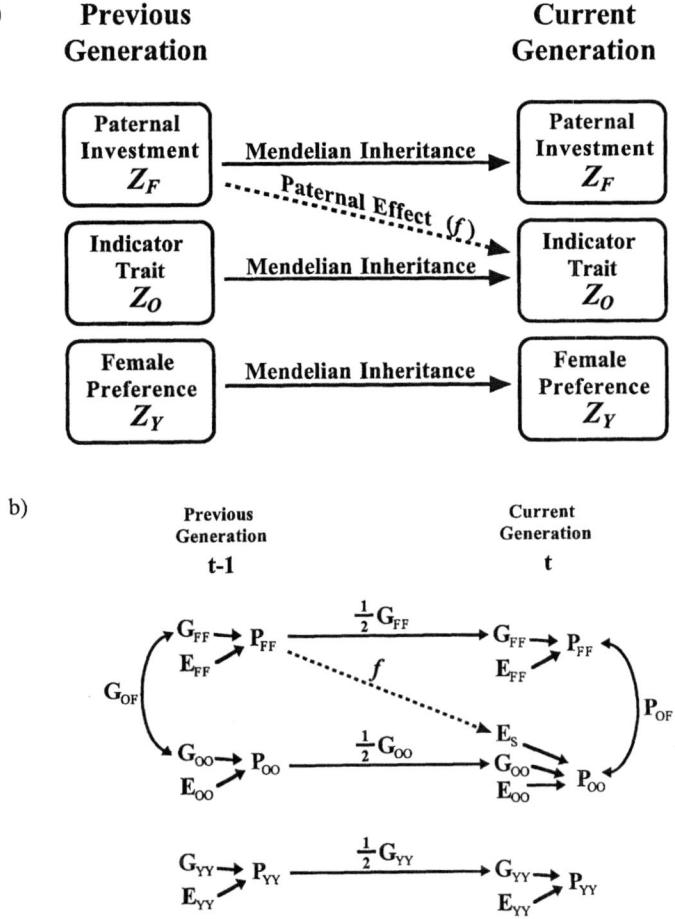

Figure 2.3 Illustration (a) and path diagram (b) of parental and offspring traits across two generations. In panel a, a verbal description of our model of male parental investment, an indicator trait, and female mate choice is presented. In panel b, paths indicating genetic (G) and environmental (E) variance components contributing to phenotypic variation (P) of both parents and offspring. Three traits are presented: male parental investment (subscript F), indicator trait (subscript O), and female mate choice (subscript Y). The strength of the effect of the parental investment on the indicator trait is indicated by f. Across generations, the dashed line indicates indirect genetic effects while solid lines indicate direct (Mendelian) inheritance.

offspring when male parental investment acts as an environmental effect on this trait. The offspring generation is indicated by t, and the parental generation is indicated by $t-1$. Following equation (2.1), the traits that are not phenotypically influenced by parental investment can be modeled (figure 2.3b) as

$$z_Y = a_Y + e_Y, \tag{2.10a}$$

$$z_F = a_F + e_F, \tag{2.10b}$$

while the indicator trait, its expression affected by the level of parental investment by the father in the previous generation (Kirkpatrick and Lande 1989), is

$$z_{O_t} = a_{O_t} + e_{O_t} + f z_{F*(t-1)}. \tag{2.10c}$$

The new term in this equation is $f z_{F*(t-1)}$, which represents the effect of the father's parental investment on the offspring's indicator trait ($z_{F*(t-1)}$) weighted by the paternal effect coefficient (f). This coefficient, f, is a factor that describes the degree to which the expression of offspring traits depends on parental investment (analogous to our earlier interaction effect coefficient, ψ, and to m, the strength of the maternal effect on the offspring phenotype). The asterisk (*) indicates that the males that affect the next generation are a subset of the original males—those males that survive and are chosen as mates. Despite appearing considerably more complicated, equation (2.10c) is completely analogous to equation (2.3) in that the phenotype of interest is influenced by additive genetic effects (a_O), environmental effects (e_O), and the environment provided by other individuals, in this case from a previous generation (in this case, $f z_{F*(t-1)}$). The major difference between equations (2.3) and (2.10) is that in the latter the individuals reside in different generations and are related by ½, and selection acts on individuals before they affect the next generation. Thus, only a subset of phenotypes remain in the population to affect the offspring phenotypes. Of course, $f z_{F*(t-1)}$ is also determined, in part, by additive genetic effects, thereby resulting in evolutionary dynamics typical of maternal effects models and others that include indirect genetic effects (see Kirkpatrick and Lande 1989; Cheverud and Moore 1994).

We can expand equations (2.10a–c) to determine the patterns of genetic and environmental variances and covariances that produce phenotypic distributions. However, when we examine the relationship between the indicator trait (which is influenced by parental investment) and any other trait, additional terms result. The total phenotypic variance in the indicator trait (P_{OO}) is

$$P_{OO} = G_{OO} + f G_{OF} + f^2 G_{FF} \tag{2.11}$$

(see Kirkpatrick and Lande 1989). The phenotypic covariance between parental investment and the indicator trait is

$$P_{OF} = G_{OF} + \frac{f}{2} G_{FF}. \tag{2.12}$$

Thus, both the phenotypic variation in the indicator trait and the phenotypic covariance between the indicator trait and parental investment are affected by indirect genetic effects. The implications of the effects on phenotypic variances and covariances bewteen traits and investment are discussed below. Other phenotypic covariances are unaffected by parental investment (and therefore indirect genetic effects) and can be derived from the paths presented in figure 2.3b (Wolf et al. 1997).

Again, following standard quantitative genetic assumptions, we can evaluate short-term evolutionary responses to selection on the male parental investment and the female mating preference as

$$\Delta \bar{z}_F = \tfrac{1}{2} [G_{FF}\beta_F + (G_{FO} + \tfrac{f}{2} G_{FF})\beta_O + G_{FY}\beta_Y], \tag{2.13a}$$

$$\Delta \bar{z}_Y = \tfrac{1}{2} [G_{YY}\beta_Y + (G_{YO} + \tfrac{f}{2} G_{YF})\beta_O + G_{YF}\beta_F]. \tag{2.13b}$$

The ½ in front of each equation reflects the sex-limited expression of the traits (z_O, z_F only in males; z_Y only in females); that is, only half of the population experiences this selection. The first term inside the brackets in each equation is direct selection acting on the heritable variation in the trait, and all of the subsequent terms define correlated responses due to selection acting on genetically correlated traits.

The expression for the response of the indicator trait to selection is somewhat more complicated, typical of traits influenced by indirect genetic effects (Kirkpatrick and Lande 1989, 1992; Lande and Kirkpatrick 1990):

$$\Delta z_O = \tfrac{1}{2}[[(G_{OO} + \tfrac{f}{2}G_{OF})\beta_O] + [G_{OF}\beta_F + G_{OY}\beta_Y]+$$
$$[f\Delta\bar{z}_{F(t-1)} + fP_{FF}(\beta_{F(t)} - \beta_{F(t-1)}) + fP_{OF}(\beta_{O(t)} - (\beta_{O(t-1)}))]]$$

(2.13c)

The first part of this equation (the terms inside the first inner brackets) reflects selection acting directly on the indicator trait. Next (the terms inside the second inner brackets), we have the effects of selection acting on genetically correlated traits. The final part of the equation (the terms inside the third inner brackets) describes the response to selection acting on the parental investment trait. Selection experienced in the previous generation (β_{t-1}) can alter the paternally inherited component. Selection on parental investment results in a genetically correlated response in the indicator and changes the magnitude of the paternally inherited portion of the trait. This occurs because of the environmental effect of parental investment. Our model thereby accounts for changes in phenotypic variances (P) resulting from changes in selection on parental investment (z_F) or the indicator (z_O). Note that if selection does not change from one generation to the next ($\beta_t - \beta_{t-1} = 0$), we are left with a pleasingly simple additional effect of male parental investment ($f\Delta\bar{z}_{F(t-1)}$), a change in the paternally inherited portion.

Mate Choice for Indicator Traits and Parental Investment: The Origination of Phenotypic Correlations

Analysis of this model provides several insights. First, we would expect indicator traits to be phenotypically variable (eq. [2.11]) when there is variation in male parental quality. Note that a lack of genetic variation underlying the indicator trait does not preclude heritable phenotypic variation because the total variance in the indicator trait is also influenced by the effects of the paternal investment (f) and indirect genetic effects (G_{FF}). Thus, additional contributions to variation exist for the indicator trait. Second, indirect genetic effects further influence phenotypic correlations. In the model we present here, there can be a phenotypic correlation between the indicator trait and parental investment simply because of, and genetic variation in, the level of investment (eq. [2.12]).

One of the more important consequences of indirect genetic effects is that mate choice based on a male indicator trait phenotypically correlated with male parental investment results in higher fitness for females. A number of conditions will lead to such a phenotypic correlation (eq. [2.12]). As expected, any genetic covariance between the two traits should result in an "honest" phenotypic correlation, that is, a trait that accurately reflects the level of parental investment that will be given. A positive correlation indicates that males with larger values of the indicator trait give more investment. Females should choose males with larger indicator traits. A negative correlation honestly indicates that males with smaller values of the indicator trait give greater investment. Females should therefore choose males with smaller indicator traits. The stronger the genetic correlation, the more "honest" the signal.

A second term, $\frac{f}{2} G_{FF}$, can also result in a phenotypic correlation. Whenever the expression of a indicator in a male is sensitive to the amount of heritable parental investment provided by his father, this term will be nonzero. The direct and indirect genetic effects set up by parental investment lead to adaptive evolution of mate choice and to honest signals.

Fisher (1915) described three phases of sexual selection: origination, elaboration, and equilibrium. In the first phase, natural selection and sexual selection work in concert, so the trait preferred by females is a "a fairly good index of natural superiority" and "associated with natural well-being" of the male (Fisher 1915, p. 187). In the second phase, sexual selection and natural selection become uncoupled, and sexual selection leads to elaboration of the trait beyond (or even counter to) that purely reflecting a natural selection advantage. In the third phase, sexual selection is eventually countered and eventually halted by natural selection. Our model provides conditions for Fisher's first phase of sexual selection. The eventual outcome of this process may be continued adaptive mate choice (Grafen 1991) or may become the runaway process envisioned by Fisher (1958; Lande 1981; Kirkpatrick 1982). To determine the outcome of evolution we would need to evaluate the equilibrium conditions for the traits.

Mate Choice for Indicator Traits and Parental Investment: Evolutionary Trajectories of Sexually Selected Traits with Indirect Genetic Effects

Indirect genetic effects resulting from the influence of parental investment on the indicator trait set up unusual evolutionary dynamics for these traits. Traits affected by paternal inheritance can evolve even when there is no selection acting directly on the trait or any genetically correlated traits (eqs. [2.13a–c]). Therefore, the net change in the mean value of the indicator trait will reflect both direct selection and correlated selection acting on parental investment in both the previous and the present generations (eq. [2.13c]). Responses to such selection can result in nonintuitive evolutionary dynamics such as time lags, evolution away from the optima, and unpredictable equilibria (Kirkpatrick and Lande 1989; Cheverud and Moore 1994).

To understand the evolution of the three traits, we have to consider the forms of selection acting on each trait as well as how the traits are inherited. Here we present a simple description of how selection might occur and leave out the mathematical details of fitness. To consider selection, we examine how each trait influences the fitness of an individual, given the value of the trait that is expressed. For each character, then, we have to consider how a given trait might influence viability, fecundity, and mating success. We do this so we can illustrate possible evolutionary outcomes.

Rather than experiencing simple directional selection, the indicator trait of males exposes them to both viability selection and sexual selection resulting from mate choice. Viability selection can be modeled as a Gaussian function (Kirkpatrick 1985; Price et al. 1993), while sexual selection is frequency dependent following the relative preference model of Lande (1981).

Male parental investment in our model affects parental fecundity and offspring survival. We assume that fecundity is an increasing function of parental investment. We further assume that there are direct costs to parents associated with increases in parental investment and therefore, as a result of trade-offs, there is net stabilizing selection on parental invest-

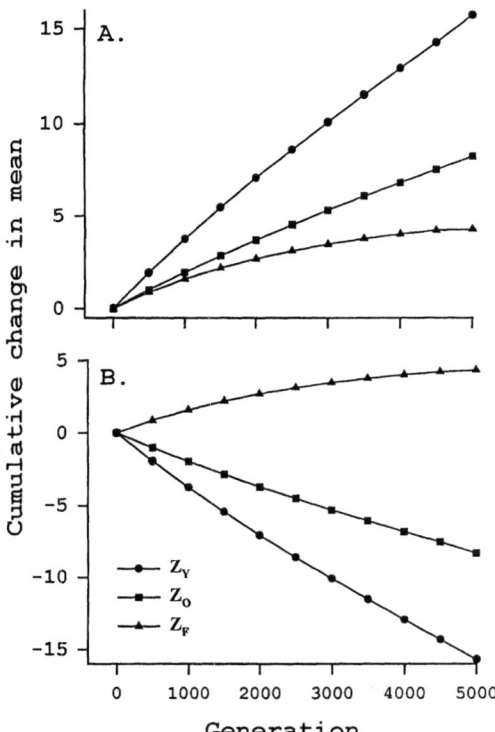

Figure 2.4 Joint evolutionary trajectories of a male indicator trait (z_O), parental investment (z_F), and mate choice for the indicator trait (z_Y). The fitness relationships and selection for each graph are as described in section 2.2. For all traits the y-axis is scaled so that the optima are arbitrarily defined to be 0. Evolution is initiated with absolute female preferences for the most common male phenotype. In panel A, genetic conditions are $f = 1$, $G_{OF} = 0.5$, $G_{FF} = G_{OO} = G_{YY} = 1$. All unspecified genetic correlations equal zero. In panel B, $f = -1$ and $G_{OF} = -0.5$. Under both negative and positive genetic covariances, parental investment continues to evolve to a value greater than the initially predicted optimum.

ment resulting from viability differences. We assume that there is no cost (i.e., no viability or mating costs) to female choice. However, selection acts on female preferences (β_Y) because the preference determines what kind of male a female mates with and males vary in the amount or quality of parental investment they provide, thereby creating variation in female fecundity (Kirkpatrick 1985). Female fitness can be defined by the probability of mating with a certain type of male multiplied by the fecundity resulting from that mating, summed over all types of males (Kirkpatrick 1985). Given these definitions of fitness and selection, we can finally define the selection coefficients (S) associated with each trait. Once these have been defined, selection gradients can be evaluated ($\beta_i = S_i/P_i$).

In our model, we find that the mean value of the indicator trait and the mean value of the female preference continue to evolve away from (or beyond) the optimum under all conditions of inheritance (figure 2.4), regardless of the starting value for the traits or the assumed optimum. This phenomenon has an intuitive explanation: while there is an optimal amount of parental investment for a male to give, females have higher fitness if they can mate with males that give more investment. The chosen males have a higher mating success that counteracts the fitness loss due to providing more than the optimal amount of investment. Thus, whenever parental investment has an effect on the expression of an indicator trait (i.e., $f \neq 0$), or there is a genetic correlation between parental investment and the indicator trait ($G_{OF} \neq 0$), both the indicator trait and the mating preference for that indicator trait will continue to evolve. Whenever f and G_{OF} have the same sign, P_{OF} will be large, resulting in more rapid evolution.

An additional outcome of our model is that the mate preference evolves more quickly than the indicator trait itself. This result occurs because we have assumed that female fitness is positively influenced by choosing males that provide increased parental investment. However, the female preference evolves away from an optimum at a faster rate than the male indicator trait (figure 2.4). This occurs because, unlike parental investment, mate choice does not experience counter or stabilizing selection.

The extent that $\bar{z}_O$ lags behind $\bar{z}_Y$ also depends on the sign of f and the genetic correlations. A nonzero genetic correlation, or phenotypic effect (f), between parental investment and the indicator causes the evolution of the indicator to slow as parental investment stops evolving (note that this is always above the specified optimum). Thus, populations with well-developed mating preferences, but less well-developed male indicator traits, should exist. As in all models where the females benefit by mate choice, the more extreme the preference, the greater the benefits to the female whenever there is direct selection on mating preferences. Thus, our model presents a potential quantitative genetic explanation for the existence of sensory biases.

As we note above, even though the expression of parental investment ($\bar{z}_F$) reaches a stable value, this is at a point greater than our specified optimum. Parental investment is not at an optimum because even when $\beta_F = 0$ (i.e., when is at the optimum), there is continued selection on the indicator trait ($\beta_O \neq 0$). The point at which the correlated selection pushing above the optimum is balanced by negative selection pulling back toward the optimum depends on the relative strength of responses to correlated and direct selection. Unlike standard multivariate selection responses, in this case a correlated response to selection does not rely solely on a genetic correlation. As long as there is genetic variation underlying parental investment ($G_{FF} \neq 0$), then $\bar{z}_F$ will evolve due to a correlated response to selection on the indicator trait (eq. [2.13a]).

Predicting even moderately long-term evolution from our analyses requires a number of assumptions including a constant genetic variance/covariance matrices, weak selection, and no selection on female preferences; many of these have been questioned and suggested to be unrealistic (Barton and Turelli 1989; Kirkpatrick and Ryan 1991). Arguments about equilibria should therefore be considered tenuous at best. In the long term, selection could begin to act against extreme female preferences, or a "true" runaway process (in the sense of Lande 1981) will be initiated as the indicator trait is decoupled from the level of parental investment that is provided. In any case, like other models of maternal effects (Cheverud 1984; Kirkpatrick and Lande 1989) and some models of sexual selection (Price et al. 1993), our model suggests that the traits could ultimately be expressed at points that differ from a predicted optimum. This result occurs because it may not be possible to simultaneously maximize both male and female fitness.

Mate Choice for Indicator Traits and Parental Investment: Empirical Support

Studies of parental care are currently driven as much by intuition as by theory (Westneat and Sargent 1996). Nonetheless, there are abundant empirical studies demonstrating parental care and parental investment in a wide variety of taxa (Clutton-Brock 1991; Reynolds 1996; Westneat and Sargent 1996). It is not surprising, then, to find empirical support for some components of our model. A phenotypic correlation between an indicator trait and the level of parental investment, plus mate choice based on that indicator trait, has been demonstrated in several species of birds and fish (Eckert and Weatherhead 1987; Hill 1991; Knapp 1991; Rogers 1995). What we lack are genetic studies of parental investment and indicator traits (but see Sakaluk and Smith 1988 for a study on the genetics of parental investment in a cricket).

Summary

We present two quantitative genetic models of phenotypic evolution, both of which provide insight into the evolutionary process and suggest that unusually rapid evolution of characteristics that influence interactions is possible. Our goal is to provide examples of the sorts of traits that might be commonly influenced by indirect genetic effects. We purposefully ignore the contributions of mothers to their offspring, as this is the form of indirect genetic effect that has been typically considered. Instead, we focus on traits expressed during other interactions.

Our first model extends the perspective of quantitative genetic models of indirect genetic effects to interacting individuals that are not related. Incorporating interactions among kin has been modeled by Cheverud (1984) and Lynch (1987) and by modifying the equations presented by Cheverud and Lynch so that $r = 0$ yields a specific form of our model. Our model of interacting phenotypes provides a genetic explanation for the rapid phenotypic evolution often seen in behavioral traits. Further, our model suggests why there might be integration of behavior expressed among interacting social partners. Evaluating the interaction effect coefficient, φ, provides an insight into the adaptive nature of traits expressed in social interactions.

Our second model is based on previous quantitative genetic models of sexual selection, especially those presented by Lande (1981) and Kirkpatrick (1985). However, to include the effects of parental investment, we again consider indirect genetic effects because parental investment could be influenced by genes and is certainly itself an environmental effect on the offspring. This model provides conditions for the evolution of honest indicator traits and the first phase of Fisher's model of sexual selection.

The influence of other individuals within an environment on the expression of almost all phenotypes is expected to be pervasive (Cheverud and Moore 1994). In addition, these influences are often expected to be manifested as behaviors (Stamps 1991; Cheverud and Moore 1994); thus, the perspective of indirect genetic effects in quantitative genetics would appear to be a useful one for studies of social interactions. Previous models have focused primarily on one form of indirect genetic effect: maternal genetic effects. This may have led to the assumption that only mother/offspring (or, occasionally, father/offspring) interactions are considered by this approach. However, as we have tried to illustrate with the modes presented in this chapter, the framework of quantitative genetics in general and indirect genetic effects in particular furnishes a powerful tool that should stimulate alternative approaches to addressing questions about parent/offspring interactions, mate choice, kin discrimination, aggression, social dominance, territoriality, sexual conflict, and intraspecific communication (see also Cheverud and Moore 1994). The creativity and contributions of behavioral ecologists are therefore not simply limited to testing what might happen in evolution but can also, in the best tradition of the application of behavior to understanding evolutionary biology (e.g., Darwin, 1859, 1871; Haldane 1932; Fisher 1958; Mayr 1963), address how evolutionary changes might occur.

Acknowledgments   We appreciate our interactions with our colleagues at the University of Kentucky who so good-naturedly tolerate our challenges but do not necessarily agree with our views. Steve Arnold, Chris Boake, Jim Cheverud, and Mary Jane West-Eberhard have provided significant discussions and critical comments on our models and ideas. Chris Boake, Chris Grill, and James Wagner provided useful comments on earlier versions of the manuscript. Our research has been supported by NSF (Grants DEB-9521821, IBN-9514063, and IBN-9616203 to A.J.M.; DEB-9509295 and IBN-9600775 to E.D.B.) and state and federal Hatch support (KAES publication 96-08-147) to A.J.M. J.B.W. was supported by an NSF graduate research training grant (DGE-9355093). E.D.B. Jr., Mingus, Maisie, Alice, Eliza, Caitlin, and Kevin-Elvis provided empirical information on the (environmental, if not genetic) effects of parental investment.

References

Andersson, M. 1994. Sexual Selection. Princeton University Press, Princeton, N.J.

Arnold, S. J. 1994. Multivariate inheritance and evolution: a review of the concepts. Pp. 17–48 in C. R. B. Boake (ed.), Quantitative Genetic Studies of Behavioral Evolution. University of Chicago Press, Chicago.

Bakker, T. C. M. 1986. Aggressiveness in sticklebacks (*Gasterosteus aculeatus*): A behaviour-genetic study. Behaviour 98:1–144.

Barton, N. H., and M. Turelli. 1989. Evolutionary quantitative genetics: How little do we know? Annu. Rev. of Genet. 23:337–370.

Bernardo, J. 1996. Maternal effects in animal ecology. Am. Zool. 36:83–105.

Brodie, E. D. III, A. J. Moore, and F. J. Janzen. 1995. Visualizing and quantifying natural selection. Trends Ecol. Evol. 10:313–318.

Cheverud, J. M. 1984. Evolution by kin selection: A quantitative genetic model of maternal performance in mice. Evolution 38:766–777.

Cheverud, J. M., and A. J. Moore. 1994. Quantitative genetics and the role of the environment provided by relatives in the evolution of behavior. Pp. 67–100 in C. R. B. Boake (ed.), Quantitative Genetic Studies of Behavioral Evolution. University of Chicago Press, Chicago.

Clutton-Brock, T. 1991. The Evolution of Parental Care. Princeton University Press, Princeton, N.J.

Craig, J. V., L. L. Ortman, and A. M. Guhl. 1965. Genetic selection for social dominance ability in chickens. Anim. Behav. 13:114–131.

Darwin, C. 1859. On the Origin of Species by Means of Natural Selection. John Murray, London.

Darwin, C. 1871. The Descent of Man and Selection in Relation to Sex. John Murray, London.

Eckert, C. G., and P. J. Weatherhead. 1987. Male characteristics, parental quality and the study of mate choice in the red-winged blackbird (*Agelaius phoeniceus*). Behav. Ecol. Sociobiol. 20:35–42.

Eisen, E. J. 1967. Mating designs for estimating direct and maternal genetic variances and direct-maternal genetic covariances. Can. J. Genet. Cytol. 9:13–22.

Falconer, D. S., and T. F. Mackay. 1996. Introduction to Quantitative Genetics. 4th ed. John Wiley, New York.

Fisher, R. A. 1915. The evolution of sexual preference. Eugen. Rev. 7:184–192.

Fisher, R. A. 1918. The correlation among relatives on the supposition of Mendelian inheritance. Trans. R. Soc. Edinburgh 52:399–433.

Fisher, R. A. 1958. The Genetical Theory of Natural Selection. 2nd rev. ed. Dover, New York.

Francis, R. C. 1984. The effects of bidirectional selection for social dominance on agonistic behavior and sex ratios in the paradise fish (*Macropodus opercularis*). Behaviour 90:25–45.

Grafen, A. 1991. Modeling in behavioral ecology. Pp. 5–31 in J. R. Krebs and N. B. Davies (eds.), Behavioural Ecology: An Evolutionary Approach. 3rd ed. Blackwell Scientific, Oxford.

Grafen, A., and R. A. Johnstone. 1993. Why we need ESS signalling theory. Philos. Trans. R. Soc. Lond. (ser. B) 340:245–250.

Griffing, B. 1967. Selection in reference to biological groups. I. Individual and group selection applied to populations of unordered groups. Austral. J. Biol. Sci. 20:127–139.

Griffing, B. 1977. Selection for populations of interacting phenotypes. Pp. 413–434 in E. Pollak, O. Kempthorne, and T. B. Bailey (eds.), Proceedings of the International Conference on Quantitative Genetics. Iowa State University Press, Ames.

Griffing, B. 1981. A theory of natural selection incorporating interaction among individuals. I. The modeling process. J. Theor. Biol. 89:635–658.

Guhl, A. M., J. V. Craig, and C. D. Mueller. 1960. Selective breeding for aggressiveness in chickens. Poult. Sci. 39:970–980.

Haldane, J. B. S. 1932. The Causes of Evolution. Longman, New York.

Hemmat, M., and P. Eggleston. 1988. Competitive interactions in *Drosophila melanogaster*: Recurrent selection for aggression and response. Heredity 60:129–137.

Hill, G. E. 1991. Plumage coloration is a sexually selected indicator of male quality. Nature 350:337–339.

Hoelzer, G. A. 1989. The good parent process of sexual selection. Anim. Behav. 38:1067–1078.

Hoffmann, A. A. 1994. Genetic analysis of territoriality in *Drosophila melanogaster*. Pp. 188–205 in C. R. B. Boake, (ed.), Quantitative Genetic Studies of Behavioral Evolution. University of Chicago Press, Chicago.

Huntingford, F., and A. K. Turner. 1987. Animal Conflict. Chapman and Hall, New York.

Kirkpatrick, M. 1982. Sexual selection and the evolution of female choice. Evolution 36:1–12.

Kirkpatrick, M. 1985. Evolution of female choice and male parental investment in polygynous species: The demise of the sexy son. Am. Nat. 125:788–810.

Kirkpatrick, M., and R. Lande. 1989. The evolution of maternal characters. Evolution 43:485–503.

Kirkpatrick, M., and R. Lande. 1992. The evolution of maternal characters: Errata. Evolution 46:284.

Kirkpatrick, M., and M. J. Ryan. 1991. The evolution of mating preferences and the paradox of the lek. Nature 350:33–38.

Knapp, R. A. 1991. Courtship as an honest indicator of male parental quality in the bicolor damselfish, *Stegastes partitus*. Behav. Ecol. 2:295–300.

Krakauer, D. C., and R. A. Johnstone. 1995. The evolution of exploitation and honesty in animal communication: A model using artificial neural networks. Philos. Trans. R. Soc. Lond. (ser. B) 348:355–361.

Krebs, J. R., and N. B. Davies (eds.). 1991. Behavioural Ecology: An Evolutionary Approach 3rd ed. Blackwell Scientific, Oxford.

Lande, R. 1979. Quantitative genetic analysis of multivariate evolution applied to brain: Body size allometry. Evolution 33:402–416.

Lande, R. 1981. Models of speciation by sexual selection on polygenic traits. Proc. Natl. Acad. Sci. USA 78:3721–3725.

Lande, R., and S. J. Arnold. 1983. The measurement of selection on correlated characters. Evolution 37:1210–1226.

Lande, R., and M. Kirkpatrick. 1990. Selection response in traits with maternal inheritance. Genet. Res. (Cambridge) 55:189–197.

Lynch, M. 1987. Evolution of intrafamilial interactions. Proc. Natl. Acad. Sci. USA 84:8507–8511.

Manning, A. 1961. The effects of artificial selection for mating speed in *Drosophila melanogaster*. Anim. Behav. 9:82–92.

Manning, A. 1963. Selection for mating speed in *Drosophila melanogaster* based on the behaviour of one sex. Anim. Behav. 11:116–120.

Mayr, E. 1963. Animal Species and Evolution. Harvard University Press, Cambridge, Mass.

Michod, R. E., and O. Hasson. 1990. On the evolution of reliable indicators of fitness. Am. Nat. 135:788–808.

Moore, A. J., E. D. Brodie III, and J. B. Wolf. 1997. Interacting phenotypes and the evolutionary process: I. Direct and indirect genetic effects of social interactions. Evolution 51:1352–1362.

Price, T., D. Schluter, and N. E. Heckman. 1993. Sexual selection when the female directly benefits. Biol. J. Linnean Soc. 48:187–211.

Reynolds, J. D. 1996. Animal breeding systems. Trends Ecol. Evol. 11:68–72.

Rice, W. R. 1996. Sexually antagonistic male adaptation triggered by experimental arrest of female evolution. Nature 381:232–234.

Riska, B., J. Rutledge, and W. R. Atchley. 1985. Covariance between direct and maternal genetic effects in mice, with a model of persistent environmental influences. Genet. Res. (Cambridge) 45:287–297.

Rogers, W. 1995. Female choice predicts the best father in a biparental fish, the Midas cichlid (*Cichlasoma citrinellum*). Ethology 100:230–241.

Sakaluk, S. K., and R. L. Smith. 1988. Inheritance of male parental investment in an insect. Am. Nat. 132:594–601.

Stamps, J. A. 1991. Why evolutionary issues are reviving an interest in proximate behavioral mechanisms. Am. Zool. 31:338–348.

Tanaka, Y. 1991. The evolution of social communication systems in a subdivided population. J. Theor. Biol. 149:145–163.

Tanaka, Y. 1996. Social selection and the evolution of animal signals. Evolution 50:512–523.

Welbergen, P., and F. R. van Dijken. 1992. Asymmetric response to directional selection for licking behavior of *Drosophila melanogaster* males. Behav. Genet. 22:113–124.

West-Eberhard, M. J. 1979. Sexual selection, social competition, and evolution. Proc. Am. Philos. Soc. 123:222–234.

West-Eberhard, M. J. 1983. Sexual selection, social competition, and speciation. Q. Rev. Biol. 58:155–183.

West-Eberhard, M. J. 1984. Sexual selection, competitive communication and species-specific signals. Pp. 283–324 in T. Lewis (ed.), Insect Communication. Academic Press, New York.

Westneat, D. F., and R. C. Sargent. 1996. Sex and parenting: The effects of sexual conflict and parentage on parental strategies. Trends Ecol. Evol. 11:87–91.

Willham, R. L. 1972. The role of maternal effects in animal breeding: III. Biometrical aspects of maternal effects in animals. J. Anim. Sci. 35:1288–1293.

Wolf, J. B., A. J. Moore, and E. D. Brodie III. 1997. The evolution of indicator traits for parental quality: The role of direct and indirect genetic effects. Am. Nat. 150:639–649.

Wolf, J. B., E. D. Brodie III, and A. J. Moore. In preparation. Interacting phenotypes and the evolutionary process: II. Social selection, kin selection and group selection.

Wolf, J. B., Brodie, E. D. III, Cheverud, J. M., Moore, A. J., Wade, M. J. 1998. Evolutionary consequences of indirect genetic effects. Trends Ecol. Evol. 13:64–69.

3

Inertial Growth

Population Dynamics Based on Maternal Effects

LEV R. GINZBURG

Maternal effects can be defined as the cross-generational transmission of individual quality such that the offspring survival, growth, and fecundity are affected by the properties of the environment in which the parents lived. Individual quality may have an uneven distribution in a population, but only the transmission of the average population quality will be discussed here since this chapter examines its influence on population dynamics. Recent reviews have shown the ecological and evolutionary importance of maternal effects in insects (Mousseau and Dingle 1991), plants (Roach and Wulf 1987), and mammals (Bernardo 1996, and references therein). Maternal effects induce a time lag in the reaction of populations to environmental changes. An environmental change may cause, for instance, a decrease in abundance. This would create better conditions for the average mother since there are less of them. Maternal quality passed to daughters (maternal effect) would therefore influence the rate of reproduction in the daughter's generation. This time lag of exactly one generation creates the need to reflect the delayed effect of the previous generation's environment on the current generation (Rossiter 1996). An approach to accounting for this effect is to introduce the concept of inertia into the population dynamics models. The inertial view of population dynamics based on maternal effects and its application to cycling populations is the subject of this chapter.

Population quality as a factor in population dynamics was first considered in the pioneer work of Wellington (1957), who showed that distinct larval morphs of the western tent caterpillar *Malacosoma pluviale* differed in their survival, behavior, dispersal, and disease resistance (i.e., the morphs had a different quality) and that the proportions of the larval types in each clutch varied with maternal quality. Chitty's (1960, 1996) ideas of intrinsic population regulation related to vole population dynamics also rely on variation in individual quality. One reason why individual quality and maternal effects have been largely omitted from population dynamics is that it is often difficult to know in advance which aspects of quality to consider and measure (Rossiter 1994, 1996). It was only recently that Wellington's and Chitty's ideas on individual quality were explicitly developed into a mathematical model.

Various authors mean different things by individual quality, and the concept must be clarified here prior to formulating a model equation for maternal effect. For most biologists, individual quality generally refers to the energetic endowment given by mothers to their offspring, and this common meaning is probably sufficient in most cases. Sometimes a more general meaning is used: quality is anything that is not quantity, and it certainly has to pass across generations to constitute the basis of maternal effects. For instance, Myers et al. (1997) found that the sex ratio in gypsy moths varies significantly with population density. Sex ratio will work as a measure of individual quality in a more general sense but may not pass the test for a narrower meaning. For the theory that is the subject of this chapter, the more general sense of individual quality will be sufficient.

The simple maternal effect model assumes nonoverlapping generations and follows just the two variables N, the population abundance, and X, the average individual quality (Ginzburg and Taneyhill 1994). In the dimensionless form, only two parameters are essential for the model: R is the maximum population growth rate of individuals with very high average quality, and M is the maximum increase in quality per generation in the most favorable conditions. Both parameters are assumed to be greater than one. If this is not the case, the population will rapidly go extinct. If t stands for time in generations, the dimensionless model has the form

$$N_{t+1}=N_t \ \frac{RX_t}{1+X_t} \ , \tag{3.1a}$$

$$X_{t+1}=X_t \ \frac{M}{1+N_{t+1}} \ . \tag{3.1b}$$

Equilibrial values of the abundance, N^*, and quality, X^*, are easily determined [$N^*= M-1$, $X^*=1/(R-1)$].

Note the argument N_{t+1}, not N_t, in the second equation of the model. The assumption is that, in the absence of the maternal effect, if X_{t+1} did not depend on X_t, the quality X_{t+1} would have been fully dictated by the abundance N_{t+1} at the concurrent generation. Thus, in the absence of maternal effects, the delay is absent and substitution of equation (3.1) into equation (3.1a) would lead to an "immediate" or direct density dependence. The presence of the maternal effect makes the model fundamentally delayed density dependent, not reducible to the traditional model of direct density-dependent growth.

Let us now assume that the average quality variable, X, is not observed. This is commonly the case; the majority of population data are just univariate time series of abundance. We can exclude X from equations (3.1a,b) and rewrite them in terms of the observed variable N. The equivalent equation has the form

$$\frac{N_{t+1}}{N_t} = \frac{N_t}{N_{t-1}} \ \frac{1}{1+\frac{1}{M}\left(1-\frac{N_t}{N_{t-1}}\frac{1}{R}\right)(N_t-N^*)} \ . \tag{3.2}$$

One way of interpreting this equation is to say that due to maternal effects, the growth rate N_{t+1}/N_t depends on both N_t and N_{t-1}. This general property has been termed "lagged" or "delayed" density dependence and has been the subject of many studies (Turchin 1990; Royama 1992; Turchin and Taylor 1992).

Another way of interpreting equation (3.2) is to focus on the change in the growth rate between consecutive generations and view this change as a function of abundance and growth rate:

$$\frac{N_{t+1}}{N_t} = \frac{N_t}{N_{t-1}} \cdot F\left[N_t, \frac{N_t}{N_{t-1}}\right]$$

(3.3)

It is this second view that this chapter tries to develop.

3.1 Evidence

Fitting published time series of forest Lepidoptera to equation (3.2) predicted cycle periods (table 3.1) that matched the observed ones well. One can also observe that the period tends to be mostly determined by R with some effect of M. Typical simulated abundance curves with two different values of R ($R = 1.3$ and $R = 3.0$) are shown in figure 3.1. Both curves show the slow rise and sharper decline that are characteristic of actual cycles. Note also that the period of the cycle increases as R decreases toward 1.0. This makes intuitive sense and fits the data well (figure 3.2) but is in total contradiction to the well-known delayed logistic model (Hutchinson 1948) in which period decreases as R decreases. This disagreement has been the subject of a recent debate (Berryman 1995). It is the position of Ginzburg and Taneyhill (1995) that the observed decline in period of the cycles with increasing R is one of the arguments in favor of maternal effects as the causative mechanism of oscillatory dynamics.

Figure 3.2 also illustrates another interesting prediction of the maternal effect model. The model predicts the lower bound for the period of cyclicity of six generations (6 years in the case of Lepidoptera). The absence of Lepidoptera cycles with a period of less than 6 years argues in favor of maternal effects as the mechanism causing the cyclicity.

Another line of evidence comes from the analysis of the northern European vole time series (Inchausti and Ginzburg 1998). In this model we assumed two "breeding periods" per year with R_1 and R_2 being the maximum reproduction rates for the spring $\rightarrow$ fall and fall $\rightarrow$ spring transitions. Breeding periods do not coincide with biological generations for voles, and the model in this case is a more distant abstraction than it is for Lepidoptera. Estimated values of R_1 and R_2 based on the published time series as well as observed periods of oscillation for five vole species are shown in table 3.2. In most cases the period of the cycle is 4 years, with a few cases of 3 years in the southern part of the cyclic region and one case of closer to 5 years in the northern part. Just as in Lepidoptera, lower values of R induce longer periods. The character of the simulated vole cycles can be observed in figure 3.3. Note

Table 3.1 Predicted periods of six species of forest moths, using the maternal effects model.

Species	R	M	T actual (year)	T predicted (year)
Choristonaura fumiferana	1.34	2.0	30–38	34
Hyphantria cunea	1.64	2.75	11.8	12
Lymantria dispar	4.05	12.0	7–8	7–8
Epirrita autumnata	4.68	3.0	9	9
Bupalus piniarius	4.80	2.5	8	8
Acleris variana	5.12	5.0	7	7

Parameters (maximum rate of population growth) and (maximum rate of increase in average individual quality) were estimated by the nonlinear regression method. Periods are mostly determined by with some effect of M. This is an abbreviated version of a similar table in Ginzburg and Taneyhill (1994).

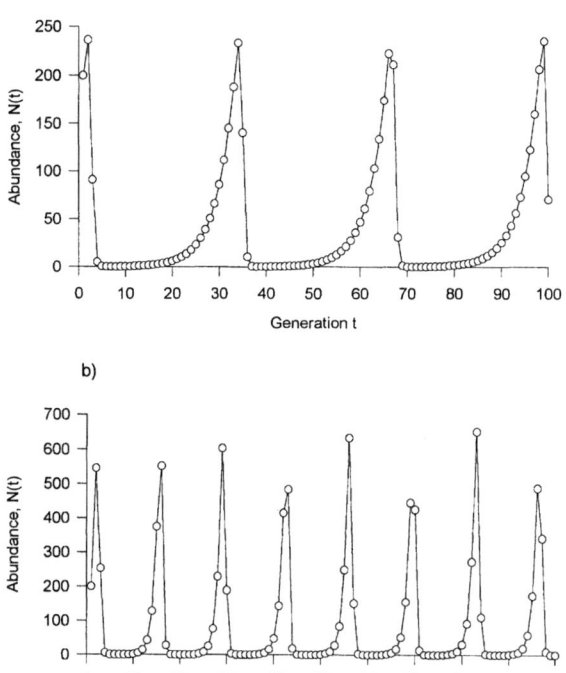

Figure 3.1 Typical behavior of the maternal effect model shown as time series plots: a, R = 1.3; b, $R = 3.0$, $M = 10$.

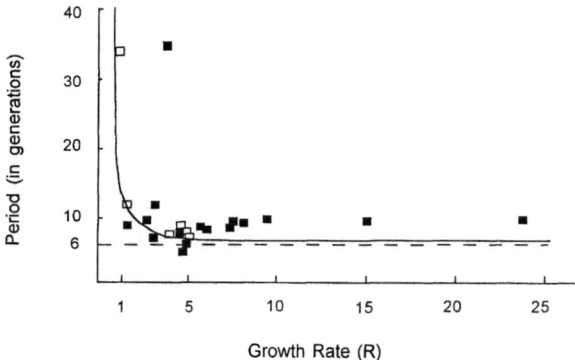

Figure 3.2 Comparison of the Berryman (1995) data (black squares), and the Ginzburg and Taneyhill (1994) data (white squares). The horizontal line corresponds to a theoretical lower limit of six generations (after Ginzburg and Taneyhill 1995).

Table 3.2 Dynamics of northern European vole time series (after Inchausti and Ginzburg 1998).

Species	Location	R_1	R_2	T actual (years)	T predicted (years)
C. rufocanus	Pallasjärvi, Finland	6.3	2.1	4–5	5
C. glareolus	Umëa, Sweden	21.8	1.1	4	4
M. agrestis	Umëa, Sweden	8.8	2.7	4	4
C. rutilus	Sotkamo, Finland	11.8	3.6	4	4
C. glareolus	Udmurt, Russia	18.4	2.7	3	3

The maximum population growth rates for the spring fall (R_1) and fall spring (R_2) transitions were calculated as the maximum of the realized growth rates (N_{fall}/N_{spring}) and (N_{spring}/N_{fall}), respectively, for a given locality and vole species. The observed periods (T actual) were determined as the dominant lag of the autocorrelogram of fall (or the trapping date closer to the fall) abundances.

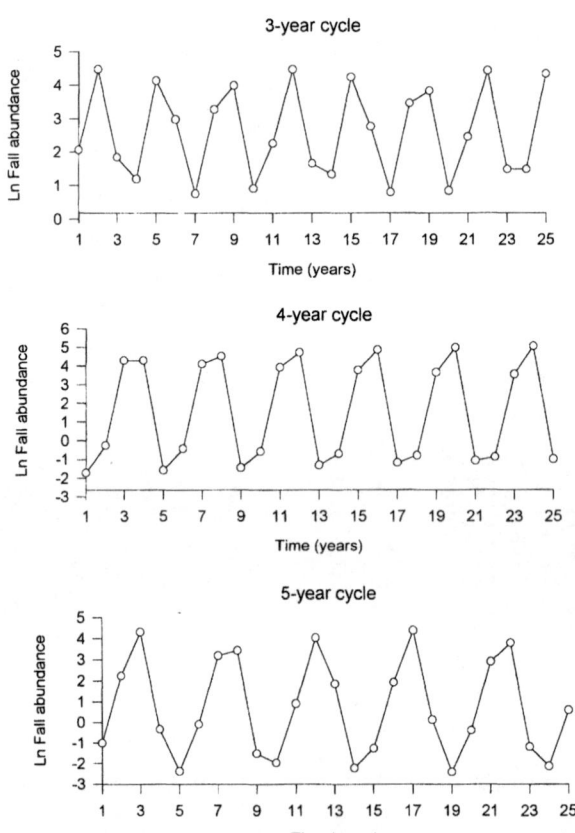

Figure 3.3 Sample trajectories for northern European vole cycles from Inchausti and Ginzburg (1998).

again the characteristic asymmetry of cycles with slower rises and sharper declines: 3 years up and 1 year down is typical of the model and agrees with observations in northern Europe.

3.2 Inertial Growth

Inertia is a tendency to continue changing in the direction and with the rate of the previous change. The proposed view captured by equation (3.3) is to consider population growth as an inertial process.

The exact parameterization of equations (3.1a,b) is not to be taken literally. The functions picked for the right sides are just a simple choice of expressions; thus, we can also view,

$$\frac{N_{t+1}}{N_t} = \frac{N_t}{N_{t-1}} \cdot F\left[N_t, \frac{N_t}{N_{t-1}}\right] \tag{3.3}$$

as a general expression of the maternal effect model. The function is defined for all positive abundances ($0 < N < \infty$) but only for a limited range of growth rates ($0 < N_t/N_{t-1} < R$). It is a general property of F that it declines as a function of N_t and that $F \to 1$ when $N_t/N_{t-1} \to R$. The latter reflects the impossibility of attaining growth rates exceeding R. This limitation is explicit in equation (3.1) but only implicit in (3.3).

Contrast the traditional view of population growth, the "kinetic" view (due to its focus on the growth rate), with the proposed, "inertial" view (due to its focus on the change in the growth rate). According to the traditional view, the Malthusian growth equation is written as $N_{t+1}/N_t = R$, and density dependence is introduced as the dependence of growth rate on population density,

$$\frac{N_{t+1}}{N_t} = F(N_t) \tag{3.4}$$

According to the inertial view, the Malthusian growth equation has the form $N_{t+1}/N_t = N_t/N_{t-1}$. Equation (3.4) says that populations grow in the next generation at the same rate as they grew in the previous one. It uses two previous population abundance points in time instead of one, has no parameters, and requires two initial conditions, the abundance and the growth rate.

The concept of density dependence has to be redefined, and the concept of growth dependence has to be defined according to the proposed view, with the dependence of F on N_t in equation (3.3) to be called "density dependence" and the dependence of F on N_t/N_{t-1} to be called "growth dependence." The inertial model in equation (3.3) recognizes the special role of geometric (exponential or Malthusian) growth and focuses on describing a deviation from it that is treated as an undisturbed behavior. Ecologically meaningful interactions modulated by the maternal effect lead to the inertial form of the population dynamics equation. The mechanism underlying the inertial view is a delayed reaction of population growth to environmental changes due to maternal effects.

Biologists often observe inertial phenomena in population growth whether or not they use the word "inertia" to describe what they see. For instance, declines in density in larch budmoth *Zeiraphera diniana* (Baltensweiler and Fischlin 1979) and Canadian lynx *Lynx canadensis* (Keith and Windberg 1978) have a tendency to continue for no apparent reason, even after population densities are very low. Recovery from densities in cyclic populations is often refractory: the initial rate of increase is generally slower than the expected short-term exponential expansion. This has been observed for hares (Keith 1983, 1990) and voles (Chitty 1996). Populations introduced into a new environment lacking the mortality agents of the original habitat reflect the growth rate attained in their original environment. Western tent caterpillars *Malacosoma pluviale* continued to cycle after being introduced into a new environment (Myers 1990). Chitty (1996) relates his many failures to founding of laboratory vole colonies from animals obtained during the decline phase of cycles.

3.3 Density Dependence and Growth Dependence

The phenomena such as those described in section 3.1 are better captured by an inertial equation (3.3) than by a kinetic, rate-based model (3.4). The use of an inertial model requires, however, a redefinition of what has been termed density dependence and the introduction of a new concept of growth dependence. To discuss these two concepts, population abundance can be expressed as the logarithmic deviation from its equilibrium value, N^*: $n = \ln(N/N^*)$. The traditional kinetic view of population growth can be expressed in terms of the new variable as $\Delta n = r$ for Malthusian growth and $\Delta n = g(n)$ for density-dependent growth. Here Δn is a logarithmic change in abundance, $\Delta n = n_{t+1} - n_t$, r is the per capita growth rate, and $g(n)$ is the density dependence growth function.

The proposed inertial view is $\Delta^2 n = 0$ for Malthusian growth and

$$\Delta^2 n = f(n, \Delta n) \tag{3.5}$$

for the density- and growth-dependent dynamic equation. Here $\Delta^2 n$ is the second difference, $\Delta(\Delta n) = (n_{t+1} - n_t) - (n_t - n_{t-1})$. Note that $f \to 0$ when $\Delta n \to r_{max}$, which constrains the growth rate to be smaller than its maximum value.

To see the usefulness of the revised concepts of density dependence and growth dependence, one should consider a specific paramaterization of f, but a linear approximation to equation (3.5) around the equilibrium $N = N^*$ ($n=0$) will suffice to illustrate the main points. In the case of pure density dependence (i.e., in the absence of growth dependence), equation (3.5),

$$\Delta^2 n \approx -\alpha n \tag{3.6}$$

describes regular oscillations. We can associate the parameter α with the resource sharing mechanism (Hassell and Varley 1969) by assuming the density dependence in the original variables is proportional to per capita resources, S, with a degree of consumer interference measured by ($0 \leqslant m \leqslant 1$): $F \sim S/N^m$.

Here $m=1$ if resources are perfectly shared, and $m < 1$ if the sharing is imperfect. Since $n = \ln N$, power function in N leads to proportionality in n. Thus, $\alpha = m$, and therefore $\alpha = 1$ in the case of perfect sharing and $\alpha < 1$ otherwise. In the case of perfect sharing of resources, $\alpha = 1$, equation (3.6) induces a period of oscillation equal to six generations, and larger periods if $\alpha < 1$. Thus, we see again from another point of view that the period of six generations is a good candidate for the theoretical minimum of biologically feasible periods. Figure 3.4 graphically represents density dependence in the inertial equation as the negative dependence of change in growth rate (second difference of n) on density.

The traditional concept of immediate direct density dependence arises from our more gen-

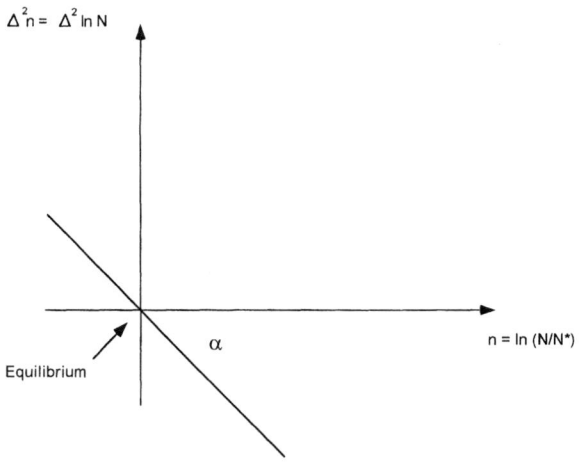

Figure 3.4 Graphical representation of the pure density-dependent component in the inertial model.

eral view, when negative growth dependence dominates on the right side of equation (3.5). If $\alpha \ll \beta$, we can write approximately

$$\Delta^2 n \approx -\beta \Delta n \text{ or } \Delta n = r - \beta n, \qquad (3.7)$$

a logistic-like equation describing direct (same generation) negative influence of abundance on the growth rate.

For the purpose of intuitive biological interpretation, we therefore have to think of the growth dependence as being caused by the direct (concurrent with growth) effects of density. Density dependence in the new sense is caused by delayed density effects of the mother's density on daughter's reproductive rate.

To include growth dependence in the linear approximation around the equilibrium, another term must be added to equation (3.6):

$$\Delta^2 n = -\alpha n \pm \beta \Delta n \qquad (3.8)$$

The equilibrium $n=0$ ($N=N^*$) is stable when $\beta < 0$ and unstable when $\beta > 0$. The population trajectory spirals either toward equilibrium or toward a limit cycle with a period greater than that of pure density dependence ($\beta > 0$), independently of the sign of β. This again argues that in the presence of growth dependence the period of the potential cycle can only exceed six generations. The qualitative character of growth dependence is shown in figure 3.5 for both $\beta > 0$ and $\beta < 0$. In both cases, change in the growth rate, $\Delta^2 n$, has to reach zero when the growth rate attains its maximum value. This essential nonlinearity of growth dependence (dashed line on figure 3.5) follows from the existence of the maximal attainable growth rate, r_{max}. While biological populations can decline unlimitedly fast (effectively instantaneously on the generation time scale), their per capita growth rate is limited by a species-specific maximum value. This limitation causes a special form of nonlinearity of the growth dependence (Ginzburg 1993) and a characteristic asymmetry of population cycles (Ginzburg and Inchausti 1997). Cyclic species typically take about 50% more time to grow than to decline, potentially

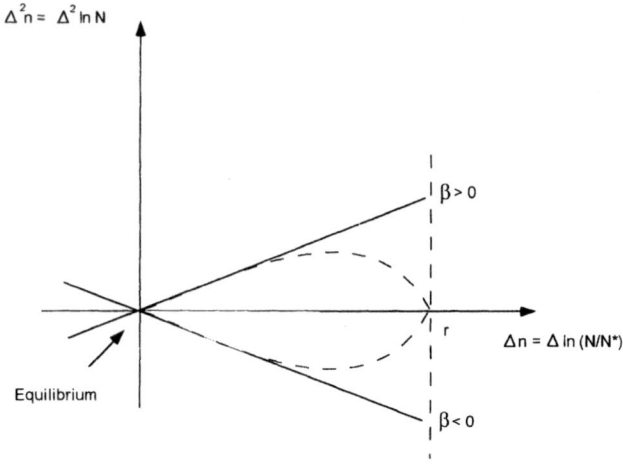

Figure 3.5 Graphical representation of growth dependence in the inertial model.

due solely to this nonlinearity induced by the existence of the upper bound in the growth rate. In the linear approximation, when $\Delta n \ll r_{max}$, this effect is unnoticeable.

Density dependence (negative effect of density on the growth) is usually thought of as responsible for equilibration of population abundance. In the inertial view this role is taken by an analogous concept of growth dependence (negative effect of growth on the change in growth). The term "density dependence," in the inertial view, is then reserved for the negative effect of the density on the change in growth, the role traditionally reserved for delayed density dependence. Thus, using new terminology (table 3.3), density dependence causes oscillations while growth dependence dampens them or, if strong enough, causes monotonic equilibration of population abundance.

It is the common strength of growth dependence in nature that may have steered us away from a more general view of population dynamics as an inertial process. In the new view, oscillations are a natural background of population dynamics, not an exception requiring a separate theory as it has been treated previously. In other words, one does not have to appeal to species interactions to explain cyclic behavior. Maternal effects compel us to generalize our approach for a single species to include both oscillatory and monotonic dynamics into one basic model.

Conclusions

To decide on the dynamic complexity of the model needed for describing population growth, consider the following thought experiment. Imagine a population equilibrated to a constant

Table 3.3 Changes in terminology required by the inertial view.

Traditional kinetic view	Density dependence	Delayed density dependence
Suggested inertial view	Growth dependence	Density dependence

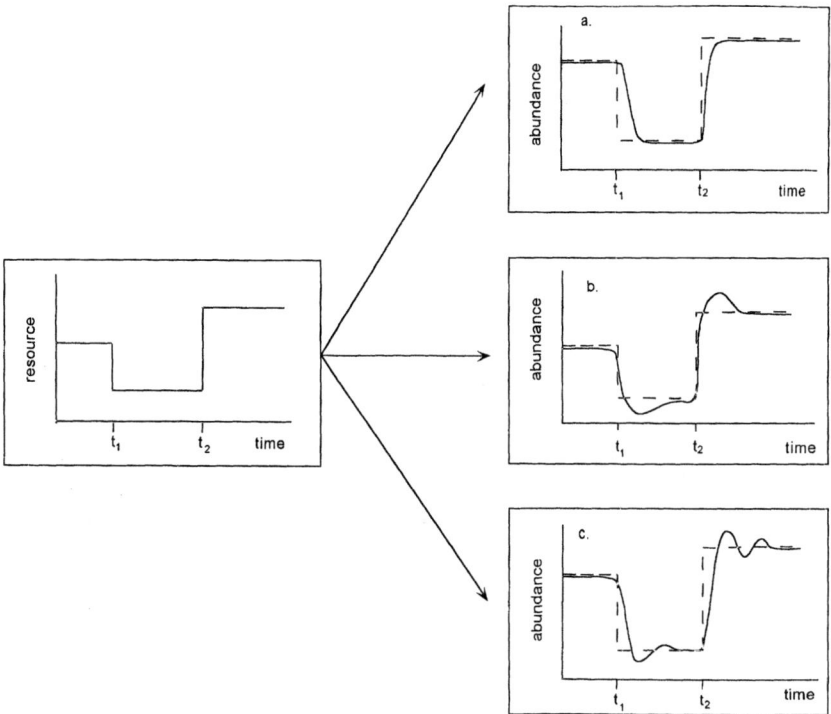

Figure 3.6 Kinetic (a) versus inertial (b and c) representation of the reaction of population abundance to environmental change. The dashed line corresponds to the abundance dictated by the resource level.

amount of a resource (figure 3.6). Assume that the resource amount had changed at time to a lower value and stayed at this value for a period of time—allowing for a new population equilibrium to be reached; then at time, resource amount increased and the population abundance followed the change.

If the only dynamic behavior allowed is the monotonic convergence to an equilibrium (figure 3.6a), the traditional kinetic approach is fully sufficient to describe this process. However, to include the possibility of overshooting the equilibrium as in figure 3.6b or oscillatory dynamics as in figure 3.6c, inertia, or delayed density dependence, must be introduced to describe the process. Whether or not such a description is needed depends on the time scale of environmental change (resource amount change in this example) compared to the generation time of the species whose dynamics is described. If environmental changes are slow (require many generations of the species in question), a kinetic approach is fully sufficient. In fact, if they are extremely slow, one can disregard population dynamics altogether and think of population abundance as following environmental change precisely. If, however, environment changes occur on a temporal scale comparable with the generation time, situations like the ones shown in figure 3.6, b and c, are more likely and delayed density dependence has to be taken into account. Of course, a monotonic approach to a new equilibrium remains a possibility. Including delayed reaction allows for dynamic complexity *and* includes simple kinetics depending on the relative strength of delayed and immediate (same generation) density dependence (density dependence and growth dependence in the new terminology).

It is the need to pay attention to temporal scales of environmental change to the generation time that focuses attention on maternal effects, the simplest mechanism of inertial growth. Other mechanisms such as predator-prey interaction may or may not be responsible for a similar delayed effect. Predator-prey interaction is more likely to be an explanation for inertial dynamics when generation times of the predator and the prey are similar, and less likely if the generation times are disparate. The passage of quality across generation has the advantage of the correct time scale even though one cannot absolutely discount species interactions as a cause of a one-generation delay. Note that in both Lepidoptera and voles, the influence of an interacting species is only hypothetical while maternal effect is an established fact. Even in the lynx-hare case where data on cycling of interacting species are unquestionable, the causality of cycling remains somewhat doubtful (Keith 1983, 1990). It seems, therefore, that before engaging into more complex species interaction modeling, we have to give serious consideration to inertial growth models for single populations in explaining complex dynamics.

The traditional kinetic view of population dynamics is so well established in ecology that some authors elevate it to the level of a fundamental principle. Nisbet and Gurney (1982) wrote:

> There is one elementary principle which underlies all population modeling, a principle which is all too easily forgotten when battling with complex mathematics, namely that the total number of individuals in a fixed region of space can only change for four reasons:
>
> (a) births,
> (b) deaths,
> (c) immigration,
> (d) emigration.
>
> In a small time interval the change in population can thus *always* be written as
>
> $$\Delta N = (B - D + 1 - E)\Delta t,$$
>
> where $B\Delta t$, $D\Delta t$ are respectively the total numbers of births and deaths during Δt, while $I\Delta t$ and $E\Delta t$ are the total numbers of individuals entering and leaving the region during the same time interval. The wide variation in the mathematical form of population equations found in the literature corresponds to different assumptions concerning the terms B, D, I, and E in this fundamental equation. (p. 19)

As described by Nisbet and Gurney, the principle sounds more like bookkeeping than an equation based on a significant biological principle. A bank account balance, water flows, and many other input-output processes are subject to this kinetic principle—which, while correct, does not refer to any phenomenon that is specifically biological. Maternal effects compel us to view population growth as inertial on a biologically meaningful temporal scale of generations. A delay in population reaction of exactly one generation suggests that it is more sensible to view ecological interactions as causing changes in growth than as causing changes in abundance (as the traditional view suggests). It may be more productive to view populations as generally growing in the direction and with the rate of a previous generation unless interactions preclude them from doing so.

Acknowledgments Pablo Inchausti provided useful discussions, Scott Ferson and Resit Akçakaya provided constructive criticism, and Matthew Spencer, Dale Taneyhill, and Karen

Kernan edited the manuscript. This is contribution 989 of the Graduate Program in Ecology and Evolution at the State University of New York at Stony Brook.

References

Baltensweiler, W., and A. Fischlin. 1988. The larch budmoth of the Alps. Pp. 332–348 in A. Berryman (ed.), Dynamics of Forest Insect Populations: Patterns, Causes, Implications. Plenum Press, New York.

Bernardo, J. 1996. Maternal effects in animal ecology. Am. Zool. 36:83–105.

Berryman, A. 1995. Population cycles: A critique of the maternal and allometric hypotheses. J. Anim. Ecol. 64:70–71.

Chitty, D. 1960. Population processes in the vole and their relevance to general theory. Can. J. Zool. 38:99–113.

Chitty, D. 1996. Do Lemmings Commit Suicide? Beautiful Hypotheses and Ugly Facts. Oxford University Press, New York.

Ginzburg, L. 1993. The logistic equation revisited: Final installment. Trends Ecol. Evol. 8:69–70.

Ginzburg, L., and P. Inchausti. 1997. Asymmetry of population cycles: Growth rate-abundance representation of hidden causes of ecological dynamics. Oikos 80:435–447.

Ginzburg, L., and D. Taneyhill. 1994. Population cycles of forest Lepidoptera: A maternal effects hypothesis. J. Anim. Ecol. 63:79–92.

Ginzburg, L., and D. Taneyhill. 1995. Higher growth rate implies shorter cycle, whatever the cause: A reply to Berryman. J. Anim. Ecol. 64:294–295.

Hassell, M., and G. Varley. 1969. New inductive population model for insect parasites and its bearing on biological control. Nature 223:1133–1137.

Hutchinson, G. E. 1948. Circular causal systems in ecology. Ann. N.Y. Acad. Sci. 50:221–246.

Inchausti, P., and L. Ginzburg. 1998. Small mammals cycles in northern Europe: Patterns and evidences for a maternal effects hypothesis. J. Anim. Ecol. 67.

Keith, L. 1983. The role of food in hare population cycles. Oikos 40:385–395.

Keith, L. 1990. Dynamics of snowshoe hare populations. Curr. Mammal. 2:119–195.

Keith, L., and L. Windberg. 1978. A demographic analysis of the snowshow hare cycle. Wildl. Monogr. 58:1–70.

Mousseau, T., and H. Dingle. 1991. Maternal effects in insect life histories. Annu. Rev. Entomol. 35:511–534.

Myers, J. 1990. Population cycles of western tent caterpillars: Experimental introduction and synchrony of populations. Ecology 71:986–995.

Myers J., G. Boettner, and J. Elkinton. In review. Maternal effects in gypsy moth: Only sex ratio varies with population density.

Nisbet, R., and W. Gurney. 1982. Modelling Fluctuating Populations. Wiley, New York.

Roach, D., and R. Wulff. 1987. Maternal effects in plants. Annu. Rev. Ecol. Syst. 18:209–236.

Rossiter, M. 1994. Maternal effects hypothesis of herbivore outbreaks. Bioscience 44:752–762.

Rossiter, M. 1996. Incidence and consequences of inherited environmental effects. Annu. Rev. Ecol. Syst. 27:451–476.

Royama T. 1992. Analytical Population Dynamics. Chapman and Hall, New York.

Turchin, P. 1990. Rarity of density dependence or population regulation with lags? Nature 334:660–663.

Turchin, P., and A. Taylor. 1992. Complex dynamics in ecological time series. Ecology 73:289–305.

Wellington, W. 1957. Individual differences as a factor in population dynamics: The development of a problem. Can. J. Zool. 35:193–323.

4

What Is an Adaptive Environmentally Induced Parental Effect?

ELIZABETH P. LACEY

Biologists working in fields as diverse as mammalian behavior, plant ecology, microbial genetics, quantitative genetics, and insect ecology have shown that environmentally induced parental effects can be found in most kingdoms of living organisms. Such effects are diverse, have multiple causes, and can be transmitted via multiple pathways. Historically, and understandably, these effects have been studied by biologists who have focused on a particular group of organisms, such as insects or plants, or who have approached the phenomenon from a particular point of view, such as quantitative genetics, ecology, or behavior. The consequence has been the development of multiple terminologies that are not used consistently across disciplines or kingdoms. I believe these inconsistencies hinder the communication among biologists studying these effects, the development of generalized models of parental effects, and the empirical testing of adaptedness of these effects.

The following essay is my attempt to provide a terminology that can be used consistently across kingdoms, regardless of approach. It is also my attempt to provide a conceptual framework for studying environmentally induced parental effects. First, I briefly discuss the diversity of environmentally induced parental effects. Then I offer general definitions for a parental effect and an environmentally induced parental effect. This is not a trivial exercise because of the present confusion about existing terms. Third, I discuss three general classes of environmentally induced parental effects. Finally, I consider what it means for an environmentally induced parental effect to be "adaptive" and consider which class(es) describes adaptive effects.

4.1 Diversity of Environmentally Induced Parental Effects

Recent books and articles have documented a variety of environmentally induced parental effects that we can observe in nature (e.g., Boyd and Richerson 1985; Roach and Wulff 1987; Clutton-Brock 1991; Mousseau and Dingle 1991a,b; Reznick 1991; Sinervo 1991; Jablonka and Lamb 1995; Bernardo 1996; Lacey 1996; Mazer and Gorchov 1996; Rossiter 1996). For example, in multicellular plants, the parental environment is known to modify such offspring traits as seed size, germination, growth, flowering time, and sexuality (see reviews by Rowe 1964; Roach and Wulff 1987; Gutterman 1992; Wulff 1995; see also Galloway 1995; Case et al. 1996). In multicellular animals, it is known to modify such traits as egg size, growth rate, resistance to pathogens, time to reproduction, sexuality, behavior, and culture (reviewed by Boyd and Richerson 1985; Mousseau and Dingle 1991a,b; Reznick 1991; Sinervo 1991; Jablonka and Lamb 1995; Bernardo 1996; Rossiter 1996). In unicellular organisms, it is known to modify such traits as cell architecture and the use of potential food sources (reviewed by Jablonka and Lamb 1995). I refer the reader to these cited references for more detailed information about these effects and their proximate causes. What I think needs to be emphasized about these effects is that they can be observed at many levels of biological organization. When the environment stimulates parents to transmit cultural beliefs and practices to a child, the environment produces an environmental effect. When an individual cell is environmentally stimulated to pass nongenetic information to daughter cells, then an environmental effect results. Environmentally induced parental effects can be observed and are being studied at many levels of biological organization, and this is only now being appreciated.

The stimulus for an environmental effect can occur at any time during a parent's lifetime and is transient in that the stimulus usually disappears before an offspring shows the effect of the stimulus. The parental response to the stimulus can be permanent or transient. If the parental response is permanent or persists for a long time, then the environmentally induced parental effect may manifest itself in a group of related individuals. All progeny, and in some cases also a parent, may exhibit the same phenotype. For example, parents may teach all offspring the same foraging techniques or cultural practices. On the other hand, if the parental response to an environmental stimulus is ephemeral, then an environmental change may result in parents producing offspring with multiple phenotypes for the trait affected. For example, diminishing maternal resources may result in variable seed sizes or birth weights within one reproductive season; changing photoperiod during the parental generation may produce offspring with variable degrees of dormancy or diapause.

The pathways by which an environmental stimulus can produce an effect are also diverse. An effect may be transmitted by an individual through word of mouth or by example, as with parental care or the teaching of cultural practices (Boyd and Richerson 1985). It may be transmitted via maternal tissue that is carried along with the offspring tissues into the next generation, for example, the coat of a seed (Roach and Wulff 1987). It may be transmitted via molecular messengers that are transferred from maternal or paternal cell to offspring cell, for example, through the cytoplasm and through proteins that regulate gene expression (e.g., Matzke and Matzke 1993; Jablonka and Lamb 1995).

4.2 What Is an Environmentally Induced Parental Effect?

Defining an environmentally induced parental effect first requires agreement on the definition of a parental effect. Parental effects have traditionally been called "maternal effects" because effects transmitted via the mother were the first to be noticed and because paternal effects, transmitted via the father, were thought to be minimal or nonexistent (e.g., see reviews: Roach and Wulff 1987; Bernardo 1996; Rossiter 1996). Recent studies suggest, however, that paternal effects may be more important than previously thought (e.g., insects: Giesel 1986, 1988; Boggs 1995; Fox et al. 1995; mammals: Clutton-Brock 1991; plants: Lacey 1996; Mazer and Gorchov 1996). Therefore, it seems time to break with tradition and use the term "parental effects" to embrace both maternally and paternally transmitted effects and those effects for which the pathway of transmission is unknown (Lacey 1996). Also, because parents can influence their young in so many ways, it seems important to define a parental effect as broadly as possible so that the term will be generally useful (Bernardo 1996).

I suggest the following definition: *A parental effect is any parental influence on offspring phenotype that cannot be attributed solely to offspring genotype, to the direct action of the nonparental components of the offspring's environment, or to their combination.* This effect is the phenotypic product of the transmission of "information" from parent to offspring above and beyond the parental contribution to offspring nuclear and cytoplasmic genes. Note that I am deliberately excluding the effects of extranuclear (maternal and paternal) inheritance in my definition of a parental effect. I am also excluding random mutations that originate with a parent and are passed to offspring. (I will discuss nonrandom mutations in section 4.2.) Maternal selection (as defined by Kirkpatrick and Lande 1989) is included. A parental effect begins when a parental genotype responds to some signal. That response induces the transmission of information along one or several transmission pathways. Both the response and transmission pathways may have genetic components (e.g., Riska et al. 1985; Kirkpatrick and Lande 1989; Cowley and Atchley 1992). The parental effect is the end product of that transmission, the phenotypic modification of an offspring trait.

There are many definitions of a parental effect in the literature, and suggesting yet another one may seem either audacious or redundant. In spite of these definitions, however, many biologists are still confused about the meaning of a parental effect. This persistent confusion, along with recent studies elucidating the possible mechanisms by which these effects are transmitted, suggest that we try to define a parent effect again. Many biologists have defined a parental effect as a parental influence on offspring phenotype that cannot be attributed to the normal Mendelian transmission of chromosomes (e.g., Mather and Jinks 1971; Riska et al. 1985; Roach and Wulff 1987; Kirkpatrick and Lande 1989; Cowley and Atchley 1992; Platenkamp and Shaw 1993; Carrière 1994; Wulff 1995; Lacey 1996; Mazer and Gorchov 1996; Rossiter, 1996). This definition assumes that a parental effect cannot be controlled by nuclear genes, an untenable assumption. Defining a parental (maternal) effect as "a part of an offspring's phenotype that does not result from the action of its own genes . . ." (Bernardo, 1996) excludes parental effects involving modifications in gene activity, which I believe should be embraced by the term "parental effects". Defining a parental (maternal) effect in terms of the effect of "parental performance" (Cheverud 1984), or "parental phenotype" (Arnold 1994; Bernardo, 1996) or the effect of "genetic and environmental differences in the maternal generation" (Mousseau and Dingle 1991a) on offspring phenotype has resulted in multiple interpretations about what can and cannot be considered a parental effect. For example, none of these definitions explicitly separates parental effects from the parental chromosomal contributions to offspring genotype; these parental chromosomes also partially determine parental phenotype. Some traits that are determined by maternal cytoplasmic genes, such as coiling in snails, are not expressed in the maternal phenotype, and yet these definitions have embraced the effects of cytoplasmic inheritance. To try to resolve the above problems, I have suggested defining a parental effect in terms of off-

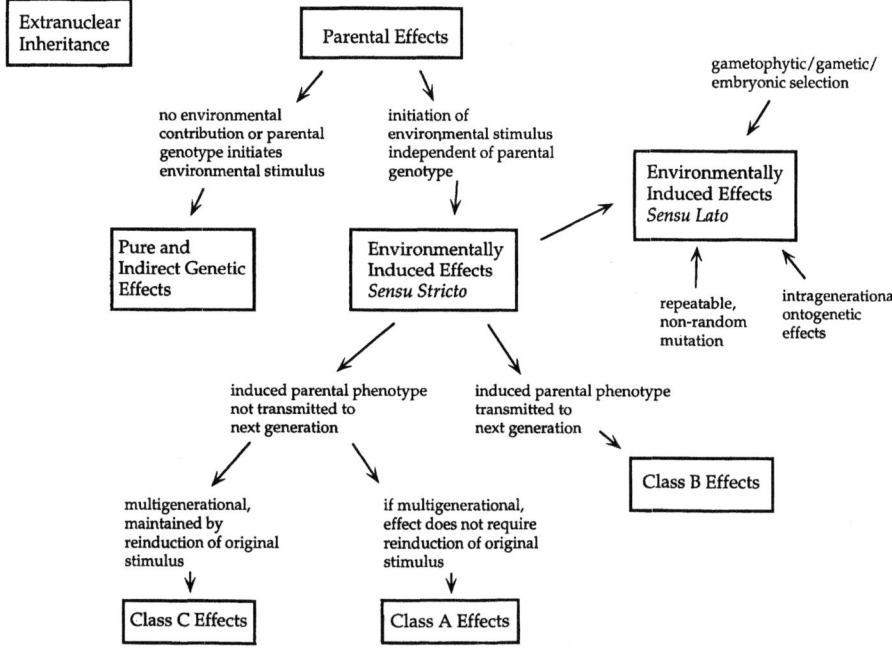

Figure 4.1 Schematic drawing of the relationships among parental effects, different classes of environmentally induced parental effects, and other phenomena that have historically been confounded with these effects. See section 4.2 for further explanation.

spring genotype and its environment. Also, I have conceptually separated parental effects from extranuclear inheritance.

A parental effect can be genetically based and/or environmentally induced. Purely genetically based effects, at least given our present understanding of parental effects, include imprinting, in which the sex of the parent determines the expression of genes passed to offspring (e.g., Solter 1988; Surani 1991; Matzke and Matzke 1993). Indirect genetic effects, as they apply to parent-offspring relationships, are contributions that parental genotypes make to offspring phenotype as mediated by parental environment (Moore et al., chap. 2, this volume.). In this case, the parental genotype determines the environmental stimulus that produces the phenotypic effect in the offspring. Environmentally induced parental effects are induced by the parental environment, but the environmental stimulus arises independently of parental genotype (Figure 4.1).

Quantitative geneticists have often looked for parental effects by measuring the asymmetry of maternal and paternal contributions to offspring phenotype in reciprocal crosses. Unequal parental contributions can only suggest that a parental effect exists, however. Parental effects may be environmentally induced in both parents and, therefore, not manifest themselves. Maternal or paternal inheritance, forms of extranuclear inheritance, will also produce asymmetrical parental contributions. Therefore, reciprocal crosses alone do not suffice to measure the existence of parental effects.

An environmentally induced parental effect is one that (1) is initiated by an environmental stimulus in the parental generation and (2) cannot be exclusively attributed to parental genotypes. In its simplest form, it is the phenotypic product of the interaction between the parental genotype and that genotype's environment as expressed in the next generation. Initiation of an environmentally induced parental effect begins when a parental genotype responds to an independently derived environmental stimulus; that parent then transmits information to its progeny via one of several possible pathways. Both the ability to respond to a stimulus and the transmission pathway may vary among genotypes. The product is the phenotypic modification of an offspring trait. The underlying genetic bases for the parental ability to respond to the environment and the process by which information is transmitted determine how long an environmentally induced effect persists across generations and how quickly and in what direction the effect evolves.

Because an environmentally induced parental effect reflects one phenotype within a range of offspring phenotypes produced by a range of parental environments, biologists have sometimes viewed these effects as manifestations of intergenerational (also called trans- and cross-generational) phenotypic plasticity (e.g., Lacey 1991; Mousseau and Dingle 1991b; Schmitt et al. 1992). Other biologists have referred to environmentally induced parental effects as intergenerational acclimation (e.g., LeRoi et al. 1994; Fox et al. 1995).

Environmentally induced parental effects are often more complex than I have described above. In reviewing the literature, Rossiter (1996) found a number of studies suggesting that in addition to genetically based differences in parental response to environmental influences, environmentally induced parental effects may manifest themselves to different degrees depending on offspring environment and offspring genotype. In quantitative genetic terms, the interaction terms $V_{G_mE_m}$, $V_{G_mE_mE_o}$, $V_{E_mE_o}$, and $V_{G_oE_m}$ could theoretically all contribute significantly to V_P, the phenotypic variance in offspring, where V_{G_m} is the genetically based maternal variance, V_{G_o} is the variance in offspring genotype, and V_{E_o} is the variance in offspring environment (Rossiter 1996). Therefore, detecting the existence and documenting the bounds of any one environmentally induced parental effect can be difficult.

Quantifying an environmentally induced parental effect is made even more difficult because of at least three confounding processes whose results mimic parental effects. The first is gametophytic/gametic/embryonic selection (Stephenson et al., 1992; Mazer and Gorchov 1996; Rossiter 1996). Although these types of selection can alter mean offspring phenotype, they involve neither the response of a parent to an environmental stimulus nor the transfer of information from parent to offspring. Therefore, gametophytic/gametic/embryonic selection should be considered a phenomenon that is distinct from an environmentally induced parental effect.

The second confounding process is environmentally induced, repeatable, nonrandom mutation. These mutations are predictable in that they are repeatedly produced by exposure to some identifiable environmental stimulus in the parental generation. The mutation is then transmitted to the offspring. Examples of this type of mutation appear to be rare, but one example appears in flax. In some genotrophs, certain parental fertilizer regimes can cause DNA amplification and RFLP alterations that are transmitted to the offspring and subsequent generations (Schneeberger and Cullis 1991). A parental effect, by definition, includes modifications in gene activity but excludes changes in gene structure. Therefore, these repeatable, nonrandom mutations should be viewed as distinct from parental effects.

The third confounding process is the direct environmental influence on offspring ontogeny. In theory, after formation of an offspring zygote, the postzygotic environment could induce the offspring's maternal parent to modify offspring phenotype. Alternatively, the environment could directly affect embryonic development. The latter is not strictly an environmentally induced parental effect because it is an intragenerational rather than an intergenerational phenomenon (Lacey 1991). Therefore, direct environmentally induced ontogenic changes also should not be treated as parental effects.

In principle, the boundaries between an environmentally induced parental effect and these three confounding processes are clear; in practice, however, they may be fuzzy. Differentiating between a parental effect and gametophytic/gametic/embryonic selection may require restricting experimental studies to homozygous lines (Mazer and Gorchov 1996), as Durrant (1962) has done with flax, or to a single genotype, as LeRoi et al. (1994) have done with *Escherichia coli*. Showing that a postzygotic effect is truly intergenerational requires demonstrating that a parent mediates the effect, for example, through hormones (Mousseau and Dingle 1991a) or oviposition site (Fox et al. 1997), which are maternal in origin, or by demonstrating that the phenotypic effect is transmitted to the offspring's offspring (e.g., Durrant 1962; Miao et al. 1991; Platenkamp and Shaw 1993; Case et al. 1996). Distinguishing between gene activity and structure may require structural analyses of DNA, and even that may not distinguish structural changes environmentally induced by modification of regulatory genes, for example, through methylation or demethylation. Therefore, differentiating among these three confounding processes and environmentally induced parental effects may be logistically difficult, costly, and time-consuming, as evidenced by the many reports that do not attempt to do so.

For these reasons, it seems useful to distinguish between environmentally induced parental effects in a broad sense and environmentally induced effects in a narrow sense (figure 4.1). Environmentally induced parental effects in the broad sense include gametophytic/gametic/embryonic selection; environmentally induced, repeatable, nonrandom mutations; and direct environmently induced changes in offspring ontogeny. Environmentally induced effects in the narrow do not. Both groups of effects may influence the course of evolution in a population; both may be adaptive or maladaptive. In the rest of this essay, however, I will focus on the narrow sense of environmentally induced parental effects.

4.3 Classes of Environmentally Induced Parental Effects

The literature suggests that there are at least three classes of environmentally induced parental effects in the narrow sense. These classes differ from each other regarding phenotypic and environmental covariances across generations (figure 4.2). Class A effects include those produced by an environmental signal that alters the phenotype j of trait B in the parental generation t; this phenotype, P_{jB_t}, then induces phenotype P_i in trait A in the offspring generation $t + 1$ (figure 4.2). Thus, the induced parental phenotype is not transmitted across generations, and the phenotype of trait A in the offspring generation does not covary with the phenotype of trait A in the parental generation; rather $P_{iA_{t+1}}$ covaries with P_{jB_t}. The covariance may be positive or negative. $P_{iA_{t+1}}$ and the covariance between traits A and B are expected to disappear at the end of the offspring generation in the absence of environmental reinforcement, that is, unless the original environmental signal reappears independently of the parental effect in subsequent generations. Examples of Class A effects are the maternal con-

GENERATION

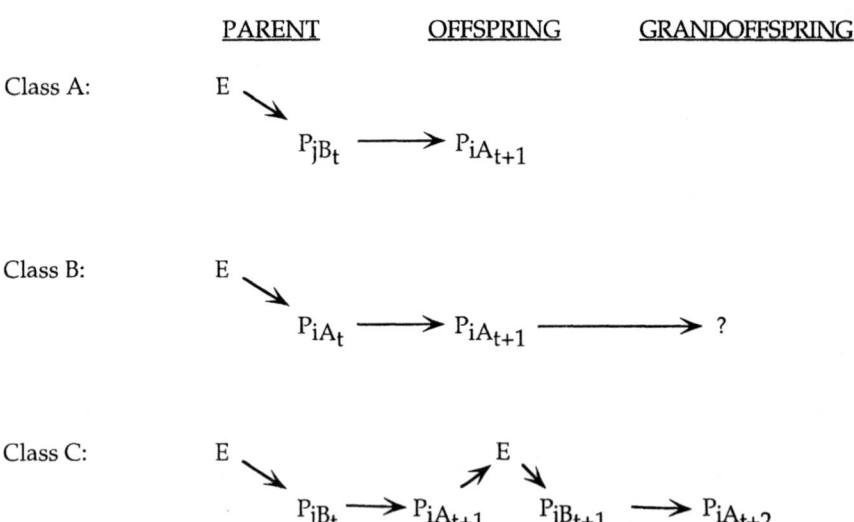

Figure 4.2 Three classes of environmentally induced parental effects, where E = the environmental stimulus, P_{iA} = phenotypic state i for trait A, P_{jB} = phenotypic state j for trait B, and t = generation. The question mark indicates that sometimes Class B effects persist past the offspring generation. See section 4.3 for further explanation.

trol of seed germination and the maternal control of offspring sex through choice of oviposition site. Regarding seed germination, the environment may alter the maternally derived seed coat, which then influences germination (e.g., Dorne 1981). In turtles, maternal decisions about where to lay eggs influence the sex ratio of her hatchlings (Roosenburg 1996).

Class B effects are those effects produced by an environmental signal that alters the phenotypic expression of trait A in the parental generation. This phenotype, P_{iA_t}, is then transmitted to and expressed in the offspring such that P_{iA_t} and $P_{iA_{t+1}}$ covary (figure 4.2). Theoretically, the phenotypes could positively or negatively covary; however, it is easier to find empirical examples showing positive covariance, which intuitively seems more likely. For example, regarding body size, abundant resources result in larger parents, who then produce larger offspring. Another example of Class B effects is the transmission of cultural beliefs. In the absence of environmental reinforcement, Class B effects should decay after the offspring generation, although the number of generations needed for total decay may vary among genotypes and likely depends on the transmission pathway(s) involved.

Class C effects are similar to Class A effects in the parental and offspring generations except that phenotype $P_{iA_{t+1}}$ causes the environmental signal that was first observed in the parental generation to reappear in the offspring generation (figure 4.2). This reappearance induces the reappearance of P_{jB} in the offspring generation, which then induces phenotype P_{iA} in the grandoffspring. Thus, the phenotypes for traits A and B and the environmental states that induced these phenotypes all positively covary across multiple generations. P_{iA} and P_{iB} both reappear in multiple generations because they perpetuate the environmental signal that induces the parental effect (figure 4.1). I must admit that I have found no examples of a Class C effect. However, it is not hard to envision one that involves the photoperiodic control of seed germination or insect diapause. For example, postzygotic photoperiod, that is, the photoperiod during seed development on the maternal parent, is known to influence subsequent seed germination in many plant species (e.g., see reviews by Roach and Wulff 1987; Gutterman 1992; Wulff 1995). Also, germination time can influence flowering time in some species (e.g., Arthur et al. 1973; McIntyre and Best 1978). Therefore, it is conceivable that seeds maturing under a particular photoperiod will germinate and consequently flower at a particular time during the next growing season, which causes the same photoperiodic regime to reappear during the development of seeds constituting the next generation. Such a scenario could lead to the evolution of populations that are polymorphic for germination and flowering times, as is *Arabidopsis thaliana* (e.g., see review by Rathcke and Lacey 1985).

In describing Classes A–C, I have assumed that the influences of offspring genotype and offspring environment are negligible, with the exception of the environmental signal in Class C effects. If these influences are not negligible, then the covariances may deviate from those discussed.

Also, I have described these classes in terms of the "transmissibility" and not the "inheritance" of phenotypes from one generation to the next. I have done this to distinguish between the passage of a phenotype across generations and the pathway by which the phenotype is passed. Words related to "inheritance" have been used in several ways in reports addressing parental effects. For example, "inherited " and "maternal inheritance" have been used very inclusively to embrace extranuclear inheritance and most environmentally induced parental effects, as I have defined them, regardless of whether or not a phenotype is transmitted across generations (e.g., Kirkpatrick and Lande 1989; Rossiter 1996). Others have used "inherited" and "heritable" to indicate the persistence of a particular phenotypic effect across generations (e.g., Hill 1965; Case et al.1996). Geneticists have severely restricted the use of "inheritance" to the transmission of gene modifications across generations, which includes structural and functional modifications. I have taken the geneticists' approach. Thus, an effect may be transmissible across generations but not necessarily heritable, or inherited. Only if the transmission pathway involves gene modification is the effect heritable.

4.4 Adaptive Environmentally Induced Parental Effects

In describing the three classes of environmentally induced parental effects, I have made no assumptions about the ecological or evolutionary consequences of the effects. The classes are purely descriptive. In principle, they can embrace adaptive and/or maladaptive effects. We can, therefore, ask if adaptive or maladaptive effects characterize the classes and if one class of effects is more likely to be adaptive than another class. To address these questions, however, we must first agree on what it means for an environmental effect to be "adaptive."

An environmentally induced parental effect may be adaptive; it may also be an adaptation. Most evolutionary biologists agree that an adaptation is a product of evolution by natural selection (reviewed by Brandon 1990). By this definition, an environmentally induced parental effect is an adaptation only after it has been subjected to natural selection and has increased in frequency in a population. Ideally, evolutionary biologists would like five types of information to establish that a trait is an adaptation: (1) evidence that selection for the trait has occurred, (2) an ecological explanation for this selection, (3) evidence that the trait is heritable, (4) information about patterns of gene flow and selective environment in which the trait is found, and (5) evidence that the trait is a derived phylogenetic character (Brandon 1990). I have found no study that provides all this information for any environmentally induced parental effect in any organism.

The term "adaptive" lacks the historical component and is, therefore, more inclusive. An adaptive environmentally induced parental effect increases the probability of reproductive success of an offspring phenotype relative to others in a population. Brandon (1990) calls this probability "adaptedness." To establish that an environmentally induced parental effect is adaptive requires only documentation that an environmental effect increases the probability of reproductive success of the phenotype exhibiting the effect over the probability of success of phenotypes not showing the effect. However, one must also establish that the observed phenotypic response in the offspring is truly an environmental effect in the strictest sense, which is not trivial. Thus far, I have found no plant study that has convincingly demonstrated adaptive environmentally induced parental effects. Studies of insect diapause provide the best body of evidence (e.g., see reviews by Mousseau and Dingle 1991a,b). Many studies have demonstrated that seasonal changes in the parental environment induce changes in parental hormone levels that subsequently affect offspring diapause, all Class A effects. These effects are most likely adaptive. Also, Fox and collaborators (1997) have recently presented strong evidence for an adaptive environmentally induced maternal effect in the seed beetle *Stator limbatus*, another Class A effect. Females lay their eggs on multiple host plants and lay eggs of differing size depending on the host plant. They increase egg size on the host species for which an increase is more likely to improve offspring survivorship and decrease egg size on the species for which an increase has little effect on survivorship. This study demonstrates the existence of an environmentally induced effect, the fitness consequence of the effect, and the transmission of information from parent to offspring that is independent of offspring genotype.

Several studies have failed to detect adaptive effects (e.g., Via 1991; LeRoi et al. 1994; Fox et al. 1995; Bernardo 1996; Donohue and Schmitt, this volume). The experiment by LeRoi and his collaborators is particularly interesting because it shows both adaptive and maladaptive consequences of parental temperature effects. LeRoi et al. used one genotype of *E. coli* that was unable to use arabinose for growth (ara-) and an ara+ mutant of that genotype. They grew both variants separately at both 32°C and 41.5°C for several generations. Then they created all combinations of the acclimated variants, grew the combinations at the two temperatures, and after some time estimated the population size of the variants in each combination. They observed that acclimation to 41.5°C reduced fitness when variants competed at 41.5°C, relative to acclimation to low temperature, regardless of the variant. However, when the variants were grown alone, acclimation to 41.5°C did improve subsequent fitness at 50°C. Thus, intergenerational acclimation, a Class B effect, was both advantageous and disadvantageous, albeit under different conditions.

Three aspects of an environmental fluctuation should influence the intensity of selection for an environmentally induced parental effect, that is, should influence its adaptiveness: amplitude, predictability, and length of the environmental fluctuation, or cycle in the case of predictable fluctuations, relative to the length of the life cycle of the organism involved. The greater the environmental fluctuation, the greater the selection for an environmentally induced effect (Rossiter 1996), regardless of type of effect (Class A or B). One might expect that where a predictable environmental cycle lasts for no more than two generations, there should be selection for short-lived effects, either Class A or B effects. The reason is that there should be selection against effects that persist into the third generation, at which time the adaptive parental information is likely to have become maladaptive. The environment will have changed. For example, any carryover effect of oviposition site in *S. limbatus* to the third generation is likely to lower the fitness advantage of the effect. If, however, the environmental cycle spans more than two generations, as it may for unicellular organisms, then we would expect that selection against the persistence of the effect would be relaxed, and persistence, or a slowing of the decay rate in Class B effects, might even be favored. For unicellular organisms that have rapid generation times, for example, bacteria, Class B effects might be common.

Conclusion

As more biologists have begun to study parental effects, the concept of a parental effect has become muddied. Here I have tried to define a parental effect and an environmentally induced parental effect in ways that are operational and generally useful. The literature indicates that there are at least two classes of environmentally induced parental effect in the narrow sense. These classes differ in the phenotypic and environmental covariances across generations. The relative abundance of classes A and B remains to be determined, as does the degree of adaptedness of these classes. Also, further research is needed to determine if one class applies to some taxonomic groups of organisms better than to others. Based on circumstantial evidence, I proposed a third class of environmentally induced parental effects (Class C). Further research will determine the reality of this class.

It has become clear that natural selection, mutation, and environmental influences on offspring ontogeny can produce effects that mimic environmentally induced parental effects, and teasing apart these processes is often difficult. Therefore, I have proposed distinguishing between narrowly and broadly defined environmentally induced parental effects. Environmentally induced parental effects in the broad sense include these confounding processes; effects narrowly defined do not. Distinguishing between narrow- and broad-sense environmentally induced parental effects serves several purposes. First, it emphasizes that one must be cautious about attributing empirically derived results to environmentally induced parental effects in the narrow sense. Second, I think that it may help us to design experiments that better address the question of the adaptiveness of environmentally induced parental effects strictly defined. Third, it may help us to design experiments that better address the evolutionary and ecological consequences of the confounding processes. Many more studies are needed to identify the environmental conditions under which environmentally induced parental effects in the narrow sense and these confounding processes are adaptive.

Acknowledgments I am very grateful to J. Antonovics, D. Byers, C. Laurie, R. Shaw, P. Thrall, and two anonymous reviewers for their comments on previous versions of the manuscript, which grew out of a presentation that I gave to the Population Biology Discussion Group at Duke University.

References

Arnold, S. J. 1994. Multivariate inheritance and evolution: A review of concepts. Pp. 17–48 in C. R. B. Boake (ed.), Quantitative Genetic Studies of Behavioral Evolution. University of Chicago Press, Chicago.

Arthur, A. E., J. S. Gale, and M. J. Lawrence. 1973. Variation in wild populations of *Papaver dubium*. VII. Germination time. Heredity 30:189–197.

Bernardo, J. 1996. Maternal effects in animal ecology. Am. Zool. 36:83–105.

Boggs, C. L. 1995. Male nuptial gifts: Phenotypic consequences and evolutionary implications. Pp. 215–242 in S. R. Leather and J. Hardie (eds.), Insect Reproduction. CRC Press, New York.

Boyd, R., and P. J. Richerson. 1985. Culture and the Evolutionary Process. University of Chicago Press, Chicago.

Brandon, R. N. 1990. Adaptation and Environment. Princeton University Press, Princton, N.J.

Carrière, Y. 1994. Evolution of phenotypic variance: Non-Mendelian parental influences on phenotypic and genotypic components of life-history traits in a generalist herbivore. Heredity 72:420–430.

Case, A. L., E. P. Lacey, and R. G. Hopkins. 1996. Parental effects in *Plantago lanceolata* L. II. Manipulation of grandparental temperature and parental flowering time. Heredity 76:287–295.

Cheverud, J. M. 1984. Evolution by kin selection: A quantitative genetic class illustrated by maternal performance in mice. Evolution 38:766–777.

Clutton-Brock, T. H. 1991. The Evolution of Parental Care. Princeton University Press, Princeton, N.J.

Cowley, D. E., and W. R. Atchley. 1992. Quantitative genetic classes for development, epigenetic selection, and phenotypic evolution. Evolution 46:495–518.

Dorne, A. (1981) Variation in seed germination inhibition of *Chenopodium bonus-henricus* in relation to altitude of plant growth. Can. J. Bot. 59:1893–1901.

Durrant, A. 1962. The environmental induction of heritable change in *Linum*. Heredity 17:27–61.

Fox, C. W., K. J. Waddell, and T. A. Mousseau. 1995. Parental host plant affects offspring life histories in a seed beetle. Ecology 76:402–411.

Fox, C. W., M. S. Thakar, and T. A. Mousseau. 1997. Egg size plasticity in a seed beetle: An adaptive maternal effect. Am. Nat. 149:149–163.

Galloway, L. F. 1995. Response to natural environmental heterogeneity: Maternal effects and selection on life-history characters and plasticities in *Mimulus guttatus*. Evolution 49:1095–1107.

Giesel, J. T. 1986. Effects of parental photoperiod regime on progeny development time in *Drosophila simulans*. Evolution 40:649–651.

Giesel, J. T. 1988. Effects of parental photoperiod on development time and density sensitivity of progeny in *Drosophila melanogaster*. Evolution 42:1348–1350.

Gutterman, Y. 1992. Maternal effects on seeds during development. Pp. 27–59 in M. Fenner (ed.), Seeds: The Ecology of Regeneration in Plant Communities. CAB International, Wallingford.

Hill, J. 1965. Environmental induction of heritable changes in *Nicotiana rustica*. Nature 207:732–734.

Jablonka, E., and M. J. Lamb. 1995. Epigenetic Inheritance and Evolution: The Lamarckian Dimension. Oxford University Press, Oxford.

Kirkpatrick, M., and R. Lande. 1989. The evolution of maternal characters. Evolution 43:485–503.

Lacey, E. P. 1991. Parental effects on life-history traits in plants. Pp. 735–744 in E. C. Dudley (ed.), The Unity of Evolutionary Biology. Dioscorides Press, Portland, Ore.

Lacey, E. P. 1996. Parental effects in *Plantago lanceolata*. L. I. A growth chamber experiment to examine pre- and post-zygotic temperature effects. Evolution 50:865–878..

LeRoi, A. M., A. F. Bennett, and R. E. Lenski. 1994. Temperature acclimation and competitive fitness: An experimental test of the beneficial acclimation assumption. Proc. Natl. Acad. Sci. USA 91:1917–1921.

Mather, K., and J. L. Jinks. 1971. *Biometrical Genetics*. Chapman and Hall, New York.

Matzke, M., and A. J. M. Matzke. 1993. Genomic imprinting in plants: Parental effects and trans-inactivation phenomena. Annu. Rev. Plant Physiol. Plant Mol. Biol. 44:53–76.

Mazer, S. J., and D. L. Gorchov. 1996. Parental effects on progeny phenotype in plants: Distinguishing genetic and environmental causes. *Evolution* 50:44–53.

McIntyre, G. I., and K. F. Best. 1978. Studies on the flowering of *Thlaspi arvense* L. IV. Genetic and ecological differences between early- and late-flowering strains. Botan. Gaz. 139:190–195.

Miao, S. L., F. A. Bazzaz, and R. B. Primack. 1991. Persistence of maternal nutrient effects in *Plantago major*: The third generation. Ecology 72:1634–1642.

Mousseau, T. A., and H. Dingle. 1991a. Maternal effects in insect life histories. Annu. Rev. Ecol. Syst. 36:511–534.

Mousseau, T. A., and H. Dingle. 1991b. Maternal effects in insects: Examples, constraints, and geographic variation. Pp. 745–761 in E. C. Dudley (ed.), The Unity of Evolutionary Biology. Dioscorides Press, Portland, Ore.

Platenkamp, G. A. J., and R. G. Shaw. 1993. Environmental and genetic maternal effects on seed characters in *Nemophila menziesii*. Evolution 47:540–555.

Rathcke, B., and E. P. Lacey. 1985. Phenological patterns of terrestrial plants. Annu. Rev. Ecol. Syst. 16:179–214.

Reznick, D. 1991. Maternal effects in fish life histories. Pp. 780–793 in E. C. Dudley (ed.), The Unity of Evolutionary Biology. Dioscorides Press, Portland, Ore.

Riska, B., J. J. Rutledge, and W. R. Atchley. 1985. Covariance between direct and maternal genetic effects in mice, with a class of persistent environmental influences. Genet. Res. 45:287–297.

Roach, D. A., and R. Wulff. 1987. Maternal effects in plants: Evidence and ecological and evolutionary significance. Annu. Rev. Ecol. Syst. 18:209–235.

Roosenburg, W. M. 1996. Maternal condition and nest site choice: An alternative for the maintenance of environmental sex determination. Am. Zool. 36:157–168.

Rossiter, M. C. 1996. Incidence and consequences of inherited environmentally induced parental effects. Annu. Rev. Ecol. Syst. 27:451–476.

Rowe, J. S. 1964. Environmental preconditioning, with special reference to forestry. Ecology 45:399–403.

Schmitt, J., J. Niles, and R. D. Wulff. 1992. Norms of reaction of seed traits to maternal environments in *Plantago lanceolata*. Am. Nat. 139:451–466.

Schneeberger, R. G., and C. A. Cullis. 1991. Specific DNA alterations associated with the environmental induction of heritable changes in flax. Genetics 128:619–630.

Sinervo, B. 1991. Experimental and comparative analyses of egg size in lizards: Constraints on the adaptive evolution of maternal investment per offspring. Pp. 725–734 in E.C. Dudley (ed.), The Unity of Evolutionary Biology. Dioscorides Press, Portland, Ore.

Solter, D. 1988. Differential imprinting and expression of maternal and paternal genomes. Annu. Rev. Genet. 22:126–148.

Stephenson, A. G., T. C. Lau, M. Quesada, and J. A. Winsor. 1992. Factors that affect pollen performance. Pp. 119–136 in R. Wyatt (ed.), Ecology and Evolution of Plant Reproduction. Chapman and Hall, New York.

Surani, M. A. 1991. Genomic imprinting: Developmental significance and molecular mechamism. Curr. Opin. Genet. Dev. 1: 241–246.

Via, S. 1991. Specialized host plant performance of pea aphid clones is not altered by experience. Ecology 72:1420–1427.

Wulff, R. D. 1995. Environmental maternal effects on seed quality and germination. Pp. 491–505 in J. Kigel and G. Galili (eds.), Seed Development and Germination. Dekker, New York.

5

Oviposition Decisions as Maternal Effects

Conundrums and Opportunities for Conservation Biologists

BERNARD D. ROITBERG

Maternal effects are ubiquitous. From maternally determined production of winged morphs in aphids (MacKay and Wellington 1977) to lactation-dependent growth rates in mammals (Young and Legates 1965), maternal effects have been demonstrated across a wide variety of taxa. Such effects are now widely accepted as important components of the phenotype (Mousseau and Dingle 1991). Less well understood, however, is the importance of maternal effects on evolutionary (e.g., Via et al. 1995) and population dynamics, in particular the latter.

The implications of ignoring maternal effects on population dynamics are potentially huge. It is this simple: if maternal effects vary in response to environment and, in turn, cause changes to environment through modification of offspring performance, nonobvious, dramatic numerical changes can occur via small alterations in one or a few parameters (e.g., Mangel and Roitberg 1992). If maternal effects are important to population and community dynamics, then perhaps they should be considered when ecological theory is applied to real world problems, for example, conservation biology. Despite the urgency of such problems, however, we know very little regarding direct and/or indirect influences of maternal effects on the evolutionary and numerical responses of inhabitants (e.g., Vogler and Desalle 1994; Werner 1994). In this chapter, I explore the issue of oviposition decisions as maternal effects on evolutionary and population dynamics in habitats undergoing change. I conclude that addition of yet another level of complexity renders these dynamics more difficult to solve but that these considerations also provide important new opportunities for both basic and applied evolutionary biologists.

5.1 The Problem

The problem that I have in mind regards the response of butterfly populations to habitat change. Suppose a conservation biologist is asked to forecast short- and long-term responses of resident butterfly populations to habitat change. Assuming that the butterfly is univoltine and polyphagous and that it deposits identically provisioned eggs in clutches of varying sizes on a range of host plants, the dynamics can be defined by

$$N_{y+1} = \sum_{i=1}^{R} c_i f(c, h), \qquad (5.1)$$

where N_{y+1} is the number of adult butterflies in year $y + 1$, i is the ith clutch deposited in year y, R is the number of clutches laid in year y, c is the size of the clutch, and f is a function that describes per capita performance of larvae deposited in clutches of size c on host plants h. Thus, according to this simple model, clutch-size decisions are integral to population dynamics.

There are several features of clutch-size decisions that are pertinent to a discussion of the link between maternal effects and population processes in butterflies:

1. An oviposition decision (i.e., clutch size and host choice) is a legitimate maternal effect (Messina, this volume) that falls well within Cheverud and Moore's (1994) parental care category of maternal effect.
2. Butterfly clutch size varies within and among females—such variation has been discussed by numerous authors including Stamp (1980) and Thompson and Pellmyr (1991).
3. Clutch size affects progeny and development—there are a number of factors that determine the success of individual larvae as a function of larval density, including (a) predators (Damman 1987), (b) parasites (e.g., Stamp 1981), (c) disease, and (d) food quantity and quality (Peterson 1987). Depending on the factor(s), that function can be positive or negative.
4. Clutches that maximize parental fitness need not maximize fitness of offspring (Godfray and Parker 1992; Roitberg and Mangel 1992; see figure 5.1). Any theory that considers evolutionary and numerical effects of clutch size decisions must take this potential conflict into account.

5.2 A Theory for Variable Clutch-Size Decisions in Evolutionarily Stable Populations

If oviposition decisions are variable, then it is important to understand the basis for this variation. As such, Marc Mangel and I (Roitberg and Mangel 1992; Mangel and Roitberg 1993) recently developed a theory to elucidate the adaptive nature of oviposition decisions (i.e., host choice and clutch size) in butterflies with mobile offspring. Specifically, we considered the situation where oviposition decisions are based on subsequent larvae making a single (final) host-acceptance decision on emergence as is common in leafrollers and other related lepidopterans (Carrière and Roitberg 1995). The larval decision to remain on or exit the plant is based on larval crowding, plant quality, and relatedness within broods; that is, it is best described as a multidimensional reaction norm (Stearns 1989).

Assuming that butterflies are under optimizing selection (Travis 1989; Orzack and Sober 1994), our theory of adaptive oviposition decisions comprises three essential components that determine the fitness payoffs to ovipositing butterflies: (1) host acceptance proclivity of individual larvae, (2) performance of individual larvae that remain on the host on which they have been deposited, and (3) performance of larvae that leave their original host plant. Using calculations of inclusive fitness, we solved the Evolutionary Stable Strategy (ESS) host ac-

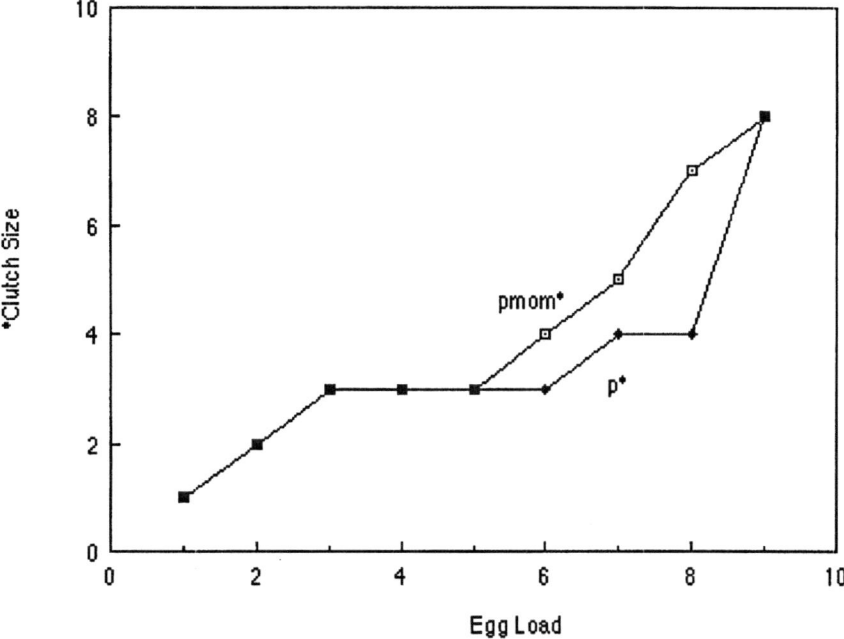

Figure 5.1 Optimal clutch size from a butterfly's (pmom*) and her offspring's (p*) perspective as a function of butterfly egg load.

ceptance rate by larvae under a variety of ecological scenarios (Roitberg and Mangel 1992). Subsequently, we used dynamic, state-variable models (Mangel and Clark 1988) to solve the optimal clutch size for butterflies whose larvae express the ESS host acceptance phenotype. Pertinent to the present discussion, we frequently found, as postulated in section 5.1, that there was an evolutionary conflict between mothers and offspring; that is, a maternal effect of clutch size and/or host choice that maximized maternal fitness frequently differed from that clutch size that maximized inclusive fitness of individual larvae (figure 5.1).

Our 1992 and 1993 reports assumed that our theoretical populations had reached evolutionary stability and thus our analysis was appropriate. In cases of instability (e.g., the habitat succession scenario discussed above), however, one must consider the evolutionary process itself (e.g., Cheverud and Moore 1994), which means that measures of the host-acceptance phenotype and genotype performance must be evaluated. That such measures are essential is exemplified by their explicit representation in equation (7) in Roitberg and Mangel (1992).

5.3 A Theory for Evolving Populations

I now consider the evolution of host acceptance behavior in evolving populations (e.g., in the face of habitat change). I assume that host acceptance is a quantitative threshold trait (Carrière and Roitberg 1996) and that my target population can be attributed as comprising 11 genotypes. Further, each genotype can be shown to express a three-dimensional, maternally directed, host acceptance reaction norm (Cheverud and Moore 1994):

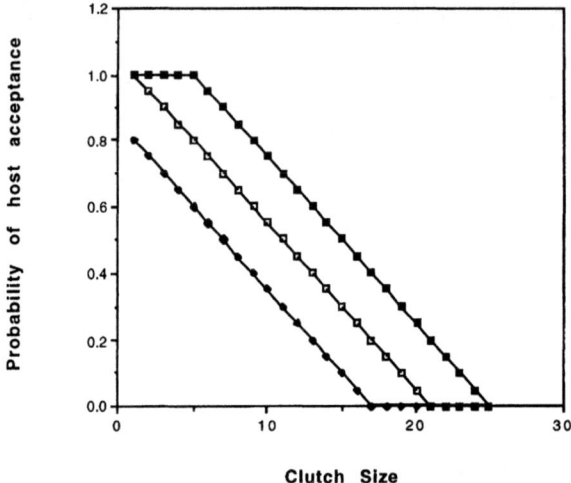

Figure 5.2 Host acceptance reaction norms for three genotypes experiencing 20 different larval crowding levels. Note the convergence in phenotype at high and low crowding.

$$P_A\,(0,\,1) = G + M + E, \tag{5.2}$$

where P_A = probability of accepting the host plant, G = additive genetic disposition to accept, M = two maternal effects C and H where C is clutch size and H is host quality, and E = environmental effects ($E = 0$ in my analysis).

Figure 5.2 shows a slice through a set of reaction norms for three genotypes on a single host plant. Several features should be noted: (1) within a single host-quality environment, reaction norms are essentially linear and parallel within the intermediate clutch-size range; (2) each genotype is equally sensitive to larval crowding; and (3) an assumption that there is little or no variation in host ranking among genotypes, that is, that within each plant, there is a repeatable deviation in the acceptance tendency of a genotype at each level of host crowding. This is equivalent to assuming a genetic correlation in host acceptance of 1. However, genetic disposition (g) could be made host specific for each genotype to accommodate different values of the genetic correlation in host acceptance (e.g., genotypes with relatively high g on one host would have relatively low g on others). Genotypes could easily have different sensitivities to crowding. I have refrained from adding such complications to keep my argument simple and to keep focus on my main question: What role do maternal effects play in the evolution of host acceptance? Finally, I also assume that there is no correlation between host acceptance disposition and larval performance at particular larval densities.

There is a feature of potential great importance that can be seen in figure 5.2: the reaction norms converge in high and low clutch-size environments; that is, there is little or no variation among genotypes in terms of phenotype (and performance). This phenotypic censorship could be an important issue if mothers tend to choose high and/or low clutches to improve their reproductive performance. Given the reaction norms described above, it is possible to evaluate the performance of different genotypes when mothers make different egg-laying decisions. Of course, how well a larva of a particular genotype performs depends not only on its tendency to accept its host but also on the number and identity of larvae with which the focal larva shares its plant. For example, a larva with a very high host acceptance disposition will likely share its host with many versus few other larvae when populations are skewed toward similar versus highly vagile individuals, respectively. Of course, the actual payoff that such a larva receives depends on the shape of the within-plant payoff function. One such function, amended from Roitberg and Mangel (1992), is

$$w = \max[0, (1 - \alpha \frac{c}{c_{max}})^\gamma], \quad (5.3)$$

where c is the clutch size, c_{max} is the maximum number of larvae that a host plant can support, α is the proportion of larvae that remain on the host plant, and γ is a shape parameter. α, of course, is the summation of genotype-specific α's weighted by frequency. Here, I suggest a host plant that can tolerate feeding with little concommitant effects on larvae until critical densities are reached, at which point host collapse occurs. As important as clutch size is, however, it is the genetic structure of the brood that determines operational larval density and thus resident larval performance.

5.4 A Dynamic Ovipositron-Decision Theory

In contrast to their relatedness to different individual larvae, all offspring are equally related to their mother by one-half; thus, the optimal egg-laying decision from the mother's perspective is determined by the summation of the within-plant payoff and off-plant payoff functions as well as several life history and ecological variables, including egg load, egg maturation rate, expectation of life, and host availability. These parameters can be amalgamated into dynamic state variable models (Mangel and Clark 1988) to solve for the optimal oviposition decision as a function of payoff functions and state. Our original theory assumed (1) that there are two host types, and the probability of encountering a host of type i during a single period is l_i; that the length of a period is chosen so that the mother matures one egg during a period, with a maximum egg complement x_m; and (3) that the probability r that the mother survives from one period to the next is independent of time, egg load, or activity. The general equation that maximizes lifetime reproductive success (our fitness surrogate) as a function of the aforementioned variables for a butterfly with x mature eggs, at time t with a maximum life expectancy T, $F(x, t, T)$, is then

$$F(x, t, T) = (1 - l_1 - l_2)\, rF(x', t + 1, T) + $$
$$\sum_{i=1}^{2} l_i \max_{c2x} \leq [f(c) + rF(x'', t + 1, T)]. \quad (5.4)$$

In this equation $x' = \min[x + 1, x_m]$, $f(c)$ is the fitness from a clutch of size c, $x'' = \min[x - c + 1, x_m]$, and $\max_{c2x}$ denotes that the maximum is taken over possible clutches ranging from 0 to x.

Equation 5.4 is solved by means of backward induction given that $F(x, T, T) = 0$ (i.e., no future fitness accrued after the butterfly is T units old; Mangel and Clark 1988). Mangel and I used this model to consider a wide range of ecological scenarios and their implications for parent-offspring conflict (Roitberg and Mangel 1992; Mangel and Roitberg 1993).

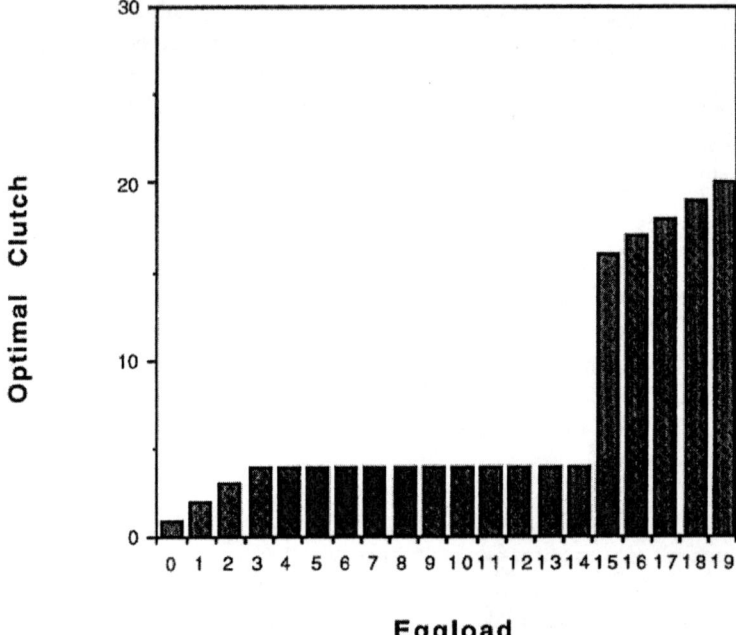

Eggload

Figure 5.3 Optimal clutch size as a function of egg load for a butterfly in a population with mean genetic disposition to accept g = 6 and sd = 1.

5.5 Clutch-Size Decisions When Broods Have Genetic Structure

As noted in section 5.2, the Roitberg-Mangel theory assumes no genetic variation in host acceptance tendency and that that tendency was determined by ESS. If such is not the case, how should females respond and what are the implications?

To answer these questions, it is necessary to modify the fitness functions in equation (5.4) by relaxing the assumption of genetic uniformity. Thus, for any given clutch, one cycles through genotypes to determine their tendency to accept the host and from there (1) the number of larvae that remain and the fitness payoff and (2) the number of larvae that leave and the fitness payoff. Figure 5.3 shows adaptive clutch-size decisions as a function of egg load for a butterfly in a population with g= 6 and sd = 1. An important feature to note is that clutch size increases as a function of egg load but that it does so in a highly nonlinear fashion (it actually declines at intermediate levels). The theory tells us that it pays mothers to either lay very few eggs or deposit their entire egg load. Further, as mothers approach the end of their life the latter behavior becomes more likely (Roitberg et al. 1993). Thus, my earlier concern regarding phenotypic censorship is well founded at least at the level of theory.

We can now address the potential impact of maternal effects on evolutionary and population dynamics. It is difficult to assign a null model regarding maternal effects. By that, I mean that there is no biologically reasonable way to remove the maternal generation from our analysis because mothers always choose sites for their offspring. Thus, there will always be maternal effects, but the critical questions are, How do maternal effects vary under different ecological conditions, and how important might these changes be to host acceptance evolution?

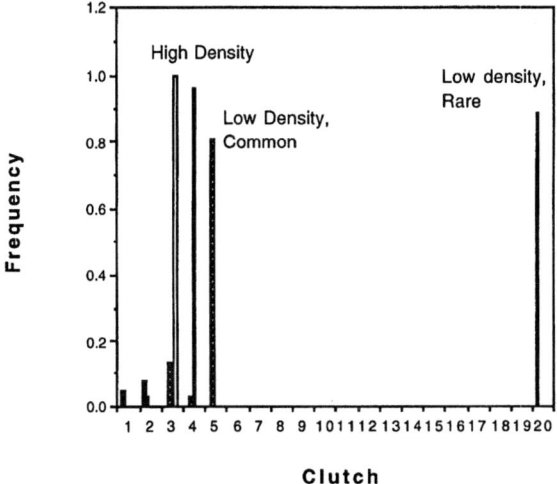

Figure 5.4 The distribution of clutch sizes varies under different densities and ratios of host plants.

5.6 Determinants of Variation in Clutch Size

First, consider how egg-laying decisions might change under different ecological conditions. Analyses I have conducted on the aforementioned clutch decision theory show that small changes in endogenous or exogenous state can lead to sudden, dramatic shifts in clutch-size and host-choice decisions. For example, figure 5.4 shows how the distribution of clutch sizes varies when one alters the perceived density and ratio of host plants. The figure demonstrates that major changes in clutch can occur with moderate changes in butterfly habitats. Further, adaptive clutches could censor phenotype to the extent that small populations would be unable to cope with habitat change (Lande 1988). Below, I deal with some specific cases and explain why such shifts occur.

Given the nature of the putative dramatic shifts in clutch and host choice, it is of interest to compare fitness functions for larval genotypes under different clutch distribution scenarios. Two "null" models with which to compare the fitness of larvae from a distribution of optimal clutches would be (1) uniform and (2) normal distribution of clutches. Figure 5.5 shows such a comparison.

Figure 5.6 shows that adaptive egg-laying behaviors of butterflies can have a significant effect on expression of larval behavior and performance among genotypes even when one assumes equal abilities to locate and exploit neighboring plants when individuals reject hosts. Equally important, egg-laying decisions (the maternal effects) are highly labile and subject to change in a complicated manner. Thus, while it might be possible to empirically derive those egg-laying decisions for conservation or management purposes, the amount of effort and number of permutations required to do so are prohibitive (see Carrière and Roitberg 1996). The theory that I have discussed provides a first step toward development of a comprehensive theory for egg laying as a maternal effects problem. Several generalizations emerge from the theory. Below, I consider some features of butterfly-caterpillar systems that can influence maternal effects and their impact on larval host acceptance dynamics: the

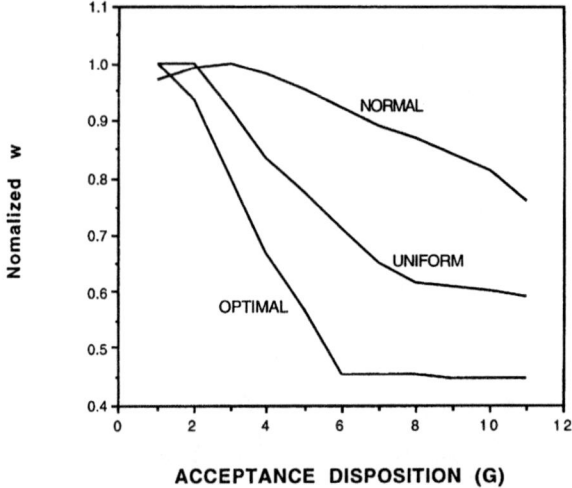

Figure 5.5 Fitness as a function of genetic disposition to accept under three different ecological scenarios: (1) butterflies deposit clutches that maximize their lifetime reproductive success, (2) butterflies deposit a uniformly distributed set of clutches (1, 20), and (3) butterflies deposit a normally distributed set of clutches $\gamma = 7$ and sd = 1.

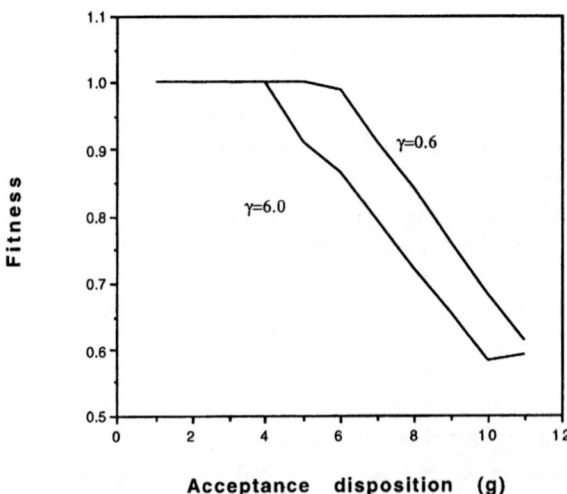

Figure 5.6 Effect of the discounting (γ in eq. [5.3]) from larval crowding on reproductive payoff on optimal clutch under varying egg loads.

within-host payoff function, butterfly egg load, host availability, expectation of life, and genetic structure.

The Fitness Function

In section 5.3 I considered the situation where increasing larval density has little effect on per capita performance, up to a particular density at which point such performance drops off precipitously. Adaptive responses by butterflies to this type of function generally will favor high-acceptance genotypes: at low clutches, larvae remain on hosts and suffer very little from competition whereas low-acceptance genotypes pay the "unnecessary" high cost of abandoning suitable plants. Of course, the degree to which this advantage is accrued depends on the genetic structure of the brood (discussed below). It is also possible, however, that per capita performance of larvae would decline exponentially (this could be the case for several reasons, including plant defense, density-dependent attack by natural enemies, and antagonistic behavior of larvae). Under such conditions, several features emerge. Figure 5.6 shows that when the function is modified as described in section 5.3 (i.e., the shape parameter γ is reduced by a magnitude), optimal clutch as a function of egg load goes from a bimodal to an increasing function; that is, there is no plateau of intermediate clutch sizes. When this is the case and when hosts are either very common or very rare, there will be little or no change in the fitness function of acceptance disposition; that is, butterflies will lay either very small or very large clutches, respectively, regardless of the payoff function. At intermediate levels of host availability, smaller clutches are favored more frequently, and this again favors high-acceptance phenotypes to the extent that a broader range of high-acceptance phenotypes experience similar if not identical fitness.

Egg Load

Relative to host availability and expectation of life, higher egg loads favor egg dumping, that is, deposition of large clutches. Egg loads are hypothesized to act through stretch receptors that somehow influence how the central nervous system responds to input of information regarding host quality and availability, though few explicit tests of egg load have been performed (see van Randen and Roitberg 1996). As noted in section 5.3, such a maternal effect can censor phenotypic differences among genotypes. Note, however, that egg load must always be considered relative to other factors such as host availability and life expectancy.

Host Availability

As host density increases, deposition of small clutches and greater acceptance of low-quality hosts become the adaptive responses by butterflies. Figure 5.4 shows how the proportion of small clutches (i.e., fewer than three eggs) varies as a function of host density. This maternal effect favors high-acceptance genotypes for reasons discussed above.

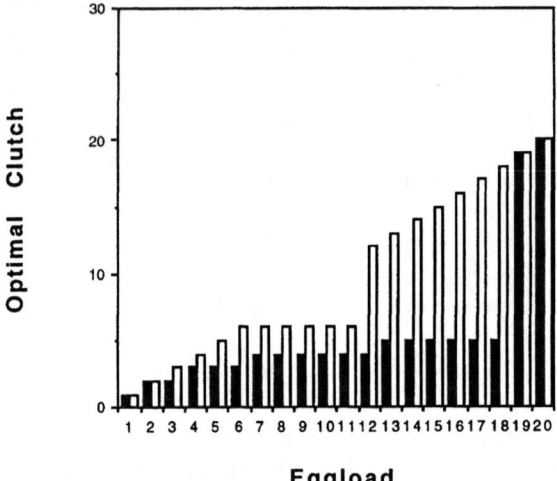

Figure 5.7 Optimal clutch size as a function of egg load for butterflies with high (solid bar) and low (open bar) expectations of life.

Expectation of Life

The theory says that butterflies that are nearing the end of their lives should be less choosy about the kinds of hosts they lay on and they should tend to lay larger clutches. In the examples discussed above, that means a shift from small to very large clutches. Figure 5.7 shows optimal clutch as a function of egg load for butterflies with high versus low expectation of life. I am not aware of any work that has explicitly tested this idea in butterflies, but it has been supported in experiments on parasitic wasps (Roitberg et al. 1993).

Genetic Structure

The payoff function for females depends on the genetic structure of the brood due to the differences in acceptance disposition among genotypes. Figure 5.8 below shows how the payoff function varies under two scenarios, when the brood is skewed toward high host acceptance versus a skew toward low host acceptance. Notice that the shape of the payoff function is not affected much but the height is. If females were able to access this information, they would alter their egg-laying decisions and rarely lay large clutches when the former distribution holds. Should females oviposit as if the population was not skewed toward host acceptance, when it actually is skewed there can be significant effects. When hosts are rare, butterflies will tend to dump eggs; when populations are skewed toward high host acceptance, individual hosts will be exhausted and most of the brood will succumb. By contrast, when populations are skewed toward low host acceptance, individuals with high host acceptance disposition will gain a frequency-dependent advantage by remaining on hosts while others leave.

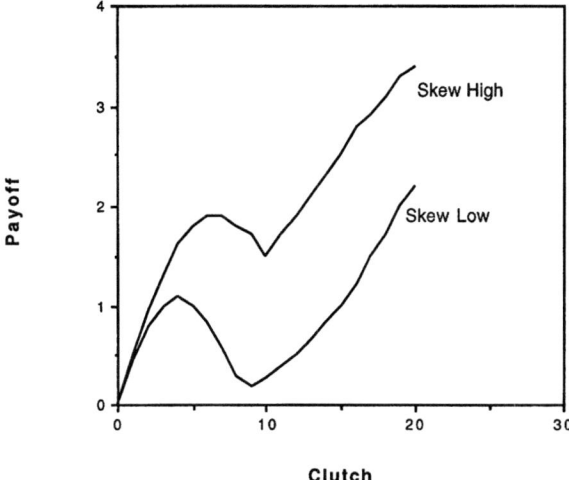

Figure 5.8 Payoff function to butterflies as a function of clutch size when clutches are skewed toward or away from high larval host acceptance.

Conclusions

Within many species of insects, mothers determine their offspring's environment. It is arguable whether such individuals act in the best interests of their offspring (see arguments regarding preference performance in herbivores, e.g., Thompson 1988; Courtney and Kibota 1990; Larsson and Ekbom 1995), but that it should be so is not surprising. What is important, however, is that the adaptive reproductive decisions by mothers can create a wide variety of environments for those offspring and thus censor or exaggerate expression of larval behavior. Abrahamson and Weiss (1997) discusses an analogous censoring in plant tolerance due to active defense; there may be many such cases across the various kingdoms.

I am not the first person to consider behavior a maternal effect, but such considerations still pale in comparison with more physiologically oriented studies. However, it should be clear that egg-laying behavior is as legitimate a maternal effect as is egg size (e.g., Braby 1994) or yolk allocation (e.g., Rossiter et al. 1993).

I have avoided many of the details and intricacies of butterfly behavior in order to make my case. I have even avoided the temptation to survey the literature for information regarding clutch-size distributions in nature because, as I pointed out in the introduction, empirical measures of such behaviors in current environments may tell us little about how they might change in response to environmental change or degradation. Singer's excellent work on the evolution of host preference in *Euphydryas* butterflies illustrates how complex changes in behavior can occur following habitat manipulation by humans (e.g., Singer et al. 1993). We do know that butterflies can respond to host plant characters in apparently adaptive ways (e.g., Pilson and Rausher 1988; but see Tatar 1989); the next step is to understand these responses in a more dynamic multidimensional framework.

We have reached a critical point in our efforts to conserve natural habitats and their inhabitants (Erwin 1991). Hard choices must be made, yet as I pointed out at the beginning of this chapter, exactly what those choices should be is not obvious; plasticity in response to environmental change might mitigate effects to some degree (Roitberg and Mangel 1997), but just how much of a buffer this provides when supplemented by transgenerational (maternal) effects should cause us to be cautious in our approach to these serious problems. Recently, Curio (1996) suggested that ethology (and its intellectual offspring, behavioral ecology) should be made an item in the conservation ecologist's toolbox. The approach that I have taken here uses behavioral ecological principles to help organize our thoughts about the adaptive nature and implications of behavior-based maternal effects.

Acknowledgments I initiated this project in response to an invitation to participate in a symposium devoted to the late Vincent Dethier. I thank E. Bowdan (a co-organizer) for that invitation as well as the organizers of the Maternal Effects as Adaptations symposium for giving me a chance to clarify my views. I also thank Art Weiss for discussions regarding phenotypic censoring and two anonymous reviewers for their excellent comments. This work is supported by an operating grant from NSERC (Canada).

References

Abrahamson, W. G. and A. E. Weis. 1997. Evolutionary ecology across three trophic levels: Goldenrods, gallmakers and natural enemies. Princeton Univ. Press, Princeton, N.J.

Braby, M. 1994. Significance of egg-size variation in relation to host plant quality. Oikos 71:119–129.

Carrière, Y., and B. Roitberg. 1995. Evolution of host preference in leaf rollers. Heredity 74:357–368.

Carrière, Y., and B. Roitberg. 1996. Behavioral ecology and quantitative genetics as alternatives for studying evolution in insect herbivores. Evol. Ecol 10:289–305.

Cheverud, J. M., and A. J. Moore. 1994. Quantitative genetics and the role of environment in behavioral evolution. Pp. 67–100 in C.R.B. Boake (ed.), Quantitative Genetic Studies of Behavioral Evolution. University of Chicago Press, Chicago.

Courtney, S. P., and T. T. Kibota. 1990. Mother doesn't always know best. Insect-Plant Interact. 2:161–188.

Curio, E. 1996. Conservation needs ethology. TREE 11:260–263.

Damman, H. 1987. Leaf quality and enemy avoidance by the larvae of a pyralid moth. Ecology 68:88–97.

Erwin, T. 1991. An evolutionary basis for conservation strategies. Science 253:750–752.

Godfray, H. J. C., and G.A. Parker. 1992. Sibling competition, parent-offspring conflict and clutch size. Anim. Behav. 43:473–490.

Lande, R. 1988. Genetics and demography in biological conservation. Science 241:1455–1460.

Larsson, S., and B. Ekbom. 1995. Oviposition mistakes in insects: Confusion or a step towards a new host plant? Oikos 72:155.

MacKay, P., and W. Wellington. 1977. Maternal age as a source of variation in the ability of an aphid to produce dispersing forms. Res. Pop. Ecol. 18:195–209.

Mangel, M., and C. W. Clark. 1988. Dynamic Modeling in Behavioral Ecology. Princeton University Press, Princeton, N.J.

Mangel, M., and B. Roitberg. 1992. Behavioral stabilization of host-parasite dynamics. Theor. Pop. Biol. 42:308–320.

Mangel, M., and B. Roitberg. 1993. Larval lifestyles and oviposition behavior of parasites and grazers. Evol. Ecol. 7:401–405.

Mousseau, T., and H. Dingle. 1991. Maternal effects in insect life histories. Annu. Rev. Entomol. 36:511–534.

Orzack, S. H., and E. Sober. 1994. Optimality models and the test of adaptationism. Am. Nat. 143:361–380.

Peterson, S. C. 1987. Communication of leaf suitability by gregarious caterpillars (*Malacosoma american*). Ecol. Entomol. 12:283–289.

Pilson, D., and M. D. Rausher. 1988. Clutch size adjustment by a swallowtail butterfly. Nature 333:361–363.

Roitberg, B., and M. Mangel. 1992. Parent-offspring conflict and life-history consequences in herbivorous insects. Am. Nat. 142:443–456.

Roitberg, B., and M. Mangel. 1997. Individuals on the landscape: Behavior can mitigate habitat differences. Oikos 80:234–240.

Roitberg, B., J. Sircom, C. Roitberg, J. van Alphen, and M. Mangel. 1993. Life expectancy and reproduction. Nature 364:108.

Rossiter, M. C., D. L. Cox-Foster, and M. Briggs. 1993. Initiation of maternal effects in *Lymantria dispar*: Genetic and ecological components of egg provisioning. J. Evol. Biol. 6:577–589.

Singer, M. C., C. D. Thomas, and C. Parmesan. 1993. Rapid human-induced evolution of insect-host associations. Nature 366:681–683.

Stamp, N. E. 1980. Egg deposition patterns in butterflies: Why do some species cluster their eggs rather than deposit them singly? Am. Nat. 115:367–381.

Stamp, N. E. 1981. Effect of group size on parasitism in a natural population of the Baltimore checkerspot *Euphydryas phaeton*. Oecologia 49:201–206.

Stearns, S. C. 1989. The evolutionary significance of phenotypic plasticity. Bioscience 39:436–445.

Tatar, M. 1989. Swallowtail clutch size reconsidered. Oikos 55:135–136.

Thompson, J. N. 1988. Evolutionary ecology of the relationship between oviposition preference and performance of offspring in phytophagous insects. Entomol. Exp. Appl. 47:3–14.

Thompson, J. N., and O. Pellmyr. 1991. Evolution of oviposition behavior and host preference in lepidoptera. Annu. Rev. Entomol. 36:65–89.

Travis, J. 1989. The role of optimizing selection in natural populations. Annu. Rev. Ecol. Syst. 20:279–296.

van Randen, E., and B. Roitberg. 1996. Eggload and superparasitism in snowberry flies. Entomol. Exp. Appl. 79:241–245.

Via, S., G. De Jong, S. M. Scheiner, C. D. Schlichting, and P. H. van Tienderen, 1995. Adaptive phenotypic plasticity: Consensus and controversy. TREE 10:215–217.

Vogler, A. P., and R. Desalle. 1994. Diagnosing units of conservation management. Conserv. Biol. 8:354–363.

Werner, E. 1994. Individual behavior and higher-order species interactions. Pp. 297–324 in L. Real (ed.), Behavioral Mechanisms in Evolutionary Ecology. University of Chicago Press, Chicago.

Young, C. W., and J. E. Legates. 1965. Genotypic, phenotypic and maternal interrelationships of growth in mice. Genetics 52:563–576.

Part II

ASSESSMENT AND MEASUREMENT

In this part, three chapters explore the conceptually difficult and technically challenging requirements for the assessment and measurement of maternal effects (readers interested in this issue are also invited to read chapters 9, 12, 17, and 19). Although the existence of a maternal effect (and parentally transmitted environmental effects in general) is not extremely difficult to demostrate, especially if experimental manipulation of maternal environment is feasible, the precise measurement of the effect often requires complex experimental designs. This is especially true if the objective is to estimate the magnitude of m, the maternal effects coefficient, and its additive genetic basis. The three chapters in this section explore the variety of methods that can be used by investigators to quantify the expression of maternal effects in plants and animals.

In chapter 6, Derek Roff surveys the maternal effects literature and evaluates the relative merits of diallel and factorial designs, the use of single locus mutants, Eisen's pedigree analysis, classical quantitative genetic designs, cross-fostering, environmental manipulation, and grandparent effects. Ruth Shaw and Diane Byers (chapter 7) explore the quantitative genetic designs necessary for estimation of V_{am}, the additive genetic variance for a maternal effect using their analysis of maternal effects in *Nemophila menziesii* as a case study. In chapter 8, MaryCarol Rossiter fully explores the necessary models for variance component dissection in complex experimental designs, especially regarding the variety of environmental sources (and their interactions) that may influence phenotypic expression in offspring. Together, these three chapters (as well as others) provide a comprehensive roadmap for future empirical studies exploring the adaptive significance of maternal effects.

6

The Detection and Measurement of Maternal Effects

DEREK A. ROFF

While the existence of maternal effects is no longer doubted, their detection and most particularly their quantitative measurement still present serious statistical and logistical problems. Maternal effects may be defined broadly as any phenotypic variation in offspring that is a consequence of the mother's phenotype rather than the genetic constitution of the offspring. Such effects may have a genetic basis or be entirely environmental; for example, offspring growth may be a function of the size of the mother, which itself may be inherited or entirely determined by environmental conditions during development (i.e., the size of the grandmother plus other environmental influences). The phenotypic response of the offspring may itself be a function of the offspring's genotype, and there may even be an interaction between the genes controlling the maternal effect and the response of the offspring.

Because of the complexity of the interaction between parental and offspring phenotypes, both theoretical and empirical studies of the significance of maternal effects have been slow in development. The general view has been that maternal (or paternal) effects are nuisances to be experimentally eliminated. However, it is now recognized, as demonstrated by the present volume, that maternal effects may play an important role in evolutionary change in natural populations and, therefore, that an analysis of their occurrence and determination is an essential component of evolutionary biology. In this chapter I review various techniques that have been used to detect or measure maternal effects, the utility of each approach typically depending critically on the biology of the organism under study. For a more complete review, see Roff (1997). Some methods (see section 6.1, 6.5, 6.6) aim to simply demonstrate the presence of maternal effects, while others assume a particular genetic model and aim to estimate the relevant parameters (sections 6.2, 6.3, 6.4, 6.6, 6.7).

6.1 Use of Single Locus Variants

This technique depends on the presence of a single locus marker that might be a visible mutation or an electophoretically detectable allele. Using such a marker one can distinguish among individuals that have the same mother but different fathers. If maternal effects are overwhelming, as might be expected, for example, in the case of the determination of egg size, then the two individuals will show no effect attributable to their male parent but there will be differences attributable to the mother. Maternal effects of a lesser magnitude can also be detected statistically. The use of a visible mutant is illustrated by the experiment of Weigensberg et al. (in press) on the maternal control of egg and hatchling size in the cricket *Gryllus firmus*. This experiment made use of a single locus eye mutant; whereas the usual eye color is red in the embryo and black in the nymph and adult, the homozygous mutant eye color is white in the embryo and orange in the nymph and adult. The mutant is recessive and does not appear to be deleterious. Triplets were set up consisting of a female homozygous for white eye and two males, one wild type and one homozygous for white eyes. Eggs laid on the same day were collected from the female and allowed to develop on moist filter paper. Measurements were made of egg length on the day they were laid and head width of the newly hatched nymphs. Differences due to the genotype of the offspring were assessed using a nested analysis of variance with sire nested within dam. If egg size is a trait of the mother then there should be no difference in egg size between sires, but there should be significant effects attributable to mother. At the time of laying, egg size was independent of sire but not of the mother, indicating maternal control of initial egg size. However, at hatching there was a significant difference in hatchling head width attributable to "sire," and hence to the genetic constitution of the offspring. Thus, by the time of hatching the genotype of the offspring had asserted itself.

The relative rarity of single locus mutants may seem to limit the general applicability of this technique (see Barnes 1984 for another example). However, the ability to separate lines based on electrophoretic variation has greatly increased the scope of potential applications (e.g., Kerver and Rotman 1987).

6.2 Reciprocal Crosses

The same principle described in section 6.1 can be applied using inbred lines, strains, or individuals (e.g., Chandraratna and Sakai 1960; Jinks and Broadhurst 1963; Smith and Fitzsimmons 1965; Fleming 1975; Corey et al. 1976; Millet and Pinthus 1980; Cadieu 1983; Garbutt and Witcombe 1986). Instead of using a single marker, several lines are crossed, typically in a diallel design, meaning that the various components of a line (genetic makeup, maternal and paternal influences) are expressed in a range of backgrounds. Differences attributable to nuclear, maternal, and paternal effects are then dissected from an analysis of variance. One model (though not the only possible model) is

$$G_{ij} = \text{Nuclear}_i + \text{Nuclear}_j + NN_{ij} + \text{Maternal}_i + \text{Paternal}_j + I_{ij}, \qquad (6.1)$$

where G_{ij} is the effect attributable to the ith and jth parent (or line), Nuclear is the nuclear contribution of the relevant parent, NN_{ij} is their interaction, Maternal_i is the maternal effect attributable to the ith mother, Paternal_j is the paternal effect due to the jth father, and I_{ij} is all other interactions. The analysis using equation (6.1) and other models is discussed by Cockerham and Weir (1977). For examples of the application of this approach using a diallel crossing design to nondomestic plant species, see Hayward and Nsowah (1969), Antonovics and Schmitt (1986), and Montalvo and Shaw (1994). The results for the last analysis demonstrate

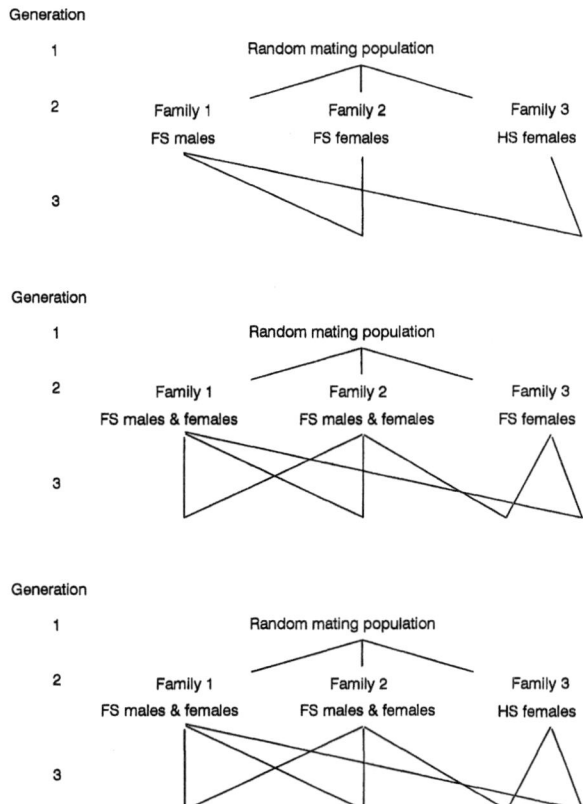

Figure 6.1 Three breeding designs suggested by Eisen (1967) for the separation of genetic and maternal effects. FS = full-sib, HS = half-sib. Modified from Thompson (1976).

the important general finding that maternal effects dissipate with age, with early traits (seed mass) determined primarily by maternal effects and later traits (leaf width) by the genotype of the offspring.

6.3 Eisen's Pedigree Analysis

For many organisms, reciprocal crosses are not easily made but several generations of crossing can be achieved. Most insects, such as *Drosophila* and *Tribolium*, and some vertebrates, such as mice, fall into this category. Eisen (1967) suggested three possible breeding designs (figure 6.1) requiring three generations of breeding to separate additive and dominance genetic variances from maternal effects in the analysis of variance model proposed by Willham (1963, 1972). In Design 1, full-sib males from family 1 are mated to females from families 2 and 3. Design 2 is more complex, with both males and females being used from families 1 and 2. The mating scheme of Design 3 is similar to Design 2 except that family 3 consists of half-sib females. The data are analyzed according to the statistical model

$$V_P = V_{AO} + V_{DO} + V_{AM} + V_{DM} + \mathrm{Cov}(A_O, A_{OM}) + \mathrm{Cov}(D_O, D_{OM}) + V_C + V_E, \quad (6.2)$$

where

V_P = phenotypic variance in the offspring;

V_{AO} = direct additive genetic variance, that is, the variation in the offspring due to the additive contribution of its own genes;

V_{DO} = direct dominance genetic variance, that is, the dominance deviation due to the genetic constitution of the offspring;

V_{AM} = maternal additive genetic variance (also called the indirect additive genetic variance), that is, variance in the trait value of the offspring due to the additive genetic variance of the mother (e.g., genetic variation in lactation, egg provisioning, and maternal care); and

V_{DM} = maternal dominance genetic variance (=indirect dominance genetic variance).

$\mathrm{Cov}(A_O, A_{OM})$ = direct maternal additive genetic covariance, that is, the additive genetic covariance between the offspring's trait value as determined by its own genes and the genes that determine the maternal performance, which are also transmitted to the offspring; and $\mathrm{Cov}(D_O, D_{OM})$ = direct maternal dominance covariance.

V_C = maternal environmental variance, that is, effect of environment provided by the mother on the offspring trait value that is not due to additive or dominant genetic effects in the mother. For example, provisioning of propagules may be a function of the mother's size, which is itself a function of genetic and environmental factors.

V_E = environmental variance, that is, that which remains after the above sources have been accounted for.

This model does not take into account possible effects that may come through the grandparent or some more distant source. Such effects can be incorporated but the model becomes very unwieldy (Willham 1963, 1972). The crosses from the three mating schemes produce a wide range of covariances and from these the variance components can be separated either by analysis of variance (Eisen 1967) or maximum likelihood (Thompson 1976), the latter being generally preferred. These designs are logistically difficult to arrange, and I have not found an example of their use, although maximum likelihood is commonly used by animal breeders to estimate the various components from general pedigrees.

6.4 Full-Sib, Half-sib and Offspring-Parent Designs

The expectations using the Willham model for these mating designs *assuming no dominance or epistasis* are shown in table 6.1. The maternal effect enters in three manners: the maternal additive variance, the direct maternal additive covariance, and the maternal environmental variance. The half-sib design gives an estimate of the additive genetic variance, but the three maternal effects cannot be separated (V_C can be estimated by using a split-family nested design).

If a trait such as propagule size is determined by the mother, then it is more rightly regarded as a trait of the mother rather than a trait of the propagule. In this case the heritability of the trait is obtained by regressing the mean trait value of the offspring on the mean trait value in the mother (e.g., mean size of offspring's eggs on mean size of the mother's

Table 6.1 Expectations of covariances obtained from the three "standard" breeding designs.

Relationship	Expected relationship
Paternal half-sibs	$\frac{1}{4}V_{AO}$
Full sibs within sire	$\frac{1}{4}V_{AO} + V_{AM} + \mathrm{Cov}(A_O, A_{OM}) + V_C$
Individuals within full-sib family	$\frac{1}{2}V_{AO} + V_E$
Offspring-sire	$\frac{1}{2}V_{AO} + \frac{1}{2}\mathrm{Cov}(A_O, A_{OM})$
Offspring-dam	$\frac{1}{2}V_{AO} + \frac{1}{2}V_{AM} + \frac{3}{4}\mathrm{Cov}(A_O, A_{OM})$

V_{AO} = direct additive genetic variance; V_{AM} = maternal additive genetic variance; $\mathrm{Cov}(A_O, A_{OM})$ = direct maternal additive genetic covariance; V_C = maternal environmental variance; V_E = environmental variance. Dominance variance is assumed to be zero. From Hanrahan and Eisen (1973).

eggs). Alternatively, some other design such as a full or half-sib could be used, again using the mean value of the character as the trait value.

If the trait is regarded strictly as a trait of the offspring, the usual offspring-parent regression is used, for example, the mean egg size produced by the offspring on the mid-parent egg size. In this design if egg size is actually a maternal trait, the offspring-dam regression will be significant but the offspring-father regression will not. This arises because if egg size were entirely maternally derived then the egg size from which the mother hatched is an estimate of the mean egg size produced by her mother (the grandmother of the offspring measured in this experiment). On the other hand, in this case, the egg size of the male parent makes no contribution to the egg size from which his offspring hatch; therefore, there should be no correlation between sire egg size and the mean egg size from which his offspring hatch. There will, however, be a correlation between the sire's egg size and the mean egg size produced by his offspring (the grandparent effect, discussed in section 6.5).

For offspring-parent data the estimation of maternal effects using the Willham model and assuming no environmental or nonadditive effects is done as follows (table 6.1). The additive genetic variance is estimated from the full-sib analysis and then subtracted from the offspring-sire component to give $\frac{1}{4}\mathrm{Cov}(A_O, A_{OM})$. The remaining component, V_{AM}, is then estimated from the offspring-dam regression. If the dominance variance cannot be assumed to be zero, the foregoing estimates will be biased. If the covariance term is positive, the slope of the mean offspring value on the mother's phenotypic value should then be higher than that on the sire (table 6.1). In particular, the "heritability" of the trait ($=2 \times$ slope) will appear higher using the offspring-on-dam regression than the offspring-on-sire regression. The standard errors associated with offspring on one-parent regressions are typically very high (Roff 1997) and hence large sample sizes are required to detect a significant difference. The absence of a difference between the two regressions does not necessarily demonstrate lack of a maternal effect since if $\mathrm{Cov}(A_O, A_{OM})$ is negative the two slopes could be very similar.

Lande and Price (1989), using the general maternal effects model proposed by Kirkpatrick and Lande (1989), developed an offspring-parent method for the estimation of the maternal components, *"providing that all of the maternal characters influencing the characters of interest are measured and included in the analysis"* (Lande and Price 1980, pp. 918; emphasis in original). To estimate the maternal-effects matrix all that is required are the two matrices of partial regression coefficients of offspring on mothers and offspring on fathers (for details see Lande and Price 1989).

6.5 The Grandparent Effect

A maternal influence that has a heritable basis will be determined by both the genes the female received from her mother and also the genes she received from her father. Therefore, the traits so influenced by the maternal effect will be visibly correlated with the phenotype of their grandfather rather than their father. Suppose, for example, initial propagule size is maternally determined but that this determination is genetically based. Let the two parents be designated $P_♀$ and $P_♂$. The propagule size produced by this mating is a function of the genotype of the mother, $G(P_♀)$, and is unrelated to that of the father. However, the phenotype of the propagules produced by the F_1 is a function of $P_♀$ and $P_♂$. Regarding these propagules, $P_♂$ is their grandfather. This provides an experimental approach to demonstrating both the presence of a maternal effect and also that it is heritable, as illustrated by Reznick (1981, 1982) for offspring weight in the mosquito fish *Gambusia affinis* and the guppy *Poecilia reticulata*. Reznick crossed mosquito fish whose grandparents had been collected from two widely separated localities in North America, North Carolina (N) and Illinois (I); crosses were made in all possible combinations giving four combinations of parents. The several generations of laboratory rearing were assumed to have removed any possible environmental effects arising directly from the natal locations. From each combination Reznick measured six offspring from six females. Analysis of variance showed that offspring weight was highly correlated with the female but not the male (figure 6.2). This experiment demonstrates the presence of a maternal effect but cannot separate environmental from genetic sources. To do this Reznick examined three further crosses involving wild-caught males from Illinois ($I_♂$) and North Carolina ($N_♂$). In one cross $I_♂$ were mated with female offspring from the Illinois × Illinois (II) cross. For the second cross $I_♂$ were mated with female offspring from the Illinois$_♀$ × North Carolina$_♂$ (IN) cross. Finally, $N_♂$ were mated with female offspring from the Illinois$_♀$ × North Carolina$_♂$ (IN). If offspring weight is due to a heritable maternal effect, then offspring from the latter two matings should not differ since they share the same grandfather stock (North Carolina). On the other hand, offspring from the first cross have different grandfathers (Illinois and North Carolina, respectively) and hence should be different. This pattern was observed (figure 6.2), confirming that the maternal effect did depend on nuclear genetic variation.

A simple variant on this scheme that demonstrates the presence of maternal (or paternal) effects in natural populations is to collect females from the field and from these raise two generations under constant conditions. Assuming that parents in the field experience conditions different from those in the laboratory (either in mean or variance), then a maternal effect in, say, early development will be generated in the offspring produced by the field-collected females but will not be present in subsequent generations (grandparental effects could be present in the first laboratory generation but would be lost by the second generation). This is illustrated by a study on the diapause propensity of the lepidopteran *Choristoneura rosaceana*: the mean proportion of larval diapausing in the first generation (i.e., from field-collected females) was 0.70 (SE = 0.24), but in the second and third generations only 0.17 (0.14) and 0.17 (0.13), respectively, of the larvae entered diapause (Carrière 1994). Note that the proportions in the second and third generations were identical, indicating the absence of any effects further back than the mother.

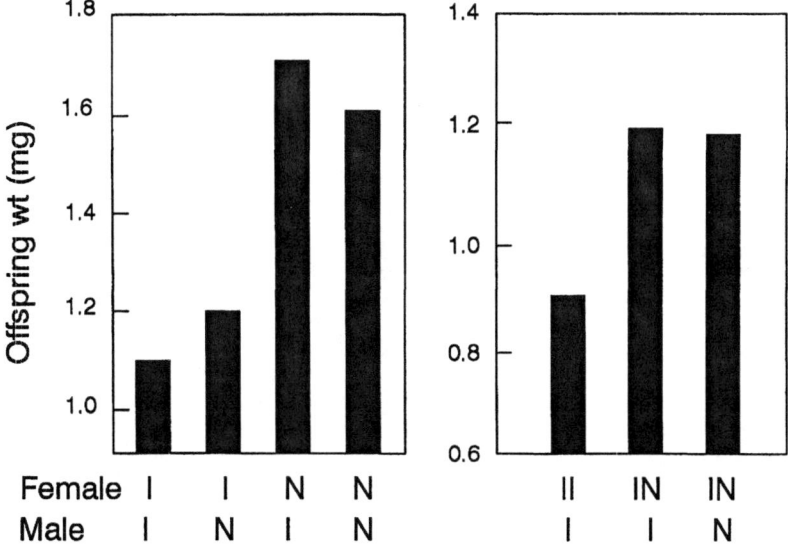

Figure 6.2 Offspring sizes of mosquito fish from different crosses. Left panel shows off-spring size from the four crosses of the stocks from Illinois (I) and North Carolina (N). Note that offspring size depends on the female but not on the male. Right panel shows crosses made from female offspring in the left panel and wild-caught males. Females that had grand-fathers from different localities (II and IN) produce offspring that differ in size, but females that had grandfathers from the same locality (IN and IN) produce the same-sized offspring regardless of the source of the father (I and N). Modified from Reznick (1981).

6.6 Environmental Manipulation

Consider a random sample of individuals drawn from a single population; one half is exposed to one set of environmental conditions during development, and the other half to an alternate environmental regime. If the offspring from the two groups, when raised in a common environment, display different phenotypes, then there must be maternal and/or paternal effects. This approach has been used in insects to demonstrate the importance of maternal influence on diapause, wing dimorphism, body size, survival rate, growth rate, lipid content, and sex ratio (reviewed in King 1987; Mousseau and Dingle 1991). Similar experiments have shown that maternal influences are important in plants, particularly regarding seed characteristics (reviewed by Roach and Wulff 1987). Alternatively, a female can be successively exposed to differing conditions. For example, suppose the environment is rotated through a sequence A, B, A, B, and so on, and the female produces eggs of size E_1, E_2, E_1, E_2, . . .; then this is evidence for a maternal effect on egg size. However, if the sequence were E_1, E_2, E_3, E_4, then this indicates variation in egg size but it is not possible to determine whether the variation is a result of the female or of the offspring genotypes (e.g., there may be physiological changes in the mother that cause selective mortality of genotypes). Several environments might be provided simultaneously. For example, King (1988) provided females of the parasitic wasp *Spalangia cameroni* with different sizes of hosts and found that the sex ratio of offspring varied according to host size, with further analyses indicating that this was not due to differen-

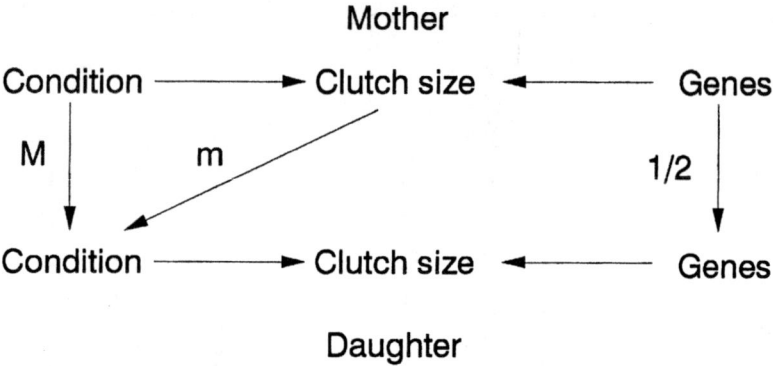

Mother

Figure 6.3 Maternal effects model suggested by Schluter and Gustafsson (1993) for clutch size in the collared flycatcher. Arrows connect dependent variables; symbols show the partial regression coefficients (=1 if not indicated). Redrawn from Schluter and Gustafsson (1993).

tial mortality. In some cases the effect of the mother may be indirect, as in *Sarcophaga crassipalpis*, in which diapause is determined by the larva as a consequence of a photoperiodic signal received through the integument of the female (Denlinger 1970, 1971, 1972).

These experiments can demonstrate the presence of a maternal effect but do not allow the separation of genetic maternal effects from environmentally based maternal effects. As discussed in sections 6.1–6.5, variation among females in the maternal effect they have on their offspring could be due at least in part to genetic variation between females (Falconer 1965; Willham 1972). Schluter and Gustafsson (1993) combined the experimental manipulation approach with a pedigree analysis to dissect these two components in the determination of clutch size in the collared flycatcher, *Ficedula albicollis*. The experimental manipulation consisted of transferring one egg from a clutch to the clutch of a neighboring female. The pedigree analysis used information on mothers and their daughters from the first-year and second-year clutches. Clutch size of the daughters was assumed to be determined by the genotype of the mother and two maternal components acting through the daughter's condition, one due to the condition of the mother (M) and one due to the clutch size that the daughter experienced (m; figure 6.3). The situation was further complicated by the fact that clutch size of a female depends on the size of the previous clutch, requiring a parameter b relating second-year clutch size to first-year clutch size.

From the presumed relationships, Schluter and Gustafsson were able to determine the expected covariance between offspring and parents. In addition to the three parameters m, M, and b, there is a parameter H, defined as the ratio of the additive genetic variance to the phenotypic variance. In the absence of maternal effects H is simply the heritability of the trait; however, in the presence of maternal effects genetic and nongenetic components of clutch size are correlated (Kirkpatrick and Lande 1989), and hence the phenotypic variance is the sum of the separate variances plus a covariance. With maternal effects the covariance between the phenotypic and additive genetic values is equal to the additive genetic variance (HV_P) plus the covariance between the additive genetic component and the condition component. The covariances between mother and daughter are similarly augmented.

The two parameters m and b were estimated from the clutch manipulation experiment. To obtain the remaining two parameters (M and H), the estimated values of m and b were substituted in the covariance equations, and the values of M and H that maximized the probability of obtaining the observed clutch sizes were found. The clutch size of a daughter was altered by approximately 0.25 eggs, giving an estimate of $m = -0.25$. The cost of the first clutch relative to the second was found to be similarly high ($b = -0.25$). From the maximum likelihood analysis, values of $M = 0.43$ (95% limits of 0.25–0.58) and $H = 0.33$ (−0.10–0.72) were obtained. Thus, if mothers in good condition produce large clutches, the negative maternal effect coming via the clutch size (m) is offset in part by the positive maternal effect coming from the condition of the mother (M). The estimate of "heritability" (H) is also quite high though the confidence region includes zero.

The clutch size model analyzed by Schluter and Gustafsson (1993) demonstrates that heritability estimates from offspring-parent regression may be confounded by maternal effects (see also Falconer 1989). Estimates of clutch size using dam-on-daughter regressions have typically given significant values (Roff in press), but because of the foregoing problem these estimates must be viewed with caution. In van Noordwijk et al.'s (1980) study of the great tit, *Parus major*, regressions based on three generations support the daughter-dam estimate (figure 6.4). All the heritabilities based on the assumption that the trait is maternally derived give similarly high estimates, while those based on the assumption of a significant paternal contribution give uniformly small and nonsignificant values. It must be stressed that this does not imply that the male does not contribute clutch-determining genes to his daughter; this is demonstrated by the daughter-on-paternal grandparent regression, which, though statistically nonsignificant (0.68, SE = 0.38), suggests a large additive genetic contribution, as does that of daughter on dam (0.37, SE = 0.12). As shown by figure 6.4, the "male" regressions amount to regression on the randomly drawn female that are predicted to give insignificant heritability estimates if the trait is a maternally expressed character. The similarity of the estimates across two generations argues for a heritable basis to the trait. Similar results have been obtained for other populations, confirming the conclusion that approximately 40% of the phenotypic variation in clutch size is due to additive genetic variation (van Noordwijk et al. 1980).

6.7 Cross-Fostering

If parental effects have no influence on the phenotype of the offspring, then switching offspring between parents should not influence, on average, the phenotypic value of the offspring. This approach has been used in the detection of maternal/paternal effects in mice and birds.

The first experiments (e.g., Brumby 1960) were able to show maternal effects but could not quantify these. Statistically more satisfactory experiments have since been carried out by Rutledge et al. (1972), Atchley and Rutledge (1980), Cheverud et al. (1983), Cheverud (1984), Leamy and Cheverud (1984), and Riska et al. (1984, 1985) using single strains of randomly bred mice. I shall use the study by Riska et al. (1984, 1985) to illustrate the approach. At birth litters were standardized to eight pups, and four (two of each sex if possible) were exchanged with four pups from another mother. There are two types of genetic effects that can be disentangled from this experiment: first, there is the usual direct genetic contribution made by both the mother and the father; second, there is an indirect contribution made by the

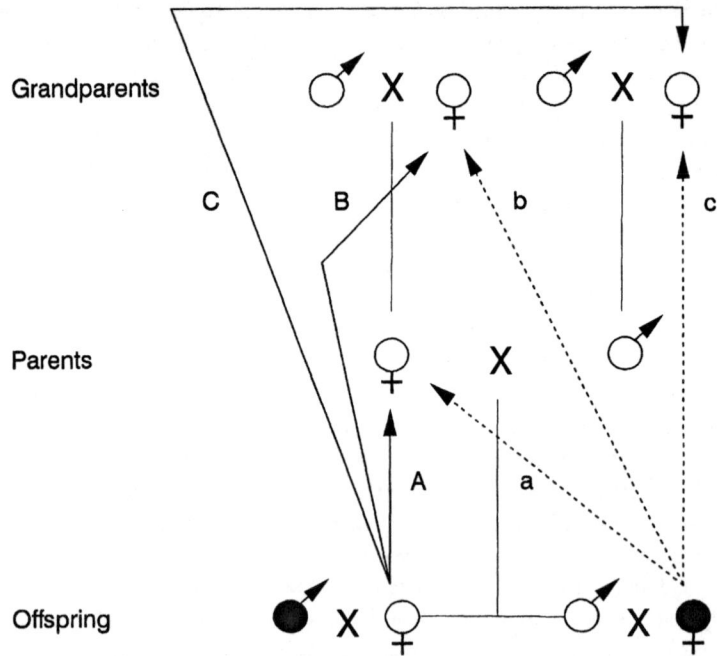

Figure 6.4 Heritability estimates of clutch size in the great tit, *Parus major*, calculated from descendent on ancestor regressions (data from van Noordwijk et al., 1980). Solid lines show pedigree relationships; solid symbols represent male and females drawn at random from the population. Arrows show the various regressions used: solid arrows are those using the female descendants; dashed arrows are those using the male descendants. Because clutch size is a sex-limited trait, it is calculated in all cases from the female, the hypothesis for the male descendants being that the male makes a contribution in some manner to the female that causes his genes to be expressed in her clutch size. *A*, daughter on dam, heritability = 0.37 (SE = 0.12); *B*, granddaughter on maternal granddam, 0.38 (0.56); *C*, granddaughter on paternal granddam, 0.68 (0.38); *a*, son on sire 0.05 (0.11); *b*, grandson on maternal grandsire, −0.04 (0.36); *c*, grandson on paternal grandsire, 0.05 (0.31).

female by virtue of the environment she provides, which may itself have a heritable component. Females may differ genetically in the type of care they provide (e.g., different physiologies or behaviors), and this will be passed on as a genetic heritage to the offspring, though they experience it as an environmental effect. These effects were separated using the Willham model.

Two types of analysis are possible with the experimental design of Riska et al. (1984, 1985), one utilizing only the cross-fostered pups and one using all the data. The linear model for the first analysis is

$$\text{Offspring weight} = c_0 + c_1 \text{Pair} + c_2 \text{Nursing mother} + c_3 \text{Genetic mother} + \text{Error}, \quad (6.3)$$

where the cs are coefficients. Assuming no cage effects and no dominance, the variance component due to the genetic mother is equal to one-half the direct-additive genetic variance (because the pups are full-sibs), and the variance due to the nursing mother is equal to the indirect-additive variance (maternal additive genetic variance) plus the common environmental variance (maternal or cage if not zero). The ratio of the maternal variance (=nursing mother) to the direct-additive genetic variance was very large at an early age, and even though it declined with age, it was still a significant ratio at age 10 weeks.

The data from both pups reared by their natural mothers and the pups reared by foster mothers can be analyzed using the variance decomposition

$$V_P = V_{AO} + V_{AM} + \mathrm{Cov}(A_O, A_{OM}) + V_C + V_E , \qquad (6.4)$$

where parameters are as defined for equation (6.2). For simplicity I have dropped the dominance components, which Riska et al. (1985) assumed to be zero. Unfortunately, while the covariance can be estimated this design does not provide a reliable estimate of V_{AM} (modification of the design such that the dams are themselves related does permit such an estimation; see Cheverud and Moore 1994 for further details). Because the cross-fostered mice are reared in an environment produced by genes that they do not share, the phenotypic variance of these mice is equal to $V_P - \mathrm{Cov}(A_O, A_{OM})$. Therefore, by two analyses of variance, one estimating the phenotypic variance of pups raised by their own mothers and one estimating the phenotypic variance of cross-fostered pups, the covariance can be estimated by subtraction. Riska et al (1985) obtained positive values of $\mathrm{Cov}(A_O, A_{OM})$, although negative values are common (Roff in press). Negative covariances could arise because of variation in litter size, an effect eliminated in the present experiment by standardization of litter size at birth. However, negative covariances have been found in experiments in which litter size either does not vary substantially (e.g., cattle) or in which litter size was standardized (e.g., Eisen et al. 1970).

The cross-fostering of mammals in natural populations is generally not feasible, although there are some species, such as squirrels, where it may be possible. In birds, however, it is relatively easy to switch eggs between nests, and thus this group is ideal for the study of maternal/paternal influences on offspring traits. Nine experiments of this kind have all shown that the offspring resemble their biological parents and not their foster parents, and that the heritabilities are independent of the rearing environment (see table 7.16 of Roff in press). Thus, at least regarding morphological traits, parental effects appear to be of minor importance in birds. But, as discussed in section 6.6, this is not the case for clutch size.

Summary and Conclusions

Experimental designs can be divided into two broad categories: those that establish the existence of a maternal effect but cannot measure its genetic basis, and those that provide estimates of some or all of the genetic variances and covariances of the maternal effect. In the first category are the use of single locus variants, comparison of offspring reared from field-collected mothers with the offspring they or their descendents produce (the grandparent effect), environmental manipulation, and simple methods of cross-fostering. To estimate genetic parameters requires the use of some type of pedigree analysis. Diallel analysis (reciprocal crosses) is a generally feasible method for plants but only partial diallels can typically be made for animals. An alternative method, Eisen's pedigree analysis, is possible for animals and plants but is logistically very difficult. Nevertheless, it does offer an important tool for the separation of both maternal and dominance components simultaneously. The more usual pedigree designs of half-sib and offspring-parent regression can only be used if nonadditive effects are assumed to be absent; whether this is a reasonable assumption remains to be demonstrated. The environmental manipulation, generally involving the cross-fostering of offspring, is potentially a very powerful method for the separation of maternal effects, although it does appear that the joint estimation of maternal and nonadditive effects may only rarely be possible. The method is also restricted to those organisms that provide parental care, which obviously greatly restricts the number of species for which it is a useful technique.

References

Antonovics, J., and J. Schmitt. 1986. Paternal and maternal effects on propagule size in *Anthoxanthum odoratum*. Oecologia 69:277–282.

Atchley, W. R., and J. J. Rutledge. 1980. Genetic components of size and shape. I. Dynamics of components of phenotypic variability and covariability during ontogeny in the laboratory rat. Evolution 34:1161–1173.

Barnes, P. T. 1984. A maternal effect influencing larval viability in *Drosophila melanogaster*. J. Hered. 75:288–292.

Brumby, P. J. 1960. The influence of the maternal environment on growth in mice. Heredity 14:1–18.

Cadieu, N. 1983. Maternal influence and the effect of heterosis on the viability of *Drosophila melanogaster* during different pre-imaginal stages. Can. J. Zool. 61:1152–1155.

Carrière, Y. 1994. Evolution of phenotypic variance: non-Mendelian parental influences on phenotypic and genetic components of life-history traits in a generalist herbivore. Heredity 72:420–430.

Chandraratna, M. F., and K. Sakai. 1960. A biometrical analysis of matriclinous inheritance of grain weight in rice. Heredity 14:365–373.

Cheverud, J. M. 1984. Evolution by kin selection: A quantitative genetic model illustrated by maternal performance in mice. Evolution 38:766–777.

Cheverud, J. M., and A. J. Moore. 1994. Quantitative genetics and the role of the environment provided by relatives in behavioral evolution. Pp. 67–100 in C. R. B. Boake (ed.), Quantitative Genetic Studies of Behavioral Evolution. University of Chicago Press, Chicago.

Cheverud, J. M., J. J. Rutledge, and W. R. Atchley. 1983. Quantitative genetics of development: Genetic correlations among age-specific trait values and the evolution of ontogeny. Evolution 37:895–905.

Cockerham, C. C., and B. S. Weir. 1977. Quadratic analyses of reciprocal crosses. Biometrics 33:187–203.

Corey, L. A., D. F. Matzinger, and C. C. Cockerham. 1976. Maternal and reciprocal effects on seedling characters in *Arabidopsis thaliana* (L.). Heynh. Genet. 82:677–683.

Denlinger, D. L. 1970. Embryonic determination of pupal diapause induction in the flesh fly *Sarcophaga crassipalpus* Macquart. Am. Zool. 10:320–321.

Denlinger, D. L. 1971. Embryonic determination of pupal diapause induction in the flesh fly *Sarcophaga crassipalpus*. J. Insect Physiol. 17:1815–1822.

Denlinger, D. L. 1972. Induction and termination of pupal diapause in *Sarcophaga* (Diptera: Sarcophagidae). Biol. Bull. 142:11–24.

Eisen, E. J. 1967. Mating designs for estimating direct and maternal genetic variances and direct-maternal genetic covariances. Can. J. Genet. Cytol. 9:13–22.

Eisen, E. J., J. E. Legates, and O. W. Robison. 1970. Selection for 12-day litter weight in mice. Genetics 64:511–532.

Falconer, D. S. 1965. Maternal effects and selection response. Pp. 763–774 in S. J. Geerts (ed.), Genetics Today. Pergamon Press, Oxford.

Falconer, D. S. 1989. Introduction to Quantitative Genetics. Longman Scientific and Technical, Essex.

Fleming, A. A. 1975. Effects of male cytoplasm on inheritance in hybrid maize. Crop Sci. 15:570–573.

Garbutt, K., and J. R. Witcombe. 1986. The inheritance of seed dormancy in *Sinapis arvensis* L. Heredity 56:25–31.

Hanrahan, J. P., and E. J. Eisen. 1973. Sexual dimorphism and direct and maternal genetic effects on body weights in mice. Theor. and Appl. Genet. 43:39–45.

Hayward, M. D., and G. F. Nsowah. 1969. The genetic organisation of natural populations of *Lolium perenne* IV. Variation within populations. Heredity 24:521–528.

Jinks, J. L., and P. L. Broadhurst. 1963. Diallel analysis of litter size and body weight in rats. Heredity 18:319–336.

Kerver, W. J. M., and G. Rotman. 1987. Development of ethanol tolerance in relation to the alcohol dehydrogenase locus in *Drosophila melanogster*. II. The influence of phenotypic adaptation and maternal effect on survival on alcohol supplemented media. Heredity 58:239–248.

King, B. H. 1987. Offspring sex ratios in parasitoid wasps. Q. Rev. Biol. 62:367–396.

King, B. H. 1988. Sex-ratio manipulation in response to host size by the parasitioid wasp *Spalangia cameroni*: A laboratory study. Evolution 42:1190–1198.

Kirkpatrick, M., and R. Lande. 1989. The evolution of maternal characters. Evolution 43:485–503.

Lande, R., and T. Price. 1989. Genetic correlations and maternal effect coefficients obtained from offspring-parent regression. Genetics 122:915–922.

Leamy, L., and J. Cheverud. 1984. Quantitative genetics and the evolution of ontogeny. II. Genetic and environmental correlations among age-specific characters in randombred mice. Growth 48:339–353.

Millet, E., and M. J. Pinthus. 1980. Genotypic effects of the maternal tissues of wheat on its grain weight. Theor. Appl. Genet. 58:247–252.

Montalvo, A. M., and R. G. Shaw . 1994. Quantitative genetics of sequential life-history and juvenile traits in the partially selfing perennial *Aquilegia caerulea*. Evolution 48:828–841.

Mousseau, T. A., and H. Dingle. 1991. Maternal effects in insect life histories. Annu. Rev. Entomol. 36:511–534.

Reznick, D. N. 1981. "Grandfather effects": The genetics of offspring size in mosquito fish *Gambusia affinis*. Evolution 35:941–953.

Reznick, D. N. 1982. Genetic determination of offspring size in the guppy (*Poecilia reticulata*). Am. Nat. 120:181–188.

Riska, B., W. R. Atchley, and J. J. Rutledge. 1984. A genetic analysis of targeted growth in mice. Genetics 107:79–101.

Riska, B., J. J. Rutledge, and W. R. Atchley. 1985. Covariance between direct and maternal genetic effects in mice, with a model of persistent environmental influences. Genet. Res. 45:287–297.

Roach, D. A., and R. D. Wulff. 1987. Maternal effects in plants. Annu. Rev. Ecol. Syst. 18:209–235.

Roff, D. A. 1997. Evolutionary Quantitative Genetics. Chapman and Hall.

Rutledge, J. J., O. W. Robison, E. J. Eisen, and J. E. Legates. 1972. Dynamics of genetic and maternal effects in mice. J. Anim. Sci. 35:911–918.

Schluter, D., and L. Gustafsson. 1993. Maternal inheritance of condition and clutch size in the collared flycatcher. Evolution 47:658–667.

Smith, W. E., and J. E. Fitzsimmons. 1965. Maternal inheritance of seed weight in flax. Can. J. Genet. and Cytol. 7:658–662.

Thompson, R. 1976. The estimation of maternal genetic variances. Biometrics 32:903–917.

van Noordwijk, A. J., J. H. van Balen, and W. Sharloo. 1980. Heritability of ecologically important traits in the great tit. Ardea 68:193–203.

Weigensberg, I., Y. Carrière , and D. A. Roff. In press. Effects of male genetic contribution and paternal investment on egg and hatchling size in the cricket *Gryllus firmus*. J. Evol. Biol.

Willham, R. L. 1963. The covariance between relatives for characters composed of components contributed by related individuals. Biometrics 19:18–27.

Willham, R. L. 1972. The role of maternal effects in animal breeding: III. Biometrical aspects of maternal effects in animals. J. Anim. Sci. 35:1288–1293.

7

Genetics of Maternal and Paternal Effects

RUTH G. SHAW & DIANE L. BYERS

The evolution of any adaptation depends primarily on additive genetic variation in the traits involved and on selection (Falconer and Mackay 1996). Thus, the adaptive nature of effects of maternal and paternal parents on their progeny involves the genetic determination of parental effects. It has long been recognized that maternal effects, in particular, may be genetically variable and hence may evolve and contribute to selection response (e.g., Eisen 1967; Van Vleck 1970, 1976). Yet consideration of maternal effects has often emphasized two possible consequences in relation to selection. (1) Environmentally induced maternal effects are typically confounded with genetic effects and may lead to overestimates of direct additive genetic variance (V_A; i.e., additive genetic variance attributable to variation in genotype of the individuals expressing the trait), and thus to erroneous prediction of rapid selection response (e.g., Mitchell-Olds and Rutledge 1986). (2) Maternal effects arising from environmental variation may impede response to selection by obscuring additive genetic effects (e.g., Stratton 1989). Kirkpatrick and Lande (1989, 1992; see also Lande and Kirkpatrick 1990) presented an approach to account for maternal effects in predicting selection responses, but this work treated as fixed the relationship between offspring traits and maternal traits and therefore does not account for evolutionary change in a maternal effect.

These emphases may have detracted from interest in the potential contribution of maternal effects to evolution of adaptive responses. Thus, in their review documenting abundant evidence of maternal effects in plants, Roach and Wulff (1987) noted that few studies of wild species had, at that time, been devoted to quantitative assessment of genetic variation for maternal effect, and though some studies addressing this issue have since appeared (e.g., Platenkamp and Shaw 1993, Schmid and Dolt 1994), they are still limited (Riska 1991). The potential for evolutionary response of a maternal effect to natural selection remains largely a matter of speculation.

After a review of potential sources of parental effects, we provide a brief overview of evidence for maternal effects. We then review studies designed to reveal evidence of genetic determination of maternal effects. Some of these address variation both in main effects of maternal parent and in the response of maternal effects to environmental conditions (i.e., plasticity of maternal effects). Focusing on parent effects per se, we illustrate the value of quantitative genetics in empirically partitioning genetic and environmental contributions to maternal and paternal variance. We present a greenhouse experiment indicating a substantial maternal additive genetic contribution to variation in seed mass of *Nemophila menziesii*. Studies that separately estimate the genetic and environmental components of maternal effects are mostly from the agricultural literature and number far fewer than studies that demonstrate maternal effects in the broad sense (Riska 1991). It is therefore not possible at present to generalize concerning the evolutionary potential of maternal effects. Our results and those of Thiede (1996), however, suggest that maternal additive genetic variance (V_{Am}), which, like V_A, can contribute to selection response, may frequently be substantial in natural populations. We conclude by considering some possible impediments to adaptive evolution of maternal effects.

Traits that are subject to parental effects are likely influenced by diverse factors, both genetic and environmental, that can affect an individual either directly, or indirectly through its parent. In our discussion of parental effects we focus on maternal effects, noting that paternal effects can be considered in an analogous way.

Variation attributable to maternal effects can result from several causes, including (1) effects of maternal, environmentally influenced condition on developing progeny (V_{Em}) and (2) variation in organelle genomes, which in many taxa are cytoplasmically transmitted to ovules (V_{Cm}). (However, biparental inheritance occurs in some angiosperm taxa, and paternal inheritance predominates in gymnosperms [Corriveau and Coleman 1988; Mogensen 1996]. Thus, the mode of organelle inheritance for a particular taxon must be taken into account in assessing the effects of this component.) Further, (3) nuclear genetic variation at loci expressed in maternal individuals and influencing traits of offspring can contribute to maternal variance (Eisen 1967; Roach and Wulff 1987). This genetic maternal variation may, in principle, have additive, dominance (V_{DM}) and epistatic (V_{IM}) components, but we focus on additive genetic maternal variance (V_{Am}) as the primary source of genetic variation expected to contribute to overall evolution of maternal effect on a given offspring trait. For example, directional selection favoring increase in seed mass could proceed through evolution of the maternal effect, given V_{Am}.

Disentangling V_{Am} from the many possible sources of variation in offspring phenotype (see figures in Riska et al. 1985; Roach and Wulff 1987) requires information on appropriate genetic relationships of maternal parents, together with measures of their progeny for the traits of interest (Eisen 1967; Shaw and Waser 1994). Precision and accuracy of estimates of V_{Am} depend on the kinds of relationships used in the experimental design and on the size of the experiment. Designs of studies to infer V_{Am} can thus be seen as closely analogous to those for inference of direct additive genetic variance (V_A; see Falconer and Mackay 1996), with the distinction that estimation of V_{Am} relies on measures of progeny of mothers whose genetic relationships are known.

Genetic variation due to maternal transmission of cytoplasmic genes (V_{Cm}) may also contribute to evolution of maternal effects. However, definitive assessment of this source of maternal genetic variation is exceptionally cumbersome (Kirkpatrick and Dentine 1988), because effects of cytoplasmic genes are typically confounded with other effects, especially (1) environmental effects that persist over multiple generations and (2) dominance effects. We discuss in section 7.4 approaches that would enable estimation of this source (V_{Cm}) of maternal effect.

Potential for adaptive evolution of maternal effect includes, in principle, evolution of environment-specific maternal effects, such that maternal effects exhibit adaptive plasticity. The potential for such evolution would be evidenced as additive genetic variation in plasticity of maternal effect. Considering, for conceptual and empirical simplicity, the case of just two environments, inference of such variation requires extending the general scheme for inferring V_{Am} to one in which pedigreed maternal parents are grown in distinct environments and their further pedigreed progeny measured. This design is analogous to those for studying direct additive genetic variation in environmental response (i.e., the variance due to interaction between breeding value and environment, $V_{A \times E}$; e.g., Rawson and Hilbish 1991; Shaw et al. 1995) but requires an additional generation of pedigreed individuals to estimate $V_{Am \times E}$. It would reveal the magnitude and nature of interactions between additive genetic maternal effect and the environment in which the maternal parent is grown, and it would permit estimation of the additive genetic correlation between maternal effects in the two environments. This information, together with that concerning selection on the progeny trait in the two environments, would provide a basis for predicting adaptive evolution of the plasticity of the maternal effect.

7.1 An Overview of Maternal and Paternal Effects

Relative to traits of mature individuals, juvenile traits (e.g., size of juveniles: seeds, larvae, or newborns) have been found to be particularly influenced by maternal effects both in plants (e.g., Schwaegerle and Levin 1990; Biere 1991; Richardson and Stephenson 1991; Finkelstein 1994; Montalvo and Shaw 1994; Waser et al. 1995; see also Roach and Wulff 1987 for review) and animals (Bernardo 1996; reviewed in Mousseau and Dingle 1991; Riska 1991). This is expected, since development and maturation of offspring occur on or in the maternal individual in many organisms. Maternal effects are less often detected for later traits (Roach and Wulff 1987), but there are exceptions (Mazer 1987; Roach and Wulff 1987; Stratton 1989; Helenurm and Schaal 1996; Lacey 1996; Byers in press). Some reported studies were designed to provide estimates of direct additive variance (V_A) and have led to the general conclusion that V_A in juvenile traits is often limited (with heritabilities in natural populations less than 10%) while V_M (variation due to all possible sources of maternal effect: maternal environmental, cytoplasmic, and maternal nuclear effects) is substantial (e.g., Mazer 1987; Waser et al. 1995). Expression of maternal effects may vary with either the maternal or offspring environment (Alexander and Wulff 1985; Stratton 1989; Miao et al. 1991a; Schmitt et al. 1992; Wulff et al. 1994; Lacey 1996), and maternal effects can last at least one generation beyond that of the initial offspring (Alexander and Wulff 1985; Miao et al. 1991b; Case et al. 1996). Interpretation of observed maternal effects has often emphasized environmental induction as a likely mechanism (e.g., Mazer 1987; Stratton 1989; Biere 1991), yet genetic contributions to a maternal effect cannot logically be excluded unless genetic differences among maternal parents are deliberately minimized or eliminated (e.g., Aarssen and Burton 1990).

Paternal effects on offspring phenotype have been sought using diallel and reciprocal factorial crossing schemes but tend not to be found (e.g., Antonovics and Schmitt 1986; Schmitt and Antonovics 1986; Biere 1991; Byers et al. 1997). Experimental manipulation has demonstrated effects of paternal environment on certain traits of offspring production (i.e., pollen quality in *Phlox*: Schlichting 1986; seed set in *Raphanus*: Young and Stanton 1990; seed germination in *Plantago*: Lacey 1996; egg size in *Drosophila melanogaster*: Crill et al. 1996). Such paternal environmental effects could have important implications for the representation of sires in the next generation but need not imply expression of environmentally induced paternal effects in offspring. For example, Crill et al. (1996) detected no effects of experimentally manipulated paternal temperature on morphological and physiological attributes of growing or mature offspring. These findings support the usual assumption that sires affect the phenotypes of their offspring predominantly through Mendelian transmission of genes determining their breeding value (value of an individual as judged by the mean value of its progeny; Falconer and Mackay 1996). This is a key assumption (also confirmed in our study, described in section 7.4) facilitating inference of components of genetic variation, including V_{Am}. Evidence that this assumption need not always hold is presented by Lacey (1996; see below).

7.2 Broad-Sense Heritability of Maternal Effect

Quantitative genetic approaches that are commonly used for studying inheritance of polygenic traits have been extended to account for maternal effect and to determine its causal basis. For a population of interest, inference of components of variance attributable to distinct causes is based on resemblance of relatives. In particular, assessment of the genetic basis of a maternal effect requires information on the relationships between maternal parents on whose progeny measurements are obtained.

For some study organisms, clonal replication is a logistically straightforward and rapid method of obtaining individuals of known relationship that can serve as maternal parents. This approach has been used in several recent studies of maternal effect in *Plantago lanceolata*. Alexander and Wulff (1985) varied temperature and CO_2 concentration in which maternal plants were grown. Apart from highly significant main effects of maternal environment on the offspring traits, seed weight, and percentage germination, they also found highly significant differences in these traits among the four maternal clones. These clonal effects could include maternal genetic variation, which could support a response of maternal effects to selection. They could also include effects of the field environment from which maternal plants were originally sampled and of systemic microorganisms, as well as effects of nuclear alleles expressed in offspring. Thus, these could equally be considered indications of maternal genetic variation or direct genetic variation "in the broad sense," as that phrase is typically applied in quantitative genetics (Falconer and Mackay 1996).

Beyond the main effects, Alexander and Wulff (1985) detected significant interactions between maternal clone and maternal environment, suggesting genetic variation among mothers in plasticity of their nonnuclear contributions to offspring. They also found significant threeway interactions between maternal clone, maternal environment, and progeny environment, indicating differences among clones in the dependence of progeny traits jointly on maternal plant and on temperatures prevailing both during seed maturation and during germination and growth of progeny. Such interactions raise the possibility of genetic variation that could support environment-specific adaptation through evolution of the maternal effect.

Schmitt et al. (1992) employed clonal replicates of *P. lanceolata* to study maternal plasticity in relation to naturally occurring environments and, in a greenhouse study, to light and nutrient availability. These experiments, too, revealed significant effects of maternal clone on seed weight and germination percentage, as well as interactions between maternal clone and maternal environment, suggesting variation among maternal sibling groups in response to maternal environment.

Lacey (1996) further extended this approach (also with *P. lanceolata*) to distinguish prezygotic effects of temperature mediated by paternal parent from those mediated by maternal parent, and both from effects of temperature prevailing during seed maturation. Effects of postzygotic temperature and of paternal prezygotic temperature on seed weight and germination percentage were detected, as were effects of parental clone. There was also evidence of interaction between parental clone and prezygotic temperature.

This collection of studies of *P. lanceolata* is consistent not only in demonstrating strong effects of parental environment on offspring traits, but also in providing evidence suggesting that the genotype expressed by parents may influence offspring phenotype. Evidence of maternal genotypic effects, in the broad sense, has also been obtained for *Nemophila menziesii* by a somewhat different approach (Platenkamp and Shaw 1993). In this case, maternal parents related as full siblings were obtained from a series of formal crosses of plants sampled from a natural population as seedlings. Half the individuals of each full-sibling group were grown singly in pots, while the remainder were grown with *Bromus diandrus*, a co-occurring grass. Each of three unrelated plants, designated as pollen donors and growing singly, was crossed with each maternal plant.

Striking effects of maternal environment (both block and competition treatment) were found for several progeny traits in *N. menziesii* (Platenkamp and Shaw 1993). A main effect of maternal pedigree was detectable only for seed dormancy, while interactions between maternal pedigree and maternal environment were detected for seed mass, time to germination, and dormancy. For seed mass, the interaction should be viewed with caution, because it includes variation among maternal individuals. As with the other studies described above, such variation among maternal pedigrees could result from an influence of nuclear genetic variation including maternal effect. It could also be due to nuclear genetic expression in the progeny, particularly in the case of dormancy, where a sire effect was also found. Finally, it could, in principle, be due to environmental effects carried over from previous generations (i.e., for the immediately previous one, variation among random positions in the greenhouse, and for the second previous, variation among locations sampled within a field).

7.3 Further Dissection of Maternal Effects

More definitive partitioning of maternal genetic effects from effects of maternal environment and offspring genotype has been accomplished to differing degrees in studies of selected or inbred lines, using either cross-fostering of young between lines or genetic crosses between lines. Alternatively, these effects can be inferred from a segregating population either by making use of known pedigrees or by implementing a formal crossing scheme spanning at least three generations (Eisen 1967; Riska 1991).

Cross-fostering studies (e.g., White et al. 1968), including embryo transfer (e.g., Moore et al. 1970; Cowley et al. 1989), have been used to distinguish between effects of maternal and offspring genotypes. For example, an investigation of prenatal and postnatal maternal influences of mice selected for high or low body weight entailed a reciprocal cross-fostering approach in which each dam nursed two of her own offspring and two offspring from each of two other females. A single individual sired all the young of a set (White et al. 1968). Significant prenatal and postnatal maternal effects on weight gain were found. Extending this study to further separate genetic and environmental components contributing to the prenatal maternal effect, Moore et al. (1970) used embryo transfer between the same lines of mice and showed that prenatal environment (uterine effect) and genetic line of the dam significantly affected birth weight. While both postnatal effect and genetic line were significant and continued for the duration of the study (10 weeks), the uterine effect on weight was not significant by one week after birth. However, the study by Cowley et al. (1989) comparing three inbred lines of mice demonstrated a persistent uterine effect.

Genetic crosses between lines must also include reciprocals in order to distinguish maternal effects and reveal their nature. Reciprocal crosses between breeds of swine demonstrated a maternal genetic effect on litter weight (Ikeobi and Ngere 1994). Güneren et al. (1996) found a strong effect of offspring genotype and weaker effect of maternal genotype on fetal size in a cross of lines of mice divergently selected for body weight. In some cases, advanced generations have been used, and these can further yield insight into the action of genes involved in parental effects. For example, Pleines and Friedt (1989), studying reciprocal F_1s, F_2s, and backcrosses from a cross of *Brassica napus*, detected an effect of maternal genotype on linoleic acid concentration of seeds, as well as a larger effect of offspring genotype. Further experiments with inbred lines in crops have revealed contributions of maternal additive effect to phenotypic variation in seed weight (*Lycopersicon esculentum:* Sanford and Hanneman 1982), tuber weight (*Solanum tuberosum*: Nieuwhof et al. 1989), and kernel growth and weight (*Triticum aestivum*: Rasyad and van Sanford 1992). In pearl millet (*Pennisetum americanum),* significant maternal effects on seed viability were expressed as an interaction between maternal genotype and the nuclear genes of the embryo (Singh et al. 1991). The analysis of inbred line experiments has most recently been considered by Beavis et al. (1987), Foolad and Jones (1992), and Zhu and Weir (1994). Zhu and Weir (1994) present a genetic model for a modified diallel crossing design using F_1s, reciprocal F_1s, and backcrosses, which provides estimates of cytoplasmic effects, maternal additive and dominance effects, and direct additive and dominance effects. For segregating populations sampled from nature, inbred lines would not typically be readily available. Investigation of the genetic basis of differences in maternal effects between populations could proceed, however, with reciprocal crosses between them and use of Lande's (1981) generalization of line cross analysis to segregating populations.

Analysis of reciprocal crosses between distinct, segregating populations can, in addition, make use of pedigree information within populations to provide estimates of genetic variation within and among breeds (e.g., Olson et al. 1985; Kahi et al. 1995). Crosses between two breeds of cattle demonstrated significant maternal heterosis effects on progeny traits and additive maternal effects of breed on pregnancy and weaning rates (Olson et al. 1985). Most studies of this kind have been undertaken to improve selection response, and Azzam and Nielsen (1987) suggest that this objective is accomplished in many cases. Kahi et al. (1995) report an exception: a diallel crossing design of several breeds of cattle found no significant maternal genetic effects on preweaning performance, and therefore the authors suggested that in this case maternal effect need not be considered in selecting (Kahi et al. 1995).

In a more general context, the basis of variation due to maternal effects within populations can be determined with reference to random mating populations according to quantitative genetic designs exemplified by Eisen (1967). These designs require at least two generations descended from crosses of individuals sampled, ideally at random, from the reference population and yield estimates of additive genetic variance of maternal effects (V_{Am}). In these designs, crosses are conducted, again, ideally at random, such that pedigrees of parents, as well as those of offspring, are known. Paternal half-sibling relationships among mothers are especially informative in inference of V_{Am}, because it is the closest relationship that avoids confounding between the additive genetic maternal effect of primary interest and other documented contributions to maternal effect. In Eisen's (1967) designs, effects of maternal environment in the preceding generation, effects of maternal environment in the present generation, and effects of organelle genomes (assuming maternal inheritance of organelles) are all associated, in principle, with maternal individual or, in the latter case, maternal lineage. Implicit in Eisen's (1967) approaches to estimating V_{Am} is an assumption that nonnuclear contributions of maternal grandsire to the phenotype of its daughter's offspring are negligible.

Estimation of V_{Am} together with other components of variation has recently become feasible for the large general pedigrees that are available from livestock breeding programs (Meyer 1992). For example, restricted maximum likelihood (REML) analysis applied to data on several breed of cattle has revealed significant V_{Am} for weaning weight (e.g., Khombe et al. 1995; Meyer in press). Estimates of V_{Am} are similar in magnitude to estimates of variance due to environment shared by maternal siblings. Meyer's (in press) analyses also revealed that effects of environment tending to increase maternal phenotype often tend to decrease offspring phenotype, as Falconer (1965) originally observed for mice.

This cursory review indicates that experimental study of the genetic basis of maternal effect has been applied to crop, livestock, and laboratory populations, often with the motivation of determining artificial selection schemes that take greatest advantage of the genetic variation available (e.g., Van Vleck 1970, 1976; Azzam and Nielsen 1987). In contrast, the additive genetic basis of maternal effect in wild populations has rarely been estimated (Riska 1991). Approaches similar to those outlined by Eisen (1967), using pedigreed individuals for two generations, have recently been applied in two studies of natural populations of plants (Thiede 1996; Byers et al. 1997). In a study of the winter annual *Collinsia verna* (Scrophulariaceae), Thiede (1996) raised progeny from three generation pedigrees in greenhouse and field. For greenhouse-grown plants, she detected significant contributions of V_{Am} to variation in several traits observed early in the life cycle, and for several traits, V_A was also found. In contrast, neither of these components made statistically significant contributions to the variance of the traits assayed on field-grown plants. In section 7.4, we consider our own study in more detail.

7.4 Seed Mass in Nemophila menziesii

Nemophila menziesii is an annual plant of the family Hydrophyllaceae that grows in coastal sage scrub communities of southern California. To examine the basis of parental effects on seed mass, we have extended the two-generation design of Shaw et al. (1995) by another generation to separate V_{Am} as well as additive genetic paternal effect (V_{Ap}) from other parental effects (Shaw and Waser 1994). This study focuses on the genetic basis of parental effect per se, without consideration of a genetic basis of response to parental environment (e.g., V_{Am*Em}).

Seedlings of *Nemophila menziesii* were sampled from a natural population on the campus

GRSIRES$_0$→			δ_A			δ_B			δ_C			δ_D			δ_E		
	GRDAMS$_0$→		$♀_1$	$♀_2$	$♀_3$	$♀_4$	$♀_5$	$♀_6$	$♀_7$	$♀_8$	$♀_9$	$♀_{10}$	$♀_{11}$	$♀_{12}$	$♀_{13}$	$♀_{14}$	$♀_{15}$
		GEN$_1$→	$\delta_{A\text{-}1}$	$\delta_{A\text{-}2}$	$\delta_{A\text{-}3}$	$\delta_{B\text{-}4}$	$\delta_{B\text{-}5}$	$\delta_{B\text{-}6}$	$\delta_{C\text{-}7}$	$\delta_{C\text{-}8}$	$\delta_{C\text{-}9}$	$\delta_{D\text{-}10}$	$\delta_{D\text{-}11}$	$\delta_{D\text{-}12}$	$\delta_{E\text{-}13}$	$\delta_{E\text{-}14}$	$\delta_{E\text{-}15}$
	$♀_{17}$	$♀_{I\text{-}17}$	X	X	X	X	X	X	X	X	X	X	X	X	X	X	X
δ_I	$♀_{18}$	$♀_{I\text{-}18}$	X	X	X	X	X	X	X	X	X	X	X	X	X	X	X
	$♀_{19}$	$♀_{I\text{-}19}$	X	X	X	X	X	X	X	X	X	X	X	X	X	X	X
	$♀_{20}$	$♀_{II\text{-}20}$	X	X	X	X	X	X	X	X	X	X	X	X	X	X	X
δ_{II}	$♀_{21}$	$♀_{II\text{-}21}$	X	X	X	X	X	X	X	X	X	X	X	X	X	X	X
	$♀_{22}$	$♀_{II\text{-}22}$	X	X	X	X	X	X	X	X	X	X	X	X	X	X	X
	$♀_{23}$	$♀_{III\text{-}23}$	X	X	X	X	X	X	X	X	X	X	X	X	X	X	X
δ_{III}	$♀_{24}$	$♀_{III\text{-}24}$	X	X	X	X	X	X	X	X	X	X	X	X	X	X	X
	$♀_{25}$	$♀_{III\text{-}25}$	X	X	X	X	X	X	X	X	X	X	X	X	X	X	X

Figure 7.1 Design of one of five blocks for the three generation design. Each X indicates a mating between plants in generation 1 and therefore also represents a full-sibling group of progeny in generation 2. Subscripts of individuals in generation 1 identify their parents drawn from the original, field-collected founders, generation 0, also shown. Plants of generation 0 that were designated as sires of plants serving as dams in generation 1 (i.e., maternal grandsires of plants in generation 2) are indicated by Roman numerals; those that served as sires of plants that would serve as sires in generation 1 (i.e., paternal grandsires) are indicated by capital letters. Maternal parents in generation 0 are indicated by Arabic numbers. The groups of maternal grandsires, paternal grandsires, maternal granddams, and paternal granddams were mutually independent. These were nested within crossing blocks, which were therefore also independent. Each cross was replicated four times.

of University of California at Riverside to establish generation 0. Relationships among these plants were unknown but assumed to be small because plants were sampled at 2-meter distances and then mated at random. These individuals were crossed in a standard nested design (each individual designated as a sire crossed with a distinct set of three plants serving as dams) to produce generation 1 (figure 7.1). From these offspring, seeds from each paternal half-sibling group (10 from each full-sibling group) were weighed and a single individual per full-sibling group grown to serve as parents in a further series of crosses. Plants were then crossed in a factorial design to produce generation 2 on which measures of seed mass were obtained. By employing paternal half-sibling relationships among paternal as well as maternal parents, this design also permitted partitioning of paternal effects on seed mass.

The data from the progeny generation are mostly complete (i.e., without missing cells) and nearly balanced (i.e., with few exceptions, there were consistently five seeds per fruit and four fruits per mating). On these grounds, observational (or design) components of variance can readily be obtained from an analysis of variance, for which we employed SAS GLM (SAS Institute 1985). More generally, estimation of variance components can be accomplished using maximum likelihood (available programs include Meyer 1992; Shaw and Shaw 1994).

The factors included in the model are maternal grandsire (MGS), paternal grandsire

Table 7.1 Results from analysis of seed weight.

Source of variation	Var. comp.	% of V_P
Maternal grandshire (MGS)	8.6*	5.3
Paternal grandshire (PGS)	0.3	0.2
Dam(MGS)	24.2***	14.1
Sire(PGS)	1.5	0.9
MGS × PGS	−1.5	0
Dam(MGS) × PGS	5.6	3.3
Sire(PGS) × MGS	5.7	3.3
Residual	124.4	73.3

*$P < 0.08$.
***$P < 0.0001$.

(PGS), dam nested within MGS, sire nested within PGS, their interactions, and fruit within mate combination. All of these factors were considered random. Employing genetic and environmental interpretations as in Eisen (1967), causal components of variance were inferred from the observational components (table 7.1). In particular, the variance components due to MGS and PGS yield an estimate of maternal additive genetic variance as $4(F^2_{MGS} - F^2_{PGS})$, giving 0.033, which accounts for 20.4% of the phenotypic variation. This estimate is consistent with the variance component for Dam(MGS) (=14.1%), which includes $3/4 V_{Am}$. These estimates suggest that, in this experiment, other causes of differences among maternal parents contribute little to differences among their seeds.

Variance due to paternal contributions is confounded, in principle, with direct additive genetic variance, V_A. However, effects of sires and paternal grandsires on seed mass are negligible in this study (table 7.1). We can obtain upper bounds for V_A, V_{Ap}, and V_{Ep}, in turn, by assuming the remaining two equal zero. Using this approach, we find that V_A accounts for at most 4% of V_P, V_{Ap} for most 0.8% of V_P, and V_{Ep} for at most 0.9%. None of these estimates are statistically significant. We further confirmed the magnitudes of V_A and V_{Am} by a joint regression of the mean of the full-sibling groups on the seed mass of the maternal and paternal parents. The estimates of the regression slope on maternal value ($b_M = 0.130$, SE = 0.042) and on paternal value ($b_P = 0.006$, SE = 0.038) agreed well with the values expected from table 7.1, supporting our interpretation.

Our results, in which the maternal effect is fully accounted for as V_{Am}, provide little motivation to separate the other sources of maternal variation (V_{Cm}, V_{Dm}, and V_{Em}). Moreover, because the latter two components do not contribute to selection response, it is not clear that knowledge of their values would repay the further substantial effort that would be necessary to estimate them. Under field conditions and in other species, however, these other sources of maternal effects may be appreciable. For completeness, we give modifications to our design that could be used to estimate them. A design modified from figure 7.1, such that a factorial cross of generation 0 produces multiple fullsibs per mating to be used in generation 1, would separate V_{Dm} from V_{Cm} and V_{Em} (and, similarly, the analogous paternal components). Further, if clonal replicates were produced for parental plants in generation 1 and each crossed as in this experiment, then environmentally induced parental effects could be distinguished from cytoplasmic effects. Reciprocal crosses of generation 1 to produce generation 2 would permit direct estimation of V_A, without the need to assume that paternal effects are negligible.

Table 7.2 Results from analysis of seed set.

Source of variation	Var. comp.	% of V_p
Maternal grandshire (MGS)	-0.31	0
Paternal grandshire (PGS)	0.03	0.3
Dam(MGS)	2.06***	23.0
Sire(PGS)	0.60*	6.7
MGS × PGS	0.10	1.1
Dam(MGS) × PGS	0.14	1.6
Sire(PGS) × MGS	-0.003	0
Dam(MGS) × Sire(PGS)	0.06	0.6
Within crosses	6.02	67.0

*$P < 0.05$.
***$P < 0.0001$.

7.5 Evolutionary Implications

Given the finding that V_{Am} accounts for a substantial fraction of the phenotypic variance (V_p), while it appears that V_A accounts for little, what is to be expected about the evolution of seed mass? In the case of selection strictly on the basis of individual seed mass, the response to selection is predicted from $R = (V_A + 3 \text{ Cov}(A, Am)/2 + V_{Am}/2) \, S/V_p$ (Willham 1963), where $\text{Cov}(A, Am)$ is the covariance between additive direct effect and additive maternal effect, S is the selection differential, and V_p is the phenotypic variance. Thus, even if V_A is quite small, as we and Platenkamp and Shaw (1993) have found, V_{Am} alone is expected to support selection response. Response could be impeded by an adverse covariance between additive direct effects and additive maternal effect. Numerous negative estimates of this component have been obtained for livestock populations (see also Thiede 1996), but Meyer (in press) provides evidence that this may be an artifact of the statistical model used. In our study, we did not reject the null hypothesis that this covariance is zero (Byers et al. 1997). Selection response could, however, be constrained due to correlations with other maternally influenced traits.

Seed mass is likely to be negatively correlated with the number of seeds per fruit due to resource constraints (Stanton 1984; Michaels et al. 1988). We used our experiment to determine the influence of maternal (and paternal) parent on the number of seeds per fruit (seed set) for a given cross and on the correlation between seed set and seed mass. We consider the number of seeds per fruit a trait of the mating between the dam and sire. We found highly significant maternal and significant paternal effects that explained 23% and 6.7% of the phenotypic variation, respectively (table 7.2). In the case of seed set, we detected no additive genetic basis for either parental effect on seed set (i.e., the effects of MGS and PGS were not significant, $P \gg 0.1$). It seems likely that these parental effects on seed set are largely attributable to environmental variation within the greenhouse, but contribution of nonadditive maternal genetic effect (e.g., V_{Dm}) and cytoplasmic genes (V_{Cm}) cannot be excluded. The greater influence of maternal parent is not surprising, because pollen tube growth, fertilization, and seed development all take place in maternal tissue.

Multivariate analysis provided estimates of the component correlations between mean seed mass per fruit and seed set. Of particular interest here is the maternal correlation, estimated as −0.21. This suggests a trade-off among maternal parents, such that variation ranges from plants tending to mature more, smaller seeds per fruit to those bearing fewer, heavier seeds. Mechanisms underlying this trade-off are unclear because, as previously noted, there are multiple possible causes of differences among dams within grandsires in this study. We have not found evidence for a genetic constraint at the level of joint additive genetic maternal effects on seed set and seed mass in this population. However, the trade-off we detected at the level of maternal phenotype could impede response to selection on seed mass. Moreover, trade-off between maternal effects on seed number and seed mass at the whole-plant level, which we have not studied, would also bear on the evolution of seed mass (Wolfe 1995).

Associations between seed mass and other traits further complicate prediction of selection response. We have estimated positive correlations between seed mass and time to germination at the level of both additive genetic effects and (overall) maternal effect (Byers et al. 1997). These results suggest that alleles tending to increase seed mass directly tend also to delay seed germination. Although the genetic basis of the joint maternal effect is unclear, together these findings suggest a further possible constraint on evolution of seed mass through evolution of the maternal effect. Thus, as with any other trait, evidence of genetic variation in maternal effect implies the potential for evolutionary response to selection, but predictions that fail to take into account correlated characters are likely to be misleading.

Conclusions

We have presented results indicating that variation in seed mass can be attributed to V_{Am} and therefore that seed mass has the potential to evolve through genetic change in maternal effect. Because few studies of the genetic basis of maternal effects in natural populations have been conducted, the prevalence and magnitude of V_{Am} are not known. Thiede's (1996) findings for *Collinsia verna*, however, suggest that substantial contributions of V_{Am} to phenotypic variation are not rare. However, Thiede's study further indicates that, as is typically true, field experiments will necessitate far larger samples than those conducted in controlled conditions to achieve comparable statistical power. Still more challenging will be field studies that include explicit consideration of genetic variation in the response of maternal effect to environmental conditions and that assess selection on progeny in relation to their own and their parents' environments. Yet these are essential to thorough understanding of the extent to which parental effects can contribute to future adaptive evolution within populations.

Acknowledgments   We thank W. G. Hill, D. S. Falconer, M. Mackinnon, S. Mbaga, and F. H. Shaw for helpful discussions; and E. Lacey, C. Fox, T. Mousseau, and anonymous reviewers for their thoughtful comments on the manuscript. This work was supported by NSF award BSR-8817756. RGS gratefully acknowledges sabbatical support from the Underwood Fund of the BBSRC (U.K.) and the Bush Sabbatical Supplement Program of the University of Minnesota.

References

Aarssen, L. W., and S. M. Burton. 1990. Maternal effects at four levels in *Senecio vulgaris* (Asteraceae) grown on a soil nutrient gradient. Am. J. Bot. 77:1231–1240.

Alexander, H. M., and R. D. Wulff. 1985. Experimental ecological genetics in *Plantago*. X. The effects of maternal temperature on seed and seedling characters in *P. lanceolata*. J. Ecol. 73:271–282.

Antonovics, J., and J. Schmitt. 1986. Paternal and maternal effects on propagule size in *Anthoxanthum odoratum*. Oecologia 69:277–282.

Azzam, S. M., and M. K. Nielsen. 1987. Expected responses to index selection for direct and maternal additive effects of gestation length or birth date in beef cattle. J. Anim. Sci. 64:357–365.

Beavis, W. D., E. Pollak, and K. J. Frey. 1987. A theoretical model for quantitatively inherited traits influenced by nuclear-cytoplasmic interactions. Theor. Appl. Genet. 74:571–578.

Bernardo, J. 1996. The particular maternal effect of propagule size, especially egg size: Patterns, models, quality of evidence and interpretation. Am. Zool. 36:216–236.

Biere, A. 1991. Parental effects in *Lychnis flos-cuculi*. I. Seed size, germination and seedling performance in a controlled environment. J. Evol. Biol. 3:447–465.

Byers, D. L. In press. Effect of cross proximity on progeny fitness in a rare and a common species of *Eupatorium*. Am. J. Bot.

Byers, D. L., G. A. J. Platenkamp, and R. G. Shaw. 1997. Variation in seed characters in *Nemophila menziesii*: Evidence of a genetic basis for maternal effect. Evolution 51:1445–1456.

Case, A. L., E. P. Lacey, and R. G. Hopkins. 1996. Parental effects in *Plantago lanceolata* L. II. Manipulation of grandparental temperature and parental flowering time. Heredity 76:287–295.

Corriveau, J. L., and A. W. Coleman. 1988. Rapid screening method to detect potential biparental inheritance of plastid DNA and results for over 200 species. Am. J. Bot. 75:1443–1458.

Cowley, D. E., D. Pomp, W. R. Atchley, E. J. Eisen, and D. Hawkins-Brown. 1989. The impact of uterine genotype on postnatal growth and adult body size in mice. Genetics 122:193–203.

Crill, W. D., R. B. Huey, and G. W. Gilchrist. 1996. Within- and between-generation effects of temperature on the morphology and physiology and *Drosophila melanogaster*. Evolution 50:1205–1218.

Eisen, E. J. 1967. Mating designs for estimating direct and maternal genetic variances and direct-maternal genetic covariances. Can. J. Genet. Cytol. 9:13–22.

Falconer, D. S. 1965. Maternal effects and selection response. Pp. 763–774 in S. J. Geerts (ed.), Genetics Today. Pergamon Press, Oxford.

Falconer, D. S., and T. F. C. Mackay. 1996. Introduction to Quantitative Genetics. 4th ed. Longman, Essex.

Finkelstein, R. R. 1994. Maternal effects govern variable dominance of two abscisic acid response mutations in *Arabidopsis thaliana*. Plant Physiol. 105:1203–1208.

Foolad, M. R., and R. A. Jones. 1992. Models to estimate maternally controlled genetic variation in quantitative seed characters. Theor. Appl. Genet. 83:360–366.

Güneren, G., L. Bünger, I. M. Hastings, and W. G. Hill. 1996. Prenatal growth in lines of mice selected for body weight. J. Anim. Breed. Gen. Zeit. Tier. Zuch. 113:535–543.

Helenurm, K., and B. A. Schaal. 1996. Genetic and maternal effects on offspring fitness in *Lupinus texensis* (Fabaceae). Am. J. Bot. 83:1596–1608.

Ikeobi, C. O. N., and L. O. Ngere. 1994. Direct genetic and additive maternal effects on swine litter size and weight in a tropical environment. Trop. Agricult. 71:77–79.

Kahi, A. K., M. J. MacKinnon, W. Thorpe, R. L. Baker, and D. Njubi. 1995. Estimation of individual and maternal additive genetic and heterotic effects for preweaning traits of crosses of Ayrshire, brown Swiss and Sahiwal cattle in the lowland tropics of Kenya. Livestock Prod. Sci. 44:139–146.

Khome, C. T., J. F. Hayes, R. I. Cue, and K. M. Wade. 1995. Estimation of direct additive and maternal additive genetic effects for weaning weight in Mashona cattle of Zimbabwe using an individual animal model. Anim. Sci. 60:41–48.

Kirkpatrick, B. W., and M. R. Dentine. 1988. An alternative model for additive and cytoplasmic genetic and maternal effects on lactation. J. Dairy Sci. 71:2502–2507.

Kirkpatrick, M., and R. Lande. 1989. The evolution of maternal characters. Evolution 43:485–503.

Kirkpatrick, M., and R. Lande. 1992. The evolution of maternal characters: Errata. Evolution 46:284.

Lacey, E. P. 1996. Parental effects in *Plantago lanceolata* L. I: A growth chamber experiment to examine pre- and postzygotic temperature effects. Evolution 50:865–878.

Lande, R. 1981. The minimum number of genes contributing to quantitative variation between and within populations. Genetics 99:541–553.

Lande, R., and M. Kirkpatrick. 1990. Selection response in traits with maternal inheritance. Genet. Res. 55:189–197.

Mazer, S. J. 1987. The quantitative genetics of life history fitness components in *Raphanus raphanistrum* L. (Brassicaceae): Ecological and evolutionary consequences of seed-weight variation. Am. Nat. 130:891–914.

Meyer, K. 1991. Estimating variances and covariances for multivariate animal models by restricted maximum likelihood. Gen. Select. Evol. 23:317–340.

Meyer, K. 1992. DFREML a set of programs to estimate variance components by restricted maximum likelihood using a derivative-free algorithm. User notes. Vers. 2.0. Animal Genetics and Breeding Unit, University of New England, Armidale.

Meyer, K. In press. Estimates of genetic parameters for weaning weight of beef cattle accounting for direct-maternal covariances. Livestock Prod. Sci.

Miao, S. L., F. A. Bazzaz, and R. B. Primack. 1991a. Effects of maternal nutrient pulse on reproduction of two colonizing *Plantago* species. Ecology 72:586–596.

Miao, S. L., F. A. Bazzaz, and R. B. Primack. 1991b. Persistence of maternal nutrient effects in *Plantago major*: The third generation. Ecology 72:1634–1642.

Michaels, H. J., B. Benner, A. P. Hartgerink, T. D. Lee, and S. Rice. 1988. Seed size variation: Magnitude, distribution, and ecological correlates. Evol. Ecol. 2:157–166.

Mitchell-Olds, T., and J. J. Rutledge. 1986. Quantitative genetics on natural plant populations: A review of the theory. Am. Nat. 127:379–402.

Mogensen, H. L. 1996. The hows and whys of cytoplasmic inheritance in seed plants. Am. J. Bot. 83:383–404.

Montalvo, A. M., and R. G. Shaw. 1994. Quantitative genetics of sequential life-history and juvenile traits in the partially selfing perennial, *Aquilegia caerulea*. Evolution 48:828–841.

Moore, R. W., E. J. Eisen, and L. C. Ulberg. 1970. Prenatal and postnatal maternal influences on growth in mice selected for body weight. Genetics 64:59–68.

Mousseau, T. A., and H. Dingle. 1991. Maternal effects in insect life histories. Annu. Rev. Entomol. 36:511–534.

Nieuwhof, M., F. Garretsen, and J. C. van Oeveren. 1989. Maternal and genetic effects on seed weight of tomato, and effects of seed weight on growth of genotypes of tomato (*Lycopersicon esculentum* Mill). Plant Breed. 102:248–254.

Olsen, T. A., A. van Dijk, M. Koger, D. D. Hargrove, and D. E. Franke. 1985. Additive and heterosis effects on preweaning traits, maternal ability and reproduction from crossing of the angus and brown Swiss breeds in Florida. J. Anim. Sci. 61:1121-1131.

Platenkamp, G. A. J., and R. G. Shaw. 1993. Environmental and genetic maternal effects on seed characters in *Nemophila menziesii*. Evolution 47:540–555.

Pleines, S., and W. Friedt. 1989. Genetic control of linolenic acid concentration in seed oil of rapeseed (*Brassica napus* L.). Theor. Appl. Genet. 78:793–797.

Rasyad, A., and D. A. van Sanford. 1992. Genetic and maternal variances and covariances of kernel growth traits in winter wheat. Crop Sci. 32:1139–1143.

Rawson, P. D., and T. J. Hilbish. 1991. Genotype-environment interaction for juvenile growth in the hard clam *Mercenaria mercenaria*. Evolution 45:1924–1935.

Richardson, T. E., and A. G. Stephenson. 1991. Effects of parentage, prior fruit set and pollen load on fruit and seed production on *Campanula americana* L. Oecologia 87:80–85.

Riska, B. 1991. Maternal effects in evolutionary biology. Pp. 719–724 *in* E. Dudley (ed.), The Unity of Evolutionary Biology. Dioscorides Press, Portland, Ore.

Riska, R., J. J. Rutledge, and W. R. Atchley. 1985. Covariance between direct and maternal genetic effects in mice, with a model of persistent environmental influences. Genet. Res. 45:287–297.

Roach, D. A., and R. D. Wulff. 1987. Maternal effects in plants. Annu. Rev. Ecol Syst. 18:209–235.

Sanford, J. C., and R. E. Hanneman Jr. 1982. Large yield differences between reciprocal families of *Solanum tuberosum*. Euphytica 31:1–12.

SAS Institute. 1985. SAS User's Guide: Statistics. Vers. 5. SAS Institute Inc., Cary, N.C.

Schlichting, C. D. 1986. Environmental stress reduces pollen quality in *Phlox*: Compounding the fitness deficit. Pp. 483–488 in D. L. Mulcahy, G. B. Mulcahy, and E. Ottaviano (eds.), Biotechnology and Ecology of Pollen. Springer, New York.

Schmid, B., and C. Dolt. 1994. Effects of maternal and paternal environment and genotype on offspring phenotype in *Solidago altissima* L. Evolution 48:1525–1549.

Schmitt, J., and J. Antonovics. 1986. Experimental studies of the evolutionary significance of sexual reproduction. III. Maternal and paternal effects during seedling establishment. Evolution 40:817–829.

Schmitt, J., J. Niles, and R. D. Wulff. 1992. Norms of reaction of seed traits to maternal environments in *Plantago lanceolata*. Am. Nat. 139:451–466.

Schwaegerle, K. E., and D. A. Levin. 1990. Quantitative genetics of seed size variation in *Phlox*. Evol. Ecol. 4:143–148.

Shaw, R. G., and F. H. Shaw. 1994. QUERCUS: Programs for quantitative genetic analysis using maximum likelihood. Available via anonymous ftp: ecology.umn.edu/pub/ftp/quercus.

Shaw, R. G., and N. M. Waser. 1994. Quantitative genetic interpretations of postpollination reproductive traits in plants. Am. Nat. 143:617–635.

Shaw, R. G., G. A. J. Platenkamp, F. H. Shaw, and R. H. Podolsky. 1995. Quantitative genetics of response to competitors in *Nemophila menziesii*: A field experiment. Genetics 139:397–406.

Singh, J., O. P. Govila, R. K. Yadav, and P. K. Agrawal. 1991. Studies on seed storability of pear millet (*Pennistum americanum* (L) Leeke) under controlled condition. 1. Maternal and genetic control of seed viability. Plant Var. Seeds 4:143–149.

Stanton, M. L. 1984. Developmental and genetic sources of seed weight: Variation in *Raphanus raphanistrum*. Am. J. Bot. 71:1090–1098.

Stratton, D. A. 1989. Competition prolongs expression of maternal effects in seedlings of *Erigeron annuus* (Asteraceae). Am. J. Bot. 76:1646–1653.

Thiede, D. A. 1996. The impact of maternal effects on multivariate evolution: Combining quantitative genetics and phenotypic selection in a natural plant population. Ph.D. dissertation, Michigan State University.

Van Vleck, L. D. 1970. Index selection for direct and maternal genetic components of economic traits. Biometrics 26:477–483.

Van Vleck, L. D. 1976. Selection for direct, maternal and grandmaternal genetic components of economic traits. Biometrics 32:173–181.

Waser, N. M., R. G. Shaw, and M. V. Price. 1995. Seed set and seed mass in *Ipomopsis aggregata:* Variance partitioning inferences about postpollination selection. Evolution 49:80–88.

White, J. M., J. E. Legates, and E. J. Eisen, 1968. Maternal effects amons lines of mice selected for body weight. Genetics 60:395–408.

Willham, R. 1963. The covariance between relatives for characters composed of components contributed by related individuals. Biometrics 19:18–27.

Wolfe, L. M. 1995. The genetics and ecology of seed size variation in a biennial plant, *Hydrophyllum appendiculatum* (Hydrophyllaceae). Oecologia 101:343–352.

Wulff, R. D., A. Céceres, and J. Schmitt. 1994. Seed and seedling responses to maternal and offspring environments in *Plantago lanceolata*. Funct. Ecol. 8:763–769.

Young, H. J., and M. L. Stanton. 1990. Influence of environmental quality on pollen competitive ability in wild radish. Science 248:1631–1633.

Zhu, J., and B. S. Weir. 1994. Analysis of cytoplasmic and maternal effects I. A genetic model for diploid plant seeds and animals. Theor. Appl. Genet. 89:153–159.

8

The Role of Environmental Variation in Parental Effects Expression

MARYCAROL ROSSITER

Despite the growing realization that parental effects are pervasive, our ability to quantify their contribution to fitness is tenuous. This is due, in part, to the fact that we have not fully appreciated the extent to which the expression of parental effects (presence/absence and magnitude) can be modified by the environment. Our ability to quantify the contribution of parental effects to fitness is also limited by the availability of experimental designs that reliably separate parental effects from genetic effects (Rossiter 1997). In this chapter, I show how the expression of maternal and paternal effects can be subject to the vagaries of environmental experience. Examples taken from the literature include parental effects whose expression, lack of expression, or intensity of expression is the result of environmental variation—unpredictable variation such as rainfall or food availability and predictable variation such as photoperiod or seasonality.

First, some definitions are in order to clarify the terminology I use here. The term "maternal effects" has a history of varied meanings, and the particular use usually depends on the biological subdiscipline practiced by the user (discussed in Rossiter 1996). To counter the limitations of this term, I use the more cumbersome but inclusive "inherited environmental effects" to define any component of phenotypic expression that is derived from either mother or father, apart from the nuclear genes. Inherited environmental effects arise as the product of parental genes, parental environment, or the interaction between parental genome and parental environment. Inherited environmental effects can include contributions owing to abiotic, nutritional, and other ecological or autecological aspects of the parental environment, such as predation intensity or density-dependent growth. The impact of inherited environmental effects is positive or negative depending on the nature of the contributions *and* the ecological context within which they are received (Rossiter 1996). This means that the environmental conditions in two (or more) succeeding generations can modify the expression of inherited environmental effects. With this definition stated, I will also, in the interest of brevity, use the terms "parental effect" to include inherited environmental effects known to arise from either the mother ("maternal effects") or the father ("paternal effects").

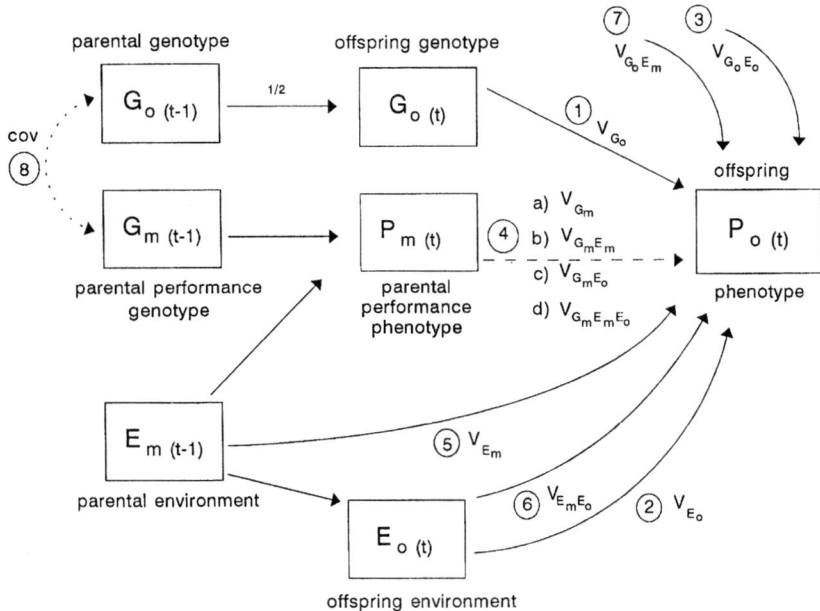

offspring environment

Figure 8.1 The components of offspring phenotype (P_o) expressed in time t, deriving from the direct contribution of nuclear genes by one parent (G_o), a time-lagged presentation of the parental environment (E_m), a time-lagged expression of parental performance genes (G_m) and their interaction with the parental environment to produce the parental performance phenotype (P_m), plus the offspring's own environment (E_o). For simplicity of presentation, G indicates additive genetic effects with dominance and epistasis (nonadditive genetic variation V_{NA}) assumed to be negligible. The numbered sources indicate possible routes of contribution to the offspring phenotype: source 4a = V_{Gm}, 4b = V_{GmEm}, and so on; see text below for a complete listing of variables. This diagram can represent contributions by either mother or father (from Rossiter 1996).

A most critical aspect of this cross-generational phenomenon is that a parental effect is received as an *environmental* (vs. a genetic) component of an offspring's phenotype, even if its source arises solely or in part from parental gene expression. Although this environmental effect is inherited (like money), it is not heritable, even if the same genes whose expression generated an inherited environmental effect are themselves inherited as part of the nuclear genetic contribution of the parent. To clarify these distinctions, consider figure 8.1 (sources of variations 1–8), which shows all potential contributing agents to offspring phenotype. This qualitative model illustrates the potentially complex construction of an inherited environmental effect. This model, based on a quantitative genetics analysis of phenotypic variation that includes parental effects and a variable offspring environment (Eisen and Saxton 1983), was extended to include the consequences of a variable parental environment (Rossiter 1996). I provide a brief description of the model here. For more extensive coverage, as well as a discussion of how inherited environmental effects influence quantitative genetic analysis, see Rossiter (1997).

Figure 8.1 shows eight potential sources of offspring phenotypic variation, $P_{o(t+1)}$, arising from genetic (G) or environmental (E) sources during the parental (m) generation in time t or the offspring (o) generation in time $t + 1$. Those components with which we are most familiar include offspring genotype derived from nuclear genes provided by the parent (source 1, $G_{o(t)}$), environmental conditions occurring contemporaneously with offspring life (source 2, $E_{o(t)}$), and their interaction (source 3, $G_{o(t)} \times E_{o(t)}$). If we aim to understand the genesis and evolution of inherited environmental effects, we must distinguish the pathways by which the parental performance phenotype adjusts offspring phenotype. Sources 4–8 represent multiple, nonexclusive ways that parental effects can contribute to offspring phenotype; each is described in turn below.

Source 4 represents any parental effect that arises from some aspect of the parental performance phenotype, $P_{m(t)}$, a potentially complex mixture of the parental genotype and parental environment. Examples of parental performance phenotypic traits that participate in parental effects include maternal body size (Cowley et al. 1989), maternal age (Bridges and Heppell 1996), maternal behavior (Roosenburg 1996), and paternal body size or behavior (Boggs 1995). The parental performance phenotype, $P_{m(t)}$, can adjust offspring phenotype through the action of any or all of its components (figure 8.1, sources of variation 4a–4d).

The first of these components is the parental performance genotype, G_m (source 4a). Here, parental performance genes (as expressed by the parent) influence the offspring phenotype. With quantitative genetic designs it has been possible to distinguish the contribution of parental genes expressed in parents to offspring phenotype (source 4a) from the contribution of genes transmitted from parents and expressed as offspring genes (source 1; e.g., Lacey 1996).

In addition to the parental performance genotype, the parental performance phenotype, $P_{m(t)}$, may include interactions between parental genes (expressed by the parent) and the environment. In other words, the environment can adjust parental gene expression. This can occur in three ways: by an interaction with the parental environment, $G_m \times E_m$ (source 4b), the offspring environment, $G_m \times E_o$ (source 4c), or both, $G_m \times E_m \times E_o$ (source 4d). For instance, Glazier (1992) found that parental genotype dictated the extent to which parental food quality influenced offspring size, an example of $G_m \times E_m$ (source 4b). Additional empirical examples of how components of $P_{m(t)}$ influence offspring phenotype include: G_m, source 4a: Atchley and Newman (1989), Carmona et al. (1994), Cowley and Atchley (1992), Lacey (1996), Plantenkamp and Shaw (1993), Reznick and Yang (1993), and Schmid and Dolt (1994); and $G_m \times E_m$, source 4b: Barton et al. (1994), Carmona et al. (1994), Case et al. (1996), Glazier (1992), Keena et al. (1995), Lacey (1996), Lin and Dunson (1995), Plantenkamp and Shaw (1993), Potvin and Charest (1991), Rogowitz (1996), and Schmid and Dolt (1994). Because sources 4c ($G_m \times E_o$) and 4d ($G_m \times E_m \times E_o$) include an environmental component from the offspring generation, pertinent examples are provided below along with a discussion of source 6, $E_m \times E_o$.

If the parental environment influences offspring phenotype independent of the parental performance phenotype, that is, beyond any influence of the parent's genotype, then this contribution is represented by E_m, source 5. For example, from the world of insects, mothers that grow up in nutritional environment A are able to sequester and transmit an antimicrobial plant compound to their offspring (e.g., Dussourd et al. 1988; Boppre and Fischer 1994). When E_m is present, the time-delayed environmental contribution to offspring phenotype is independent of or beyond the control of parental genotype. A hypothetical example, based on the known transmission of antimicrobial agents to offspring, would be that the quantity of the antimicrobial plant compound transmitted to offspring will depend on some attribute of the parent's environment such as plant quality. Examples of source 5 contributions to offspring phenotype can be found in Barton et al. (1994), Islam et al. (1994a), and Schmid and Dolt (1994).

If there is an interaction between the influence of the parental and offspring environments on offspring phenotype, then this contribution is represented by $E_m \times E_o$, source 6. For example, in the gypsy moth, the *sequence* of host plants used in successive generations (parent, then offspring) can adjust offspring phenotype (Rossiter 1991). There are many experiments showing that the quality of the offspring environment influences the intensity of the parental effect (Boonstra and Boag 1987; Groeters and Dingle 1987, 1988; Gould 1988; Stratton 1989; Aarssen and Burton 1990; Kobayashi 1990; Miao et al. 1991a, b; Rossiter 1991; Kaplan 1992; Schmitt et al. 1992; Wulff and Bazzaz 1992; Brett 1993; Futuyma et al. 1993; Fox et al. 1995; Galloway 1995; Lin and Dunson 1995 [here, the presence of *source 4c* was specifically verified]; Watson and Hoffmann 1995). Unfortunately, the experimental designs employed in most of these experiments cannot distinguish among $G_m \times E_o$, $G_m \times E_m \times E_o$, and $E_m \times E_o$ (sources 4c, 4d, and 6). The experiments do, however, suggest excellent model systems for further work to distinguish among these three avenues of parental effect influence.

If the genotype of the offspring adjusts the impact of the environmental contribution from the parents, then this contribution to offspring phenotype is represented by $G_o \times E_m$, source 7. For example, in the herbaceous perennial *Plantago lanceolata*, the impact of the thermal parental environment (prezygotic) on germination and growth in the next generation varied among offspring genotypes (Lacey 1996).

Finally, if there is a genetic correlation between a parental performance trait and the same or different trait in offspring, then this contribution to offspring phenotype is represented by $\text{Cov}G_mG_o$, source 8a. The contribution of this covariance to offspring phenotype is generally assumed to be absent as a simplifying assumption in studying the genetics of natural populations, although there is good empirical work to support its existence (e.g., Dickerson 1947; Willham 1963; Bondari et al. 1978; for a discussion of the consequences of this assumption to genetic analysis of natural populations, see Rossiter (1997).) It is of special interest to note that this type of cross-generational genetic correlation can also involve gene-environment interactions expressed in the context of either parental or offspring environments. These are designated as $\text{Cov}(G_m \times E_m)(G_o \times E_m)$ (Rossiter 1996) and $\text{Cov}(G_m \times E_o)(G_o \times E_o)$ (Eisen and Saxton 1983), referred to here as sources 8b and 8c. Simple genetic cross-generational covariance, $\text{Cov}G_mG_o$, has been documented (e.g., Willham 1972; Bondari et al. 1978; Riska et al. 1985; Atchley and Newman 1989; Southwood and Kennedy 1990), but, to my knowledge, measurement of complex cross-generational covariance, $\text{Cov}(G_m \times E_m)(G_o \times E_m)$ and $\text{Cov}(G_m \times E_o)(G_o \times E_o)$, is seldom addressed outside the study of animal husbandry. Nonetheless, it is another way in which the environment can alter the intensity of parental effects expression. It should now be clear that there are many paths by which the environment can participate in adjusting parental effects expression. If a species is capable of generating inherited environmental effects in any of these ways (except exclusively via source 4a, G_m, or source 8a, $\text{Cov}G_mG_o$), then the intensity of the parental effect expression, ranging from "not expressed" to maximum impact, will be dictated by the condition of the external environment in (at least) two consecutive generations. In sections 8.1 and 8.2, I discuss empirical work that shows that the intensity of parental effects expression can be adjusted by the environmental experience of the parents, as well as the environmental experience of the offspring.

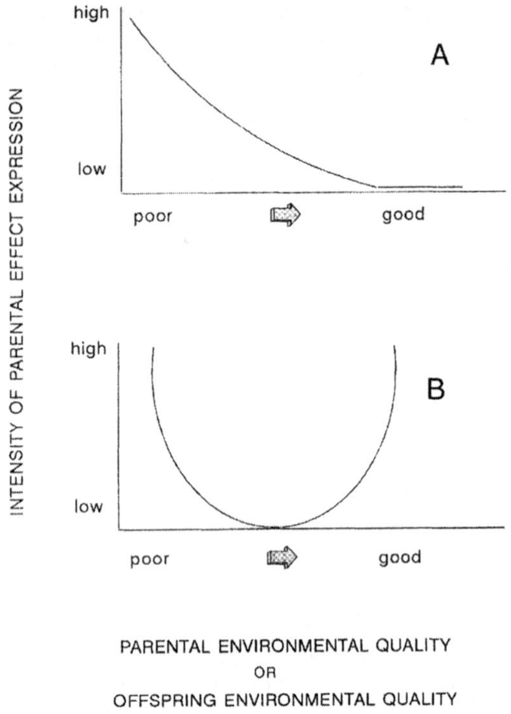

Figure 8.2 Response of two hypothetical species (A and B): the impact of the environmental condition on the intensity of their parental effect expression. The x-axis can be read as either parental or offspring environmental quality, depending on when the environment influences the parental effect: at its initiation or completion.

8.1 Parental Environment Influences Parental Effects Expression

When the condition of the parental environment influences parental effect expression, off-spring phenotype includes one to all of these components: $G_m \times E_m$ (source 4b), $G_m \times E_m \times E_o$ (source 4d), E_m (source 5), $E_m \times E_o$ (source 6), $G_o \times E_m$ (source 7). Regardless of the path of influence, the intensity of parental effects expression is a function of the environmental quality experienced by the parents (figure 8.2). The shape of this response will be species dependent. Since there are a number of possible functions to describe the relationship between parental environmental quality and the intensity of parental effect expression, I describe two hypothetical cases for illustration. In figure 8.2A, the intensity of parental effect expression decreases as environmental conditions improve, such that beyond some environmental quality threshold, parental effects are not manifested. In figure 8.2B, parental effects expression is dampened or absent as environmental quality approaches some average value; as environmental quality moves toward the extremes, parental effects expression intensifies. This response curve is what I would expect for species that exhibit cycling or irruptive population dynamics (Rossiter 1994).

Below I describe just over a dozen examples demonstrating that shifts in the parental environment alter parental effects expression (see table 8.1 for summary). These examples rarely include a quantitative measure of the contribution of any specific parental effect source (i.e., sources 4b, 4d, 5, 6, and 7 in figure 8.1), but their presence can be established by inference.

Table 8.1 Environmental variation experienced by the parents influences the the intensity of parental effects expression.

Organism	Source of variation in parental environment	Offspring trait influenced by parental environment	Reference
Rat	Toxin dose	Developmental time, juvenile mortality, shock performance	Barton et al. 1994
Vole	Food quality	Gonadal development	Nelson 1991
Cattle	Food quality	Hip height at weaning varies by parental forage	Brown et al. 1993
Gastropod	Egg size	Correlation structure in offspring	Shibata and Rollo 1988
Rotifer	Density	Propensity for mictic response	Carmona et al. 1994
Daphnia	Food quantity	Egg size, but mediated by genotype	Glazier 1992
Cricket	Temperature and photoperiod	Propensity to diapause	Bradford and Roff 1995
Moth	Dietary iron	Survival, development time, developmental stability	Keena et al. 1995
Moth	Foliage quality	Pupal weight, length of prefeeding stage	Rossiter 1991
Fly	Temperature	Cold tolerance	Watson and Hoffmann 1995
Plant	Temperature	Seed weight, germination, growth rate, time to reproduce	Lacey 1996
Plant	Competition	Seed weight, time to germination, dormancy	Plantenkamp and Shaw 1993
Plant	Temperature	Seed weight, nutrient content, cold acclimation	Potvin and Charest
Plant	Temperature $(t - 2)$	Leaf area and allometry, but mediated by parental flowering time	Case et al. 1996
	Flowering time $(t - 1)$	Germination, leaf area	

Animals

In their work on lab rats, Barton et al. (1994) treated a parental generation with naftopidil (a receptor-blocking agent) using treatment groups differing by dose (0, 10, 40, and 160 mg/kg). They found a dose-dependent influence on parents' salivation, body weight, and reproductive behavior. A dose-dependent, time-delayed influence of naftopidil was seen in the (untreated) offspring in terms of survival, development schedule, and ability to respond to shock. At the lowest dose of naftopidil (10 mg/kg), there were no toxic effects on mothers or offspring, compared to the highest dose (160 mg/kg), where about 30% of the mothers lost their entire litters. From these results I conclude that the expression of parental effects is dependent on parental environmental quality. The results also suggest the presence of $G_m \times E_m$ (source 4b), given the differential response among offspring of mothers exposed to the highest dose of naftopidil.

The study of Brown et al. (1993) provides an example from animal husbandry research. Four cattle strains (angus [A], brahman [B], and reciprocal crosses, AB and BA) were managed under different forage types: common Bermuda grass and an endophyte-infected tall fescue. They found that the nutritional regime experienced by parents influenced the strength of maternal effects component of offspring phenotype. Specifically, they found that maternal effects for hip height at weaning favored Angus managed on Bermuda grass but not fescue. For another offspring trait, ratio of weight to height at weaning, there was a significant maternal effects component, but unlike hip height, its value did not differ across parental forage environments. In other words, the maternal effect on the second trait included no aspect of E_m. These results do support the presence of a $G_m \times E_m$ (source 4b) for hip height at weaning and G_m (source 4a) for ratio of weight to height at weaning.

In a study of the interaction between diet and reproduction, Nelson (1991) reared prairie voles (*Microtus ochrogaster*) on specially milled food that varied in its concentration of 6-MBOA, a naturally occurring compound from grasses that is known to stimulate reproduction in rodents. Maternal exposure to 6-MBOA ranged from prenatal period only to postnatal, nursing period only. Nelson found that only after exposure of mothers to 6-MBOA during the prenatal period was a maternal effect manifested—enhanced gonadal development in sons. In other words, the parental environment determines whether the parental effect is expressed.

In a study of *Daphnia magna*, Glazier (1992) examined the effect of maternal nutrition (low vs. high food rations) on offspring size in two clones differing in metabolic rate. He found that the clone with a lower metabolic rate produced more and larger offspring as parental food resources dropped. The clone with a higher metabolic rate showed no difference size and number of offspring as parental food resources dropped. The among-clone difference suggests the presence of $G_m \times E_m$ (source 4b). In addition, Glazier found that the impact of the parental effect on subsequent offspring fitness depends on food resources in the offspring generation. Under good conditions, larger eggs become larger adults. Under bad conditions (e.g., starvation), there is no detectable maternal effect. This suggests the presence of $E_m \times E_o$ (source 6) or $G_m \times E_m \times E_o$ (source 4d), that is, genetic variation for the impact of the environment on expression of parental effects.

In a study of the rotifer *Brachionus plicatilis*, Carmona et al. (1994) measured the effects of population density on mictic response. They compared the response of isolated females from 13 clones, each of which was represented in two rearing densities. They found that, on average, there was a higher proportion of mictic daughters when parents grew under high-density conditions. Moreover, this higher average was owing to the elevated response of only half the clones (6 of 13) compared to their counterparts reared under low density, suggesting a maternal effect with a $G_m \times E_m$ (source 4b) component of offspring phenotype.

In a study of the wild slug *Deroceras laeve*, Shibata and Rollo (1988) found that the weight of eggs taken from the field was highly correlated with later fitness traits, juvenile growth rate, and time to maturity when slugs were reared on a high-quality diet. The correlations disappeared with low-quality diets. These same individuals served as the parental generation in a second experiment to evaluate the consequences of parental nutrition (low vs. high quality) on offspring traits. However, members of the F_1 generation were selected on the basis of egg weight in an attempt to equilibrate the starting size of offspring regardless of parental diet. (Parents on high-quality diets produced eggs that were, on average, larger.) Despite this equilibration, Shibata and Rollo found that parental dietary quality influenced offspring maturation size and egg production rates; that is, offspring of mothers with a high-quality diet grew larger and produced more eggs than their counterparts from parents on a low-quality diet.

In a study of voltinism in an egg-diapausing cricket (*Allonemobius socius*), Bradford and Roff (1995) used a common garden setting to subject crickets from several populations (differing in their proclivity for diapause) to a range of photoperiod and temperature conditions that mimicked a range of natural field conditions. These treatments generated a reaction norm for each group tested. They found that the propensity of offspring to break diapause, although genetically based, was influenced significantly by the environmental conditions experienced by parents.

In an investigation of "poor-quality" lab colonies of gypsy moths (*Lymantria dispar*), Keena et al. (1995) found that parents experiencing a dietary deficit in iron produced offspring with a lower probability of hatch and survival and slower, more variable development times. Moreover, they found differences among families in the degree to which parental dietary deficits influenced offspring fitness, suggesting the presence of $G_m \times E_m$ (source 4b).

In another study of *L. dispar*, Rossiter (1991) reared a parental generation on red oak trees of known defoliation and phenolic levels. All offspring were reared on synthetic diet. After familial effects were taken into account, daughters from mothers who fed on trees with higher damage levels had greater pupal weights and a shorter prefeeding (dispersal) stage compared to offspring from mothers who fed on trees with lower damage levels. Sons from mothers who fed on trees with greater condensed tannin levels had lower pupal weights compared to sons from mothers who fed on trees with lower condensed tannin levels.

Watson and Hoffmann (1996) subjected two *Drosophila* species to selection for increased cold resistance. They found that the response to this selection regime depended, in part, on the fly's environmental experience prior to the selection event. Specifically, flies that were cold hardened (achieved by holding the fly at 4°C for 1 hour) had a milder response to selection for cold resistance compared to flies that were not cold hardened. After selected lines achieved a plateau in cold resistance, continued selection for cold resistance yielded *decreasing* levels of difference. Specifically, offspring of cold-stressed parents in selected lines of both species were less able to handle the same level of thermal stress experienced by their parents. From this, Watson and Hoffmann inferred the presence of inherited environmental effects and suggested that the mechanism involves a change in egg provision quality (glycogen, lipids) due to cold exposure.

Plants

Lacey (1996) measured the effects of parental temperature (pre- and postzygotic) on off-spring traits in the herbaceous perennial *Plantago lanceolata*. She found that when prezy-gotic temperature (i.e., prior to fertilization and seed set) was lower, seeds were heavier but less germinable and offspring had lower growth rates but faster time to reproduction. Addi-tionally, she found that postzygotic temperature (an environmental condition experienced simultaneously by mother and offspring) influenced offspring phenotype more than prezy-gotic temperature. This result highlights a difficulty in studying parental effects in plants: the contribution of an inherited environmental effect to offspring phenotype can be con-founded with the contribution of the external environment during embryonic development because plants hold their seeds during development. Since Lacey was not able to distinguish whether response to the parental thermal environment is intra- or intergenerational in her two-generation design, another experiment (Case et al. 1996) was done.

Case et al. (1996) extended the previous experiment to a third generation of *P. lanceolata* to investigate the impact of temperature regime during a first generation (t) on the quality of grandoffspring. They found that grandparental (t) temperature influenced seed weight, ger-mination, leaf area and allometry, flowering time, and male sterility in grandoffspring ($t + 2$), although some of these inherited environmental effects were mediated by flowering time in the parental generation ($t + 1$). Finally, they found a significant interaction between mater-nal family (clone) and grandparental temperature for seed weight, leaf area, flowering time, and male sterility, suggesting the presence of genetic variation in the parental effect and a $G_m \times E_m$ (source 4b) component of offspring phenotype.

Plantenkamp and Shaw (1993) estimated the relative contribution of genetic maternal effects and parental environment on seed characters in an annual plant, *Neomophila men-ziesii*. They found that the parental environmental experience (growing under competition or not) had a significant influence on seeds; under competition, resulting seeds were smaller, took longer to germinate, and exhibited a higher incidence of dormancy. Moreover, there was a significant interaction between the maternal genotype and the maternal environmental experience associated with competition, suggesting the presence of $G_m \times E_m$ (source 4b).

Potvin and Charest (1991) studied the physiological response of a grass species, *Echi-nochloa crus-galli*, under several temperature regimes. During the parental generation, par-ticipating genotypes were represented in both a low- and high-temperature treatment. Off-spring of parents reared under low temperatures had greater seed weight, and higher germinability, and higher seedling leaf reserves (proteins, reducing sugars, starch) and were better able to acclimatize to cold compared to parents. In addition, there was an interaction between family (genotype) and parental temperature experience, suggesting the presence of $G_m \times E_m$ (source 4b).

8.2 Offspring Environment Influences Parental Effects Expression

When the condition of the external offspring environment influences parental effect expres-sion, offspring phenotype includes one to all of these components: $G_m \times E_o$ (source 4c), $G_m \times E_m \times E_o$ (source 4d), $E_m \times E_o$ (source 6), $Cov(G_m \times E_o)(G_o \times E_o)$ (source 8). The intensity of parental effects expression is a function of the environmental quality experienced by the offspring. As was the case for the impact of the parental environment on parental effects expression, the relationship between offspring environmental quality and intensity of parental effect expression will be species dependent (figure 8.2). The following empirical studies demonstrate how shifts in the external offspring environment alter parental effects expression (summarized in table 8.2). These examples rarely include a quantitative measure of the contribution of any specific parental effect source (e.g., sources 4c, 4d, 6, 8), but their presence can often be inferred.

Animals

Parichy and Kaplan (1992) equated maternal investment with egg size in the frog *Bombina orientalis*. They reared tadpoles from eggs representing the entire egg size range under each of two environmental quality conditions: limited versus unlimited food. They found that offspring environmental quality modified the impact of egg size on offspring success (development time and body size) only when the environment was stressful, that is, under food limitation. Likewise, in seed beetles and slugs, offspring environmental quality modified the impact of egg size on subsequent fitness (Shibata and Rollo 1988; Fox and Mousseau 1996). Fox and Mousseau (1996) also noted that for some insects (e.g., butterflies; Braby 1994), egg size affects progeny growth only under harsh conditions. Overall, these studies demonstrate that the impact of egg size (or egg-size variation) on offspring phenotype (or offspring phenotypic variation) can be modified by the offspring's environmental experience.

Brett (1993) reared members of a parental generation of *Daphnia longispina* on one of two prey items (*Microcystis* and *Rodomonas*), as were their offspring for all possible combinations of successive parent-offspring nutritional environments. For the offspring traits of time to and size at maturity, maternal food source explained 7% of the total variance while offspring food source explained about 25%. For three other offspring traits—mean clutch size, total fecundity, and body size—a maternal effect was seen only when the offspring environment included *Microcystis*.

In the seed beetle *Stator limbatus*, Fox et al. (1995) found a paternal effect that arose from the response of fathers to host quality and was further modified by the host quality experience of the offspring. In particular, fathers reared on *Cercidium florida* produced offspring that developed at a slower rate when they grew up on *C. florida* compared to *Acacia greggii*. Offspring of fathers reared on *A. greggii* showed a weaker version of this same relationship.

In a study of a leaf beetle (*Ophraella notulata*) by Futuyma et al. (1993), parents and offspring were reared on all sequential combinations of host A (*Iva fructescens*) and host B (*Ambrosia artemisiifolia*). When offspring were reared on host B, there were no detectable parental effects. By contrast, when offspring were reared on host A, a paternal effect was detectable such that daughters from host A fathers were more fecund than daughters from host B fathers.

Gould (1988) reared a parental generation of the tobacco budworm, *Heliothis virescens*, on either a plain synthetic diet or one amended with quercitin, a phenolic known to inhibit growth. The offspring from each parental treatment group were reared on plain diet and three phenolic-amended diets holding either quercitin, rutin, or gramine. The parental diet influenced growth rate when the offspring diet held quercitin or rutin but not when the offspring diet was plain or held gramine.

Groeters and Dingle (1987) reared a parental generation of the milkweed bug, *Oncopel-*

Table 8.2 Environmental variation experienced by the offspring influences the intensity of parental effects expression.

Organism	Source of variation in parental environment	Offspring trait affected	Parental effects were differentially expressed when offspring experienced	Reference
Toad	Competition for food	Body length	Greater vs. lesser competition for food	Parichy and Kaplan 1992
Daphnia	Food quality	Clutch size, body size	One prey species over another	Brett 1993
Beetle	Paternal host quality	Development time	One host species over another	Fox et al. 1995
Beetle	Paternal host quality	Fecundity	One host species over another	Futuyma et al. 1993
Moth	Dietary phenolics	Growth rate	Some phenolics over others	Gould 1988
Bug	Photoperiod	Age to reproduction	Photoperiod of LD 11:13 vs. LD 14:10	Groeters and Dingle 1987
Bug	Temperature	Age to reproduction	Thermal regime of 27°C vs. 23°C	Groeters and Dingle 1988
Plant	Soil nutrients	Plant size	Greater vs. lesser competition for resources	Stratton 1989
Plant	Soil nutrients	Seed size, early and late plant size traits	Low vs. high soil nutrient levels	Wulff and Bazzaz 1992
Plant	Habitat quality	Germination	Lower vs. higher light intensity	Schmitt et al. 1992

tus fasciatus, under each of two photoperiod regimes (light:dark [LD] 14:10 and LD 11:13). When offspring were reared under the photoperiod LD 11:13, there was a pronounced parental effect. The age at onset of reproduction was 2 weeks earlier for offspring from parents reared under LD 11:13 compared to offspring from parents reared under LD 14:10. However, if offspring were reared under the photoperiod of LD 14:10, no parental effect was detected. In another experiment on milkweed bugs, Groeters and Dingle (1988) reared members of a parental generation at either 23°C or 27°C. Offspring from each family were reared at each of these temperatures. Although a parental effect was seen regardless of the offspring thermal environment, its magnitude depended on the rearing temperature of the offspring. A greater parental effect on development time and age at first reproduction was seen when offspring grew up at 27°C.

Plants

Competitive circumstance in the offspring generation can influence the intensity of parental effect expression. For example, Stratton (1989) grew the parental generation of aster clones (*Erigeron annuus*) under both high- and low-nutrient conditions; seeds from high-nutrient parents were larger. When offspring from each parental treatment group were grown under different competitive conditions, offspring under high-competition conditions not only maintained their initial relative size advantage but the relative advantage increased through development. This postseedling difference was not manifested when offspring were grown in the absence of competition. Stratton suggested that under the specified offspring environment, persistent competition, the recurring parental effects might retard evolution.

Wulff and Bazzaz (1992) showed that response to the parental nutritional environment can depend on both offspring environment and offspring genotype. A parental generation of velvetleaf clones (*Abutilon theophrasti*) and their offspring were grown at each of two nutrient levels for all possible combinations of parent-offspring environmental condition. The authors found significant interactions between parent nutritional environment, offspring nutrition, and offspring genotype, indicating the presence of $E_m \times E_o$ (source 6) and $G_o \times E_m$ (source 7).

In a study of *P. lanceolata*, a grandparental generation was derived from two populations, one inhabiting mowed lawn, the other an abandoned field (Schmitt et al. 1992). From these starting stocks, clones were made and transplanted in a reciprocal design across both sites. Offspring from each clone/field treatment group were reared in each of four greenhouse treatments that varied in light and fertilizer levels. This design enabled the separation of parental environmental components (E_m) from parental genetic components (G_m) in offspring phenotype. The response of offspring depended on the light level they experienced (E_o), suggesting the presence of $G_m \times E_o$ (source 4c), $G_m \times E_m \times E_o$ (source 4d), or $E_m \times E_o$ (source 6). The authors concluded that the potential for natural selection to discriminate among maternal genotypes (source of the parental effect) can differ according to offspring environment.

8.3 Studies That Highlight the Interactive Effect of Parent and Offspring Environments

Table 8.3 lists examples of both parental and offspring environment influencing parental effects expression, thereby generating a component of offspring phenotype that includes E_m

Table 8.3 Environmental variation experienced by both parents and offspring influences the intensity of parental effects expression.

Organism	Source of variation in parental environment	Offspring trait affected	Parental effects were differentiality expressed when offspring experienced	Reference
Vole	Density	Development time	Field vs. lab rearing conditions	Boonstra and Boag 1987
Fish	Food rations	Development time, size at maturity	Low vs. high food rations	Lin & Dunson 1995
Locust	Density	Gregarization	Different degrees of crowding	Islam et al. 1994a,b
Moth	Host quality	Development time	One host species over others	Rossiter 1991
Plant	Soil quality	Plant height	Glasshouse and garden rearing	Schmidt and Dolt 1994
Plant	Soil nutrients	Longevity	Low vs. high soil nutrient levels	Aarssen and Burton 1990
Plant	Soil nutrients	Spike biomass	Greater vs. lesser competition for resources	Miao et al. 1991a
Plant	Soil nutrients in $t-2$ and $t-1$	Germination	Greater vs. lesser competition for resources	Miao et al. 1991b

$\times E_o$ (source 6). When variation among families for the form of the interactive effect can be detected, the presence of $G_m \times E_m \times E_o$ (source 4d) is also indicated.

In a study of the meadow vole (*M. pennsylvanicus*), Boonstra and Boag (1987) evaluated the impact of population density changes on the expression of genetic variation in life history traits. The parental generation was derived from a population at different density states (increasing and peak) by sampling from the same population in consecutive years. Parents were bred in field enclosures according to a half-sibling mating design; offspring were reared in the lab. The results indicated that the proportion of variation owing to maternal effects changed with both parental and offspring density experience. When parents came from the population at increasing density, weight at sexual maturity included a significant maternal effects component. There was no maternal effects component of offspring phenotype when parents came from the population at peak density. This indicates that offspring phenotype included some form of E_m (sources 4b, 4d, 5, 7) and E_o (sources 4c, 4d, 6, 8). When the offspring environment was "field," offspring from parents in the "increasing" population grew more rapidly than offspring from parents in the "peak" population. By contrast, when the offspring environment was "lab," the opposite pattern obtained. In combination, these results indicate the presence of $E_m \times E_o$ (source 6).

Lin and Dunson (1995) studied the response of five hermaphroditic strains of an estuarine fish (*Rivulus marmoratus*) raised under either high or low availability of food rations in the parental generation. Likewise, offspring were divided among these same two treatments of environmental conditions. There was a significant interaction between parental and offspring environmental components for offspring reproductive traits (time to maturity, body length at maturity). For some lines, offspring encountering "good" conditions (higher food rations) were able to utilize the resource more efficiently when their parents grew up under a low versus a high food level. These results demonstrate the presence of an $E_m \times E_o$ component of offspring phenotype. Since the hermaphroditic lines differed from one another in their two-generation norm of reaction lines, the presence of $G_m \times E_m \times E_o$ (source 4d) is indicated.

In a study of gregarization in desert locusts (*Schistocerca gregaria*), Islam et al. (1994a) subjected a parental generation of a laboratory colony to varying degrees of density ranging from a lifetime in isolation to a lifetime under crowding. They found that the density experience of parents dictated the degree of gregarization displayed by offspring. For example, a prereproductive life in isolation followed by crowding during the period of oviposition resulted in offspring that were gregarious. When the order of parental treatments was reversed (crowded, then isolated), offspring were solitary in behavior. Moreover, the impact of this parental effect was modified further by the density experience of the offspring (Islam et al. 1994b), suggesting the presence of $E_m \times E_o$ (source 6).

In a study of the gypsy moth, *L. dispar*, Rossiter (1991) reared members of a parental generation on either red or black oak trees. In the laboratory, offspring of each mother were reared on each of four diet treatments: red or chestnut oak foliage, regular or low-protein synthetic diet. While parental host species accounted for 24% of the total variation in development time, offspring host experience modified the impact of parental diet on offspring phenotype: when the offspring diet was chestnut oak, offspring of mothers reared on black oak grew faster than offspring of mothers reared on red oak. This pattern was not detected under the conditions of other offspring host environments, indicating the presence of $E_m \times E_o$ (source 6).

In experiments with the tall goldenrod, *Solidago altissima*, Schmid and Dolt (1995) made clonal replicates of 15 genotypes, grew them in each of two parental environments (sand and soil), then used a diallel crossing design to produce offspring representing every combination of maternal/paternal pairing of each environmental type (four parental treatment groups in all). Independent of genotype, the type of soil in which parents grew influenced seed size, likelihood of germination, and early seedling mass. These early differences were likely responsible for later correlated effects wherein the parental environment made a significant contribution to height and stem mass at harvest. The quality of the offspring environment modified the impact of the parental environment (i.e., generated $E_m \times E_o$) as evidenced by the fact that (1) maternal effects were less persistent in offspring reared in the greenhouse compared to those reared in the garden and (2) a ranking of crosses by mean offspring phenotype yielded a different order for offspring reared in field versus the greenhouse. Moreover, there was variation among clones in the form of this response, suggesting the presence of $G_m \times E_m \times E_o$ (source 4d).

In a study of the groundsel *Senecio vulgaris*, Aarssen and Burton (1990) reared genetically similar parents under high or low nutrients and then split their offspring between the same treatment groups for a total of four combinations of paired sequential environments. They found that mothers grown in low-nutrient soil produced offspring with lower seed mass and later germination time. The impact of maternal soil quality was also seen on juvenile size and survival: when reared under low- (but not high-) nutrient conditions, offspring of mothers from low-nutrient environments grew to be smaller juveniles with a greater likelihood of survival. Their discussion of the possible adaptive value of this response highlights the role of $E_m \times E_o$ (source 6) contributions to offspring fitness.

In studies of *Plantago major* and *P. rugelii*, Miao et al. (1991a, b) reared members of a parental generation under four different treatments that were combinations of nutrient level (low/high) and timing of nutrient delivery (with or without pulses). Offspring from each parental treatment were themselves divided among four different treatment groups that were combinations of competition level (present or not) and timing of nutrient delivery for a total of eight offspring groups. They found that the effect of maternal nutrient timing was apparent only when parents were grown at low nutrient levels *and* offspring were grown under competition (Miao et al. 1991a), indicating the importance of inherited environmental effects in the form of $E_m \times E_o$. In another experiment on *P. major*, alone, Miao et al. (1991b) considered the impact of two generations of environmental experience (grandparental and parental environments) on offspring. Their treatment groups included stock that received nutrient pulses in generations 1 and 2, 1 only, 2 only, or not at all. Fitness characters were measured on offspring (generation 3) growing with and without competition. Overall, maternal and grandmaternal environmental factors influenced offspring phenotype and the intensity of maternal effect expression varied with the external environmental conditions experienced by offspring during their lifetime. For example, in the absence of competition, offspring from parents and grandparents not exposed to competition germinated earlier and in higher proportion compared to all other "parental" treatment groups. By contrast, when offspring were reared under competition, offspring from parents and grandparents that had received nutrient pulses germinated earlier and in higher proportion compared to all other "parental" treatment groups. Examination of other traits yielded results demonstrating that the influence of the environment in previous generations differed among plant characters.

8.4   Negative Parental Effects and Cross-Generational Covariance

The last category of environmental impact on parental effects expression involves the influence of the environment on cross-generational genetic correlations, that is, the genetic correlation between genes expressed in successive generations. The impact of a $G_m G_o$ correlation, $\text{Cov} G_m G_o$ (source 8), on evolutionary processes has been well studied and discussed, particularly regarding mammals (e.g., Dickerson 1947; Eisen 1967; Bondari et al. 1978; Riska et al. 1985; Atchley and Newman 1989; Cowley et al. 1989; Kirkpatrick and Lande 1989). We know that it is possible for the environment to interact with cross-generational genetic correlations to produce components of offspring phenotype such as $\text{cov}(G_m \times E_o)(G_o \times E_o)$ (Eisen and Saxton 1983). However, in general, such phenomena are assumed absent, not surprising since their measurement poses feasibility problems for appropriate experimental designs. To include these pesky considerations in studies of how the environment adjusts parental effects expression, it is possible to use the presence of negative parental effects as a premise for further work to verify the presence of complex cross-generational covariance.

A negative parental effect occurs when there is an inversion of phenotype across generations, with any contributing effects of the offspring environment on offspring phenotype accounted for (see Rossiter 1992). While easier to detect than positive parental effects, most examples of negative parental effects arise as an inadvertent by-product of an experiment focused on another topic. Some examples include the work of Falconer (1965), who found that female mice who are smaller and less fecund than average produce offspring that are larger and more fecund than average, presumably related to resource partitioning among progeny. Lin and Dunson (1995) found that low (vs. high) food rations resulted in slower time to maturity and smaller size at maturity in a river fish. By contrast, offspring of these smaller parents grew to be relatively larger in a shorter period of time compared to offspring of the better fed, larger parents. Rossiter (1991) found that gypsy moth parents who grew up on foliage with greater defoliation levels ended up smaller and less fecund than parents who grew up on foliage with relatively less defoliation. By contrast, offspring of the parents on the poorer quality food were more robust (greater pupal weights) than offspring from parents with the "better" food. The chance that the results were influenced by selection was minimized by the fact that parents and their offspring were measured rather than sampling members of two successive generations and by controlling the environmental resources during the experiment.

Generational inversions of phenotype, collectively referred to as negative parental effects, can arise in several ways. First, they arise from negative cross-generational genetic covariance, $-\text{Cov} G_m G_o$ (source 8), between parental performance genes expressed in two time periods (Riska et al. 1985; Atchley and Newman 1989). Second, they arise when interactions between temporally correlated genes and one environment yield a different response, producing $-\text{Cov}(G_m \times E_o)(G_o \times E_o)$ or $-\text{Cov}(G_m \times E_m)(G_o \times E_m)$. These latter terms may be difficult to envision unless we keep in mind that (1) the expression of parental performance genes can occur in both parental and offspring environments (e.g., body size of parent and offspring) and (2) components of the parental environment can be physically experienced by both generations (e.g., enzymes, antimicrobial plant compounds). Finally, it is possible that negative parental effects are related not to genetic correlation but to how the environment influences a developmental pathway involved in an inherited environmental effect.

It may be difficult or impractical to distinguish the underlying cause(es) of a negative parental effect. Despite this, evidence of their occurrence suggests likely places to investigate

Table 8.4 Examples of negative parental effects.

Organism	Parental environmental influence, if studied	Characters with inverse value for parents and offspring	Reference
Cattle	Unknown	Growth, muscular and skeletal development	Shi et al. 1993
Mice	Clutch size	Body weight	Riska et al. 1985
Voles	Unknown	Pupal weight, reproductive parameters	Hansson 1988
Fish	Food rations	Time to and size at maturation	Lin and Dunson 1995
Springtails	Seasonal	Age at maturity	Janssen et al. 1988
Drosophilids	Temperature	Ability to tolerate cold stress	Watson and Hoffmann 1996
Moths	Foliar defoliation	Pupal weight, fecundity	Rossiter 1991
Beetles	Unknown	Pupal weight, family size	Bondari et al. 1978

These effects suggest the presence of $Cov(G_mG_o)$, which is due, strictly, to genetic covariance, or the presence of $Cov(G_mE_m)(G_oE_m)$ or $Cov(G_mE_o)(G_oE_o)$, which include effects of the external parental (E_m) or offspring (E_o) environment.

$Cov(G_m \times E_o)(G_o \times E_o)$ or $Cov(G_m \times E_m)(G_o \times E_m)$ (sources 8b and 8c). Table 8.4 lists some examples of negative parental effects, a few of which are described here (also see beginning of this section). For two *Drosophila* species, Watson and Hoffmann (1996; experiment described in section 8.1) found that offspring of cold-stressed parents were less able to handle the same level of stress. In a study of wild springtail populations (*Orchesella cincta*) raised under homogeneous lab conditions, Janssen et al. (1988) found a negative relationship between age at reproduction for mothers and daughters but a positive one between grandmothers and daughters. From this result they speculated that a generational oscillation for age to reproductive maturity phenotypic was mediated by a maternal effect owing to $-Cov-G_mG_o$. Since the populations under study have two generations per year, each encountering very different environments (summer vs. winter conditions), the authors suggested the maternal effects were of adaptive value, resulting in phenotypes better suited for two very different sets of environmental conditions. Finally, an example from the work of Bondari et al. (1978) illustrates a negative maternal effect, its inclusion of some type of $-CovG_mG_o$, and the consequences of its expression on evolution. Their quantitative genetics study of *Tribolium castaneum* found that relatively large mothers produced daughters that were relatively smaller; the same pattern held for family size. The authors assumed the underlying mechanism was associated with nutritional provisioning of the eggs such that changes in family size resulted in changes in the amount or quality of egg provisions, leading to the reduced likelihood of success in offspring from larger clutches. They suggested that genetic antagonism between two components of a characteristic (i.e., their temporal expressions: a maternal trait in generation t and a maternal effects trait in generation $t + 1$) reduces the opportunity for selection to improve one without causing an unfavorable change in the other. It is likely that egg-provisioning levels will also be influenced by the external environment in the parental generation and that the impact of egg-provisioning levels will be influenced by the external environment in the offspring generation. Consequently, it is reasonable to consider that some type of negative cross-generational covariance, $Cov(G_m \times E_o)(G_o \times E_o)$ or $Cov(G_m \times E_m)(G_o \times E_m)$, may also be adjusting offspring phenotype.

8.5 Parental Effects as Adaptations

Groeters and Dingle (1987) studied the impact of photoperiod on the genetics of life history in multivoltine milkweed beetles, *Oncopeltus fasciatus*, and found strong photoperiod-mediated maternal effects for propensity to diapause. Short-day photoperiod (approaching winter) resulted in the production of offspring with greater likelihood of diapause while long-day photoperiod (approaching summer) resulted in the production of offspring less likely to diapause. They concluded that the maternal effect (adjustment of diapause propensity) conferred flexibility in the life history by allowing greater realization of a phenotypic optimum (by entering diapause only when beneficial), despite the action of natural selection under two very different sets of environmental conditions (late spring and fall). Studies such as this one offer the opportunity to make inferences about the adaptive nature of parental effects. However, the subject is complex and often empirically intractable, as are many topics in evolutionary biology. Consequently, I focus on information at hand to suggest some avenues of thinking and further investigation.

It is necessary to answer two different questions when discussing the subject of adaptive parental effects. The first is the issue of adaptive value. A parental effect has adaptive value whenever its expression enhances the probability of survival and reproduction. And to estimate the adaptive value of a parental effect, one must assess the contribution of the parental effect to fitness, relative to contributions from other sources (e.g., nuclear genes). The second issue regards the genesis and maintenance of an adaptive parental effect: Did the parental effect arise specifically because it conveyed adaptive advantage? Or, did the parental effect comprise traits established (by selection) for other reasons, but maintained by selection, ad hoc, because of an emerging adaptive value?

In the first case (parental effect arose specifically as an adaptation), natural selection favored a particular set of character states (i.e., established a mechanism) that spanned at least two points in the life history, for example, one expressed in the parental stage and one expressed in the juvenile stage. And because the expression of a parental effect crosses generations (e.g., parent stage expression in time $t - 1$ and offspring stage expression in time t), natural selection will usually act under two very different sets of environmental conditions.

In the second case (parental effect arose for other reasons but is maintained now by natural selection), the mechanisms/character states involved in parental effects arose as a by-product of selection on other traits (e.g., ability to sequester dietary components to deter pathogens during parental life; when present in high titer, the defense compounds are inadvertently transmitted to eggs/fetus). Fortuitously, some mechanisms have adaptive value in a subsequent generation and are maintained (and possibly improved) by natural selection.

For selection to establish or be responsible for the maintenance of an adaptive parental effect, there must be genetic variation for the parental effect trait (i.e., some to all of its mechanistic components). These sources of genetic variation (shown in figure 8.1) include the parental performance genotype, G_m (source 4a); genetic variation for plasticity in parental effect expression, $G_m E_m$ (source 4b), $G_m E_o$ (source 4b), $G_m E_m E_o$ (source 4c), $G_o E_m$ (source 7); and genetic covariation, $\mathrm{Cov}G_m G_o$ (source 8). The latter type of variation is important for it can influence the rate and direction of evolutionary change, sometimes in an unpredictable manner (Riska et al. 1985; Atchley and Newman 1989; Kirkpatrick and Lande 1989; Rossiter 1997). Whenever genetic variation exists for the sources listed immediately above, there is potential for the development of an adaptive parental effect.

Beyond the issue of genetic variation, the evolution of an adaptive parental effect will be very sensitive to both the intensity and timing of selection. If selection occurs on the parent, evolution of a parental effect is possible if one of two criteria are met: parental performance genes, which eventually influence offspring, (1) have pleitropic effects important to the vigor of the parent itself or (2) are linked to genes important to the vigor of the parent. If selection occurs on the offspring (i.e., when phenotypic expression of the parental performance gene is manifest), the establishment or maintenance of a parental effect is possible only if there is covariance between the parental effect genes expressed in successive generations. However, the impact of such cross-generational genetic covariance on the evolutionary trajectory of the parental effect is unpredictable (Kirkpatrick and Lande 1989). This is due to a time lag in the impact of natural selection on gene expression. In light of these constraints, the evolution of an adaptive parental effect may be more likely when the parental performance genes are subjected to natural selection during the parental generation, rather than or in addition to the offspring generation.

We have seen in this chapter that the environment can adjust the intensity of parental effect expression. Whether this interaction influences the evolution of parental effects is a topic of great interest and little, if any, empirical work. Consequently, I offer some speculation that may be of use to future empirical and theoretical efforts. The examples in this chapter show that parental effects expression is sensitive to environmental input, suggesting that parental effects provide a vehicle for short-term phenotypic adjustment that may act to preserve a genotype from the ravages of natural selection in a variable environment. The very complexity of these effects (often involving cross-generation phenotypic coordination) suggests that they arose as adaptations; that is, natural selection favored their development—or at least their maintenance. I am inclined to think that the predictability of local environmental quality will be an important criterion for the evolution of adaptive parental effects. When the pattern of variation in a critical environmental factor is recurrent, for example, photoperiod and seasonality (guaranteed experience of known variation) or water column quality as experienced by the juveniles of many marine invertebrates (guaranteed experience of uncertain variation), it is more likely that natural selection will favor the development of parental effects that can "prepare" offspring for the expected environment—even if the expected environment is one of uncertainty (e.g., Kaplan and Cooper 1984; Rossiter 1995). I also think that adaptive parental effects are more likely to develop or be maintained as the discrepancy between parental and offspring environmental experience increases.

Acknowledgments   I thank Tim Mousseau and Chuck Fox for the opportunity to add my *two sense* to an emerging topic of great complexity and repercussions. Comments from two anonymous reviewers helped me clarify some ideas. I also thank Mark Hunter, Brian and Viv Netherwood, Sally Sherman, and Jack Schultz for thought-fulls. This work was supported by a grant from the National Science Foundation, DEB-9629735.

References

Aarssen, L. W., and S. M. Burton. 1990. Maternal effects at four levels in *Senecio vulgaris* (Asteraceae) grown in a soil nutrient gradient. Am. J. Bot. 77:1231–1240.
Atchley, W. R., and S. Newman. 1989. A quantitative-genetics perspective on mammalian development. Am. Nat. 134:486–512.

Barton, S. J., G. Bode, H. G. Sterz, F. Fukunishi, and Y. Kobayashi. 1994. Reproductive and developmental toxicity study: Effect of naftopidil on fertility and general reproductive performance in rats. Pharmacometrics 48:17–30 (in Japanese).

Boggs, C. L. 1995. Male nuptial gifts: Phenotypic consequences and evolutionary implications. Pp. 215–242 in S. R. Leather and J. Hardie (ed.), Insect Reproduction. CRC Press, New York.

Bondari, K. R., L. Willham, and A. E. Freeman. 1978. Estimates of direct and maternal genetics correlations for pupa weight and family size of Tribolium. J. Anim. Sci. 47:358–365.

Boonstra, R., and P. T. Boag. 1987. A test of the Chitty hypothesis: Inheritance of life-history traits in meadow voles *Microtus pennsylvanicus*. Evolution 41:929–947.

Boppre, M., and O. W. Fischer. 1994. Zonocerus and Chromolaena in West Africa. Pp. 108–126 in S. Krall and H. Wilps (ed.), *New Trends in Locust Control*. GTZ, D-Eschborn.

Braby, M. F. 1994. The significance of egg size variation in butterflies in relation to host plant quality. Oikos 71:119–129.

Bradford, M. J., and D. A. Roff. 1995. Genetic and phenotypic sources of life history variation along a cline in voltinism in the cricket *Allonemobius socius*. Oecologia 103:319–326.

Brett, M. T. 1993. Resource quality effects on *Daphnia longispina* offspring fitness. *J. Plant. Res.* 15:403–412.

Bridges, T. S., and S. Heppell. 1996. Fitness consequences of maternal effects in *Streblospio benedicti* (Annelida: Polychaeta). Am. Zool. 36:132–146.

Brown, M. A., L. M. Tharel, A. H. Brown Jr., W. G. Jackson, and J. R. Miesner. 1993. Genotype × environment interactions in preweaning traits of purebred and reciprocal cross angus and Brahman calves on common Bermuda grass and endophyte-infected tall fescue pastures. J. Anim. Sci. 71:326–333.

Carmona, M. J., M. Serra, and M. R. Miracle. 1994. Effect of population density and genotype on life-history traits in the rotifer *Brachionus plicatilis* O.F. Mueller. J. Exp. Mar. Biol. Ecol. 182:223–235.

Case, A. L., E. P. Lacey, and R. G. Hopkins. 1996. Parental effects in *Plantago lanceolata* L. II. Manipulation of grandparental temperature and parental flowering time. Heredity 76:287–295.

Cowley, D. E., and W. R. Atchley. 1992. Quantitative genetic models for development epigenetic selection and phenotypic evolution. Evolution 46:495–518.

Cowley, D. E., D. Pomp, W. R. Atchley, E. J Eisen, and D. Hawkins-Brown. 1989. The impact of maternal uterine genotype on postnatal growth and adult body size in mice. Genetics 122:193–204.

Dickerson, G. E. 1947. Composition of hopg carcasses as influenced by heritable differences in rate and economy of gain. Iowa Agricult. Exp. Station Res. Bull. 354:492–524.

Dussourd, D. E., K. Ubik, C. Harvis, J. Resch, J. Meinwald, and T. Eisner. 1988. Biparental defensive endowment of eggs with acquired plant alkaloid in the moth *Utetheisa ornatrix*. Proc. Natl. Acad. Sci. USA 85:5992–5996.

Eisen, E. J. 1967. Mating designs for estimating direct and maternal genetic variances and direct-maternal genetic covariances. Can. J. Genet. Cytol. 9:13–22.

Eisen, E. J., and A. M. Saxton. 1983. Genotype by environment interactions and genetic correlations involving two environmental factors. Theor. Appl. Genet. 67:75–86.

Falconer, D. S. 1965. Maternal effects and selection response. Pp. 763–774 in S. J. Geerts (ed.), Genetics Today. Pergamon Press, Oxford.

Fox, C. W., and T. A. Mousseau. 1996. Larval host plant affects the fitness consequences of egg size in the seed beetle *Stator limbatus*. Oecologia 107:541–548.

Fox, C. W., K. J. Waddell, and T. A. Mousseau. 1995. Parental host plant affects offspring life histories in a seed beetle. Ecology 76:402–411.

Futuyma, D. J., C. Herrmann, S. Milstein, and M. C. Keese. 1993. Apparent transgenerational effects of host plant in the leaf beetle *Ophraella notulata* (Coleoptera: Chrysomelidae). Oecologia 96:365–372.

Galloway, L.F. 1995. Response to natural environmental heterogeneity: Maternal efects and selection on life-history characters and plasticities in *Mimulus guttatus*. Evolution 49:1095–1107.

Glazier, D. S. 1992. Effects of food genotype and maternal size and age on offspring investment in *Daphnia magna*. Ecology 73:910–926.

Gould, F. 1988. Stress specificity of maternal effects in *Heliothis virescens* (Lepidoptera: Noctuidae) larvae. Mem. Entomol. Soc. Can. 146:191–197.

Groeters, F. R., and H. Dingle. 1987. Genetic and maternal influences on life history plasticity in response to photoperiod by milkweed bugs (*Oncopeltus fasciatus*). Am. Nat. 129:332–346.

Groeters, F. R., and H. Dingle. 1988. Genetic and maternal influences on life history plasticity in milkweed bugs (*Oncopeltus*): Response to temperature. J. Evol. Biol. 1:317–333.

Hansson, L. 1988. Parent-offspring correlations for growth and reproduction in the vole *Clethrionomys glareolus* in relation to the Chitty hypothesis. Z. Saeugetierk 53:7–10.

Islam, M. S., P. Roessingh, S. J. Simpson, and A. R. McCaffery. 1994a. Effects of population density experienced by parents during mating and oviposition on the phase of hatchling desert locusts, Schistocerca gregaria. Proc. R. Soc. Lond. (ser. B) 257:93–98.

Islam, M. S., P. Roessingh, S. J. Simpson, and A. R. McCaffery. 1994b. Parental effects on the behaviour and colouration of nymphs of the desert locust *Schistocerca gregaria*. J. Insect Physiol. 40:173–181.

Janssen, G. M., G. DeJong, E. N. G. Joose, and W. Scharloo. 1988. A negative maternal effect in springtails. Evolution 42:828–834.

Kaplan, R. H. 1992. Greater maternal investment can decrease offspring survival in the frog *Bombina orientalis*. Ecology 73:280–288.

Kaplan, R. H., and W. S. Cooper. 1984. The evolution of developmental plasticity in reproductive characteristics: An application of the "adaptive coin-flipping" principle. Am. Nat. 123:393–410.

Keena, M. A., T. M. O'Dell, and J. A. Tanner. 1995. Phenotypic response of two successive gypsy moth (Lepidoptera: Lymantriidae) generations to environment and diet in the laboratory. Ann. Entomol. Soc. Am. 88:680–689.

Kirkpatrick, M., and R. Lande. 1989. The evolution of maternal effects. Evolution 43:485–503.

Kobayashi, J. 1990. Effects of photoperiod on the induction of egg diapause of tropical races of the domestic silkworm, *Bombyx mori*, and the wild silkworm, *B. mandarina*. Jap. Agricult. Res. Q. 23:202–205.

Lacey, E. P. 1996. Parental effects in *Plantago lanceolata* L. I. A growth chamber experiment to examine pre- and post-zygotic temperature effects. Evolution 50:865–878.

Lin, H.-C., and W. A. Dunson. 1995. An explanation of the high strain diversity of a self-fertilizing hermaphroditic fish. Ecology 76:593–605.

Miao, S. L., F. A. Bazzaz, and R. B. Primack. 1991a. Effects of maternal nutrient pulse on reproduction of two colonizing *Plantago* spp. Ecology 72:586–596.

Miao, S. L., F. A. Bazzaz, and R. B. Primack. 1991b. Persistence of maternal nutrient effects in *Plantago major* the third generation. Ecology 72:1634–1642.

Nelson, R. L. 1991. Maternal diet influences reproductive development in male prairie vole offspring. Physiol. Behav. 50:1063–1066.

Parichy, D. M, and R. H. Kaplan. 1992. Maternal effects on offspring growth and development depend on environmental quality in the frog *Bombina orientalis*. Oecologia 91:579–586.

Plantenkamp, G. A. J., and R. G. Shaw. 1993. Environmental and genetic maternal effects on seed characters in *Neomophila menziesii*. Evolution 47:540–555.

Potvin, C., and C. Charest. 1991. Maternal effects of temperature on metabolism in the C-4 weed *Echinochloa crus galli*. Ecology 72:1973–1979.

Reznick, D., and A. P. Yang. 1993. The influence of fluctuating resources on life history: Patterns of allocation and plasticity in female guppies. Ecology 74:2011–2019.

Riska, B., J .J. Rutledge, and W. R. Atchley. 1985. Covariance between direct and maternal genetic effects in mice, with a model of persistent environmental influences. Genet. Res. (Cambridge) 45:287–297.

Rogowitz, G. L. 1996. Trade-offs in energy allocation during lactation. Am. Zool. 36:197–204.

Roosenburg, W. M. 1996. Maternal condition and nest site choice: An alternative for the maintenance of environmental sex determination? Am. Zool. 36:157–168.

Rossiter, M. C. 1991. Environmentally-based maternal effects: A hidden force in insect population dynamics. Oecologia 87:288–94.

Rossiter, M. C. 1992. The impact of resource variation on population quality in herbivorous insects: A critical component of population dynamics. Pp. 13–42 in M. D. Hunter, T. Ohgushi, and P. W. Price (ed.), Resource Distribution and Animal-Plant Interactions. Academic Press, San Diego.

Rossiter, M. C. 1994. Maternal effects hypothesis of herbivore outbreak. Bioscience 44:752–763.

Rossiter, M. C. 1995. Impact of life history evolution on population dynamics: Predicting the presence of maternal effects. Pp. 251–275 in N. Cappuccino and P. W. Price (eds.), Population Dynamics: New Approaches and Synthesis. Academic Press, San Diego.

Rossiter, M. C. 1996. Incidence and consequences of inherited environmental effects. Annu. Rev. Ecol. Syst. 27:451–476.

Rossiter, M.C. 1997. Assessment of genetic variation in the presence of maternal or paternal effects in herbivorous insects. *In* S. Mopper and S. Strauss (eds.), *Genetic Variation and Local Adaptation in Natural insect Populations: Effects of Ecology, Life History and Behavior.* Pp. 113–138. Chapman & Hall, New York.

Schmid, B., and C. Dolt. 1994. Effects of maternal and paternal environment and genotype on offspring phenotype in *Solidago altissima* L. Evolution 48:1525–1549.

Schmitt, J., J. Niles, and R. D. Wulff. 1992. Norms of reaction of seed traits to maternal environments in *Plantago lanceolata*. Amer. Nat. 139:451–466.

Shi, M. J., D. Laloe, F. Menissier, and G. Renand. 1993. Estimation of genetic parameters in the French Limousin cattle breed. Genet. Select. Evol. 25:177–189.

Shibata, D. M., and C. D. Rollo. 1988. Intraspecific variation in the growth rate of gastropods: Five hypotheses. Mem. Entomol. Soc. Can. 146:199–213.

Southwood, O. I., and B. W. Kennedy. 1990. Estimation of direct and maternal genetic variance for litter size in Canadian Yorkshire and Landrace swine using an animal model. J. Anim. Sci. 68:1841–1847.

Stratton, D. A. 1989. Competition prolongs expression of maternal effects in seedlings of *Erigeron annuus* (Asteraceae). Am. J. Bot. 76:1646–1653.

Watson, M. J. O., and A. A. Hoffmann. 1995. Cross-generation effects for cold resistance in tropical populations of *Drosophila melanogaster* and *D. simulans*. Austral. J. Zool. 43:51–58.

Watson, M. J. O., and A. A. Hoffmann. 1996. Acclimation, cross-generation effects, and the response to selection for increased cold resistance in *Drosophila*. Evolution 50:1182–1192.

Willham, R. L. 1963. The covariance between relatives for characters composed of components contributed by related individuals. Biometrics 19:18–27.

Willham, R. L. 1972. The role of maternal effects in animal breeding: III. Biometrical aspects of maternal effects in animals. J. Anim. Sci. 35:1288–1293.

Wulff, R. D., and F. A. Bazzaz. 1992. Effect of the parental nutrient regime on growth of the progeny in *Abutilon theophrasti* malvaceae. Am. J. Bot. 79:1102–1107.

PART III

REVIEWS OF MATERNAL
EFFECTS EXPRESSION

The ubiquity of maternal effects in plants and animals is not surprising given the intimate relationship between mothers and offspring. We are, after all, what our mothers make us. What is surprising is the fact that, until recently, many maternal effects were not recognized or characterized as such. As Michael Wade points out in chapter 1, few current textbooks make more than passing reference to maternal effects. A principle objective of this volume is to tie together many disparate fields of biology that bear on the evolution of maternal effects. In Part III, we explore the taxonomic and phenomenological range of adaptive maternal effects. The overwhelming conclusion to be drawn from these reviews is that maternal effects are ubiquitous in their distribution and are often an integral component of organismal life history variation.

Maternal effects have been reported for a wide variety of organisms. In chapter 9, Kathleen Donohue and Annie Schmitt report many examples of maternal effects in a variety of plants (see also chapters 7 and 19). Chuck Fox and Tim Mousseau (chapter 10, and references therein) report over 70 species of insects that show maternal effects, most of which involve adaptive plastic responses to environmnetal cues (see also chapters 5, 13 and 16 for insect examples). Dan Heath and Max Blouw survey fish (chapter 11), and report 20 species that show maternal effects. Most examples involve the relationship between maternal size and egg size or quality, and the resultant effects on offspring growth and survival, although maternal effects on resistance to disease and environmental contaminants, hormone titers, and behavior are also reported. Trevor Price tackles birds in chapter 12 where he reviews the large literature concerning the trade-offs birds exhibit between size, number, and quality of eggs, which have been demonstrated in many species to have profound effects on offspring fitness. In chapter 13, Frank Messina reviews some of the insect literature concerning the consequences of oviposition behavior, and also provides a detailed review of the impact maternal oviposition behavior can have on larval competition in a seed beetle. Bob Kaplan (chapter 14, and references contained therein) dissects many adaptive maternal effects in frogs and salamanders, most often relating to maternal effects on egg size or quality, and their impact on offspring growth and survival. The final chapter in this section (chapter 15) by Mertice Clark and Jeff Galef reviews some of the growing literature concerning maternal effects on behavior in mammals. Of particular note is their finding that interuterine position of embryos can profoundly influence reproductive behavior of offspring later in life.

9

Maternal Environmental Effects in Plants

Adaptive Plasticity?

KATHLEEN DONOHUE & JOHANNA SCHMITT

There is now considerable evidence that, in plants, progeny phenotype can be profoundly influenced by the parental environment. Such parental effects are of particular interest to evolutionary biologists for two reasons. First, parental effects determine the level at which selection acts (Schmitt and Antonovics 1986; Wade, this volume) and thus may strongly affect patterns of selection response (Kirkpatrick and Lande 1989; Wade, this volume). Second, parental environmental effects may be viewed as a form of phenotypic plasticity, which may in themselves evolve in response to natural selection (Lacey 1991, 1996; Schmitt et al. 1992; Platenkamp and Shaw 1993; Schmitt 1995; Mousseau and Fox, this volume). In particular, if the cue eliciting the parental effect also predicts the offspring environment, then parental effects that enhance progeny fitness in that environment may be selectively advantageous. It is therefore important to ask whether parental effects are the product of adaptive evolution.

To understand how adaptive parental effects may evolve, it is critical to examine selective events across at least two generations. Parental effects are unlikely to enhance progeny fitness unless the progeny environment is predictable from events in the parental generation; thus, information is needed concerning intergenerational environmental correlations and predictability. The effect of a parentally influenced trait on progeny fitness will be determined by selection in the offspring environment; however, the impact on parental fitness will be determined by the number of progeny produced in the parental environment and the mean fitness of those progeny across the range of offspring environments. If there are trade-offs between offspring quality and quantity, selection at the parental level may differ in direction from individual selection among offspring. Thus, multilevel studies of selection on parents and progeny are needed across the range of environments that they experience. Moreover, genetic studies need to examine not only genetic variation for a progeny character under selection, but genetic variation for the plastic maternal response that determines that character.

Although considerable evidence now exists for a wide range of parental environmental effects in plants (for reviews, see Roach and Wulff 1987; Gutterman 1992; Schmitt 1995;

Wulff 1995), surprisingly little information is available about the ecological context, fitness consequences, or genetic basis of such effects. We therefore focus on selected examples of the types of information needed to test the hypothesis that parental effects are adaptive. In particular, we review the data that can be brought to bear on three important questions. First, how well does the parental environment predict the progeny environment? Second, how do parental effects influence the fitness of both progeny and parent in the relevant environments? We argue that for a parental environmental effect to be considered adaptive, it must increase the relative fitness of the parental genotype in the offspring environment, that is, the product of offspring fitness in the offspring environment and the number of offspring produced in the maternal environment. Finally, is there genetic variation for parental environmental effects, that is, for reaction norms of offspring phenotype to parental environment? In other words, does the genetic potential exist for evolutionary response to selection? We also present two case studies to illustrate how these questions may be approached.

9.1 How Well Does the Parental Environment Predict the Progeny Enivironment?

Heterogeneity of environments in time and space is common in natural plant populations. Consequently, the environment experienced by progeny may often differ from that experienced by the maternal parent. If the environment were truly unpredictable between generations, then a maternal response to its environment would be unlikely to produce a progeny phenotype that would be favored by natural selection in the progeny environment. However, even if environments vary in space and time, plastic maternal responses may be beneficial to the progeny if the environment that elicits the maternal effect provides predictive information about the environment the progeny will experience. If temporal variation in the environment is predictable, if maternally controlled habitat selection is possible, or if environmental cues are accurate, then the maternal environment need not be the same as the progeny environment in order for the progeny environment to be predictable.

Several recent studies provide evidence for heterogeneous natural selection on a spatial scale well within the distance of seed dispersal in natural plant populations (Hartgerink and Bazzaz 1984; Kalisz 1986; Schmitt and Antonovics 1986; Antonovics et al. 1987; Stewart and Schoen 1987; Schmitt and Gamble 1990; Bell et al. 1991; Stratton 1994). Spatial autocorrelation analyses of edaphic factors (Lechowicz and Bell 1991) and plant performance (Bell and Lechowicz 1991; Bell et al. 1991; Stratton 1994, 1995) indicate that environmental similarity between sites declines markedly over short distances. Thus, the farther progeny are dispersed, the greater the difference from the maternal parent in the selective environment they will experience, even in apparently spatially homogeneous and temporally stable environments (Stratton 1994). Nevertheless, it is important to note that seed shadows are normally leptokurtic, often with a majority of the progeny dispersing in close proximity to the maternal parent (Levin and Kerster 1974; Fenner 1985; Schmitt et al. 1985). A spatially stable environment can be considered to be one in which the scale of spatial variation in selective environments is greater than the dispersal distance of the progeny. Similarly, a temporally stable environment is that for which variance between successive generations is low. In stable environments, progeny are likely to experience environments similar to the maternal environment, and thus maternal effects conferring a selective advantage in that environment will enhance their fitness. The few progeny dispersed farther than the scale of environmental autocorrelation may be either advantaged or disadvantaged by the maternal effect depending on the environments they encounter, which cannot be predicted. For these progeny, fitness may depend strongly on their ability to assess their environment through environmental cues.

The progeny environment may be predictable from the maternal environment even in temporally variable environments. One condition that promotes predictability occurs when the environment is changing over time in a predictable manner. For example, competitive environments can change predictably during succession (Platt and Weiss 1977, 1987). Light environment changes predictably for gap species during canopy closure, and habitat degradation causes predictable reduction in nutrients, water, or safe sites.

Just as insects choose oviposition sites, maternal effects in plants may exert a form of habitat selection and thus increase the predictability of the progeny environment. For example, the structures involved in attraction and reward of seed dispersal agents are largely maternally controlled. Directed dispersal of seeds via a particular guild of animal dispersers is a type of maternal "oviposition site preference" that may increase the probability that progeny will reach a favorable microsite (Beattie and Lyons 1975; Howe and Smallwood 1982). Maternally controlled dormancy and germination cues may also increase the predictability of the seedling environment and can be considered a form of maternal habitat selection for progeny development. Germination requirements such as cold stratification, light, rainfall, or nutrient availability are commonly observed (e.g., Fenner 1985; Ballaré 1994; Vazquez-Yanes and Orozco-Segovia 1994), and these requirements are largely determined by the seed coat, a product of the maternal genotype. Responses to such cues can have a strong effect on the environment in which progeny experience natural selection. The influence of these cues, moreover, can vary according to the maternal environment. For example, exposure of developing seeds to a low ratio of red:far red wavelengths (R:FR) characteristic of foliage shade may induce a phytochrome-mediated light requirement for germination (McCullough and Shropshire 1970; Hayes and Klein 1974; Cresswell and Grime 1981). Here, the plasticity in germination response of the offspring to light is itself a maternal effect.

Environmental cues present in the maternal environment may provide information about the probable progeny environment. For example, the reduction in R:FR of light transmitted or reflected by nearby vegetation may provide an accurate cue to the maternal parent of both present and future competition (Smith 1982; Smith et al. 1990; Schmitt and Wulff 1993; Ballaré 1994). The size of the parent may predict the density of its seed shadow and thus the level of sibling competition experienced by its progeny (Donohue 1993; Philippi 1993a). Seasonal cues may also predict the progeny environment. For example, the temperature or photoperiod experienced by the maternal plant during seed maturation may be an accurate predictor of the seasonal environment the progeny will experience.

Variation in the environment between maternal and progeny generations influences the adaptive value of maternal environmental effects in two important ways. First, it determines the selective environments of both maternal plants and their progeny. To measure the adaptive value of a maternal response, selection in both environments needs to be estimated (Bell and Lechowicz 1994). When the progeny environment cannot be predicted from the maternal environment through any of the mechanisms discussed above, then maternal environmental effects are less likely to be adaptive. Second, the degree of environmental variation experienced by a genotype influences whether phenotypic plasticity—or more precisely in this context, maternal environmental effects—will be advantageous, or whether specialization to a consistent environment is more adaptive (Van Tienderen 1991). The relative advantage of plasticity as opposed to specialization depends strongly on possible costs associated with responding to the environment. In the case of maternal environmental effects, these potential costs should be measured on maternal plants and within the range of environments experienced by them. Therefore, knowing not only the progeny environment, but also the maternal environment and the correlation between them, is necessary for predicting the adaptive value of maternal environmental effects. Explicit studies of intergenerational environmental correlations are clearly needed.

9.2 Parental Effects in Plants and Their Fitness Consequences

There are several ways in which parental genotype and environment may influence offspring phenotype and fitness. During seed development, the maternal parent provides resources and regulatory influences to the developing embryo and triploid endosperm; such maternal effects may strongly influence seed and seedling phenotypes and may also carry over later into the progeny life cycle. In addition, the seed coat and surrounding ovary, perianth, and bract tissues are entirely maternal in origin and may have an important impact on dormancy, germination, dispersal, and resistance of seeds to predators or pathogens. The maternal parent also may control the number of progeny, which can also be viewed as a maternal effect since trade-offs may exist between parental fecundity and offspring quality (Smith and Fretwell 1974; Lloyd 1987; Haig and Westoby 1988). It is also important to remember that the expression of maternal environmental effects may depend on the offspring environment (Alexander and Wulff 1985; Parrish and Bazzaz 1985; Stratton 1989; Miao et al. 1991a,b; Potvin and Charest 1991; Schmitt et al. 1992; Wulff and Bazzaz 1992; Wulff et al. 1994). For example, the effects of parental resource provisioning may have greater impact if the progeny experience a low-resource environment, or a maternally induced light requirement for germination may only be apparent if progeny are shaded.

The question is whether maternal environmental effects are adaptive or simply the result of resource limitation or developmental constraints (Sultan 1996; Mazer and Wolfe, this volume). Adaptive effects of maternal environment should result in increased fitness in the progeny environment. However, if maternal effects are simply a reflection of resource availability, then poor parental environments should result in reduced fitness of both parent and offspring. Below we review three maternally influenced offspring traits that are good candidates for testing the adaptive plasticity hypothesis, and the limited information that is available concerning their effects on fitness.

Offspring Provisioning

Parental investment in seed provisioning should be expected to increase the fitness of individual progeny, and directional selection for increased seed mass has frequently been observed at this level (e.g., Stanton 1984; Morse and Schmitt 1985; Wulff 1986; Haig and Westoby 1988; Kalisz 1989; Schmitt and Ehrhardt 1990; Gross and Smith 1991; Argyres and Schmitt 1992; Mazer and Wolfe, this volume). However, the evolution of offspring provisioning may be constrained by several trade-offs. Larger seeds may have reduced dispersal

(Morse and Schmitt 1985), and increased seed provisioning may increase vulnerability to seed predators or pathogens (Mitchell 1977). Most important, resource allocation to larger seeds may result in a decrease in offspring number due to a trade-off between number and quality of progeny (Smith and Fretwell 1974; Haig and Westoby 1988; Westoby et al. 1992), although this trade-off is not always observed (Mazer and Wolfe, this volume). Thus, from the fitness viewpoint of the parent, the optimal seed size may be smaller than from the fitness viewpoint of the progeny (Haig and Westoby 1988). Unfortunately, studies of selection at both levels are extremely rare. Winn (1988) examined the benefit:cost ratio (Smith and Fretwell 1974) of seed size in *Prunella vulgaris*, but her data do not permit explicit analysis of selection on parental phenotypes or genotypes. To examine selection at the maternal level, it is necessary to examine the effects of different maternal environments on both offspring size and offspring number, and to determine the fitness consequences of variation in allocation to offspring size and number on maternal fitness in different offspring environments. If maternal environmental effects are adaptive, we would predict that the optimal seed size (from the maternal genotype perspective) differs among offspring environments and that maternal effects on seed size are in the direction of the predicted optimum in each environment.

There is growing evidence that the strength of individual selection on seed mass may differ among offspring environments (Stanton 1984; Wulff 1986; Haig and Westoby 1988; Winn 1988; Schmitt and Ehrhardt 1990; Gross and Smith 1991; Winn and Miller 1995; Donohue 1997). In particular, comparative studies suggest that large seeds may be more strongly favored in shaded environments (Mazer 1989; Westoby et al. 1992) or in environments with greater abiotic stress (Gross and Smith 1991). There is also experimental evidence that seed mass is more strongly selected under highly competitive conditions (Stanton 1984; Waller 1985; Stratton 1989; Schmitt and Ehrhardt 1990), although a few studies demonstrate stronger selection for increased seed mass at low density (Donohue 1997; Mazer and Wolfe, this volume). Thus, maternal effects that result in increased seed provisioning to offspring in shaded, stressful, or competitive environments may be hypothesized to be adaptive.

Is there evidence for such effects? Numerous studies have demonstrated effects of parental environment on seed mass (Baker 1972; Marshall 1986; Nakamura and Stanton 1986; Roach and Wulff 1987; Mazer and Wolfe 1992; Schmitt et al. 1992; Schmid and Dolt 1994; Wulff 1995; Lacey 1996; Sultan 1996), although it is important to note that such effects may be due to maternal plasticity in allocation to seed coat tissues in addition to variation in provisioning of endosperm or embryo (Lacey 1996; Sultan 1996). Contrary to the adaptive prediction, many of these cases involve a reduction in seed provisioning by parents exposed to low-resource environments. However, there is also evidence for increased seed mass in response to low-resource parental environments. In *Polygonum persicaria*, achene mass and offspring provisioning increased for maternal plants exposed to drought stress (Sultan 1996), and *Solidago altissima* plants grown in sand produced fewer, larger seeds than plants grown in soil (Schmid and Dolt 1994). In a reciprocal transplant experiment, *Prunella vulgaris* plants grown in woodland habitats produced larger seeds than those grown in old fields, regardless of the population of origin (Winn 1988); however, there was no evidence for stronger selection on seed mass in woodland sites, contrary to the adaptive hypothesis. Similarly, clonal replicates of *Plantago lanceolata* planted reciprocally into the shaded environment of an abandoned hayfield produced larger seeds than those planted into an open lawn (Schmitt et al. 1992; see section 9.4).

These studies show that maternal environmental effects exist for seed mass, that seed mass influences progeny fitness differently in different environments, and that maternal responses that alter seed size sometimes result in progeny with phenotypes that are expected to be favorable in the maternal environment and sometimes result in progeny with unfavorable phenotypes. Still unknown, however, is the magnitude of selection on offspring that matured in different maternal environments, measured within a range of probable environments. Also unknown is maternal fitness in conjunction with progeny fitness. Such information is necessary to estimate the predicted optimum seed mass in any environment.

Dormancy and Germination Cues

Seed dormancy can be considered a form of dispersal through time, and it is widely viewed as an adaptive mechanism enabling seeds to germinate under appropriate environmental conditions. Usually dormancy is broken by specific cues associated with favorable microsites or climatic conditions (e.g., Fenner 1985), and in a few cases there is direct evidence that germination in response to such cues may increase fitness (Rice 1985; Pake and Venable 1996). Within-year dormancy ensures that germination will occur during the appropriate season for seedling establishment. Between-year dormancy is often viewed as a bet-hedging strategy in temporally heterogeneous environments; by reducing variation in fitness between years, it can increase long-term relative fitness. Like spatial dispersal, dormancy can reduce the risk of extinction of a genetic line, increase the environmental sampling by progeny, and increase the probability of progeny germinating under favorable environmental conditions—particularly if environmental cues are reliable (Venable and Lawlor 1980; Venable 1985; Venable and Levin 1985; Venable et al. 1987; Venable and Brown 1988; Rees and Long 1992; Pake and Venable 1996). Examples of such cues include rainfall (Went 1949; Gutterman 1980–81, 1992), nutrient availability (Fenner 1985; Adler et al. 1993), or "gap detection" signals such as fluctuating temperatures (Fenner 1985; Rice 1985) or light quality (Gorski 1975; Fenner 1985; Rees and Brown 1991; Adler et al. 1993).

If the maternal environment can predict the quality of the environment that germinating seedlings may experience in the near future, then maternal environmental effects controlling the fraction of dormant progeny may be adaptive. Seed dormancy is largely maternally controlled (Gutterman 1980–81, 1992; Roach and Wulff 1987; Fenner 1991), although the embryo may also have an influence (Garbutt and Witcombe 1986; Adler et al. 1993). In particular, seeds matured at different positions on the maternal plant often differ in dormancy (Gutterman 1980–81, 1992; Venable and Levin 1985; Philippi 1993b). Consequently, plasticity of maternal architecture may result in correlated plastic responses in the fraction of progeny that are dormant. Maternal environmental effects on dormancy are common (Gutterman 1980–81, 1992), but little is known about their impact on progeny fitness in natural populations (Fenner 1985). Nevertheless, many cases suggest plausible hypotheses for the adaptive significance of plastic maternal control.

One potential example of adaptive maternal effects on dormancy is a plastic response to seasonal cues. Such cues provide important predictive information to the maternal plant about the environment that will be experienced by germinating seeds. For example, day length during seed maturation frequently influences the fraction of seeds that are dormant, a possible mechanism for preventing germination in an inappropriate season (Gutterman 1980–81, 1992). In certain Negev Desert annuals, maternal exposure to long days results in increased seed dormancy (Gutterman 1980–81), which may prevent germination in the spring, when seedlings would not be able to complete their life cycle before the end of the rainy season.

The maternal environment may also provide information about the competitive environment the progeny may experience. For example, exposure to a low ratio of R:FR light during seed maturation on the maternal plant may induce a light requirement for germination (McCullough and Shropshire 1970; Hayes and Klein 1974; Cresswell and Grime 1981; Van Hinsberg 1996). Reduced R:FR is an accurate signal of vegetation shade (Smith 1982; Smith et al. 1990; Schmitt and Wulff 1993) and may provide a cue that the maternal plant is growing under a closed canopy. For many fugitive species in such conditions, the fitness of the progeny may be increased by delaying germination until a suitable gap opens. Similarly, dormancy is increased in seeds produced by maternal plants of the California annual *Nemophila menziesii* grown in competition with grass, a maternal effect that may increase the chance of persistence in patches dominated by a superior competitor (Platenkamp and Shaw 1993). In the desert annual *Lepidium_lasiocarpum*, dormancy is positively correlated with size of the maternal parent, a response hypothesized to be a mechanism for reducing sibling competition in dense seed shadows (Philippi 1993a).

These studies show that maternal effects on dormancy duration and germination requirements do exist, and many of these studies suggest that maternal environmental effects on seed dormancy can be adaptive. To test this hypotheses, however, it will be necessary to determine experimentally the fate of seeds produced under different parental conditions across the range of possible seedling environments (e.g., Rice 1985).

Dispersal

Seed dispersal is maternally determined. Traits that influence seed dispersion patterns could be traits of the seeds and fruits or architectural traits of the maternal plant. When the maternal plant responds to its environment in ways that influence dispersal, then dispersal can be said to exhibit maternal environmental effects.

Direct evidence of maternal environmental effects on seed dispersion patterns is scarce, but circumstantial evidence is provocative. Many studies have shown that the environment influences the proportion of different seed types that a plant produces, for example, cleistogamous (Brown 1952; Levin 1972; Waller 1980; Clay 1982), amphicarpous (Weiss 1980; Cheplick and Quinn 1982), or heteromorphic seeds (Baker and O'Dowd 1982; Venable and Levin 1985). Quite often, the different seed types are known or suspected to differ in their dispersal ability (Schmitt et al. 1985; Venable et al. 1987). More generally, environmental conditions often influence plant architecture, either allometrically (Gerakis et al. 1975; Silander 1978; Thompson and Beattie 1981; Smith 1983) or through a specific developmental response, such as photomorphogenic shade avoidance (Smith 1982; Schmitt and Wulff 1993). Plant architectural traits such as height, branchiness, or fruit abundance often influence seed dispersion patterns (McCanny and Cavers 1989; Shipley and Dion 1992; Donohue 1993; Theide and Augspurger 1996). Consequently, the maternal environment is expected to influence seed dispersal indirectly through its influence on maternal plant architecture and fruit maturation patterns. The ubiquity of such responses to environmental conditions and the abundant evidence that characters that respond to the environment also influence dispersal suggest that maternal environmental effects on dispersal could be quite common. The next question is whether variation in dispersal in response to the maternal environment influences fitness and, if so, whether it is adaptive.

Nearly every published study on the selective value of dispersal concludes that dispersal greatly influences progeny fitness (reviewed in Howe and Smallwood 1982); thus, maternally determined variation in dispersal is very likely to have a fitness impact on progeny. Dispersal can be advantageous when the maternal home site is deteriorating in quality (Platt 1976; Horvitz and Schemske 1986) or when dispersal decreases sibling competition (Van Valen 1971; Liew and Wong 1973; Hamilton and May 1977; Comins et al. 1980; Donohue 1997) or other density-dependent processes such as pathogen attack or predation (Janzen 1971, 1972; Wilson and Janzen 1972; Burdon and Chivers 1975; Augspurger 1983; Augspurger and Kitajima 1992). Dispersal can also reduce the probability of extinction of a genetic line by increasing the likelihood that at least some of the progeny inhabit a site suitable for reproduction (Venable and Lawlor 1980; Hastings 1983; Venable 1985; Horvitz and Schemske 1986; Venable et al. 1987; Venable and Brown 1988). On the other hand, dispersal can be disadvantageous when local adaptation occurs at small spatial scales or when the maternal plant site is predictably better than average in a spatially heterogeneous environment (Balkau and Feldman 1973; Roff 1975; Levin et al. 1984; Holt 1985). The selective advantage of dispersal, therefore, depends on the degree of spatial heterogeneity, temporal variation in environmental conditions, and local genetic structure of the population in the form of locally adapted lines or potentially competing kin groups. Whatever the selective agent on dispersal may be, evidence that progeny dispersion patterns influence progeny fitness is ubiquitous.

No studies, however, have documented fitness consequences of dispersal for the maternal plant, although circumstantial evidence suggests that such effects should be expected. One example is the potential trade-off between seed size, a character that often influences dispersal ability (Sheldon and Burrows 1973; Cremer 1977; Rabinowitz 1978; Green 1980; Augspurger and Hogan 1983; Dolan 1984; Morse and Schmitt 1985; Matlack 1987), and seed number, an important component of maternal fitness (Haig and Westoby 1988; Westoby et al. 1992). In addition, increasing evidence exists that plant architectural traits that could influence dispersal also influence the ability of a maternal plant to intercept light (Menges 1987; Berntson and Weiner 1991; Weiner and Thomas 1992) and have important fitness consequences (Dudley and Schmitt 1996). Therefore, the suggestion is strong that maternal traits that influence progeny dispersion patterns may also influence maternal fitness.

For maternal environmental effects on dispersal to be adaptive, they must increase the product of maternal fecundity and offspring fitness. Some responses, such as increased height under competitive conditions, may enhance both maternal fitness, through shade avoidance, and progeny fitness, through dispersal out of a competitive environment. High density has been shown to induce phenotypes that would promote dispersal (e.g., in *Hypochoeris glabra*; Baker and O'Dowd 1982) and very likely increase progeny fitness, but the fitness consequences to the maternal plant remain unknown. Maternal plants in water-stressed environments also have been shown to express characters that enhance seed dispersal (Peroni 1994), again possibly increasing progeny fitness, but again the fitness consequences to the maternal plant are not known. Other responses may increase fitness in one generation while reducing it in the other. For example, a response that increases branch production may simultaneously be associated with increased fruit production and impeded dispersal (Donohue 1993). Responses of many traits that influence dispersal, such as branch production and height, may be allometric responses, not adaptive in themselves but nonetheless having significant fitness consequences to progeny. In fact, because so many maternal traits that influence dispersal are likely to vary allometrically and reflect maternal plant vigor, it seems clear that the real selective value of dispersal traits cannot be estimated by measuring the postdispersal fitness of progeny alone.

Although there is circumstantial evidence that maternal environments may influence progeny dispersal, studies that explicitly investigate this possibility are very few (Donohue 1993; Theide and Augspurger 1996). Moreover, although abundant evidence exists that dispersal influences the fitness of offspring, very little information is available about how it may influence maternal fitness. Such information is needed before we can assess whether maternal effects on dispersal are adaptive.

It is of interest to note that much of the literature on the selective advantage of seed size, dormancy, and dispersal presents mechanisms of evolution that are based not on individual selection but on selection at higher levels. Differential extinction of genetic lines and selection for competition reduction within kin groups are examples of kin selection, not individual selection. This intuitive approach to addressing the evolution of seed mass, dormancy, and dispersal makes two points clear. First, seed mass, dormancy, and dispersal are maternal characters, controlled not by the progeny but by maternal phenotype and genotype. Second, the evolution of maternal characters is most effectively addressed by considering mechanisms of selection at the kin rather than the individual level (Wade, this volume). For empirical researchers, this means that fitness consequences at least to the maternal parent, if not to the entire kin group, should be investigated in conjunction with fitness consequences to the progeny. With increasing theoretical and methodological work on the mechanisms of higher level selection processes (Wade 1985 and this volume; Goodnight et al. 1992), the ecological models of Venable and others may be adaptable to models of the genetic evolution of such complex characters as seed mass, dormancy, and dispersal.

9.3 Genetic Variation for Parental Effects

If maternal environmental effects are considered a form of intergenerational plasticity, it becomes important to ask whether they can evolve in response to selection. For adaptive maternal effects to evolve, there must be genetic variation within populations for the response of the progeny phenotype to the maternal environment, detectable as a significant maternal genotype by maternal environment interaction effect for the progeny trait of interest (Schmitt et al. 1992; Platenkemp and Shaw 1993; Schmitt 1995). Such variation can be measured by replicating a set of maternal genotypes across different environments and measuring progeny produced by each maternal genotype in each environment. By taking the mean phenotypic value of the progeny in each environment, norms of reaction of the progeny trait to the maternal environment can be characterized.

Several recent studies have detected significant variation in reaction norms of progeny characters to parental environments in a variety of wild plant species (Alexander and Wulff 1985; Lacey 1991, 1996; Schmitt et al. 1992; Wulff and Bazzaz 1992; Platenkemp and Shaw 1993; Schmid and Dolt 1994; Wulff et al. 1994; Sultan 1996; Van Hinsberg 1996; Mazer and Wolfe, this volume). Although genetic variation in parental environmental effects is not observed for all traits or environments (e.g., Stratton 1989; Schmitt et al. 1992; Schmid and Dolt 1994; Lacey 1996; Sultan 1996), it appears to be relatively common in natural plant populations. Many of these studies were conducted using open-pollinated or self-fertilized progeny, so maternal and paternal contributions to the observed parental effects could not be distinguished. Only a few studies have controlled paternity across maternal environments so that plastic responses of maternal genotypes could be distinguished from paternal effects such as nonrandom fertilization or gametophytic selection (Platenkemp and Shaw 1993; Schmid and Dolt 1994; Lacey 1996; Mazer and Wolfe, this volume; Shaw and Byers, this volume). More such studies are needed, but nevertheless the growing evidence for genetic variation in environmental parental effects suggests considerable evolutionary potential for response to selection on parental reaction norms.

9.4 Case Studies

Seed Mass in Plantago lanceolata

To support the hypothesis that a maternal environmental effect is adaptive, it is necessary to demonstrate that it results in increased relative fitness of the maternal genotype in the relevant offspring environment. Here we describe an example of such a test for environmental maternal effects on seed mass from a large field experiment with *Plantago lanceolata* (Schmitt and Donohue, in preparation). This species, a widespread perennial rosette herb of pastures, lawns, and hayfields, has been well studied in an ecological genetic context (for a review, see Kuiper and Bos 1992), and maternal effects have been observed in several studies (Alexander and Wulff 1985; Lacey 1991, 1996; Schmitt et al. 1992; Wulff et al. 1994; Van Hinsberg 1996). In particular, seed mass has been shown to increase when parents are exposed to grass canopy shade in the field (Schmitt et al. 1992), a potentially adaptive response if selection on seed mass is stronger in shaded environments. In addition, genotypes from shaded hayfield habitats tend to produce larger, fewer seeds than populations of open habitats (Van der Toorn and Van Tienderen 1992), suggesting adaptive differentiation among populations. Seed dispersal in *P. lanceolata* is usually quite localized, so it is reasonable to assume that progeny are likely to experience the same canopy environment as the maternal parent.

Our study was therefore designed to address the following questions: (1) Do maternal environmental effects increase progeny fitness in the same canopy environment as the parent? (2) Do maternal environmental effects increase the relative fitness of the maternal genotype when progeny experience the same environment as the parent? (3) If maternal environmental effects on fitness are observed, can they be attributed to differential selection on seed mass in different canopy environments? (4) Is there genetic variation for environmental maternal effects, that is, for response of seed size to the maternal canopy environment?

To address these questions, we performed a large factorial field experiment, in which clonal replicates of 18 genotypes (9 from a lawn population and 9 from an adjacent abandoned hayfield) were grown in two maternal environments within the lawn population at Brown University's Haffenreffer Reserve (Bristol, R.I.). In spring 1991, a total of 576 plants were planted into four randomized blocks; in two of these blocks grass was allowed to grow throughout the season (unmowed treatment), while in the other blocks it was clipped periodically to create an open, high-light environment (mowed treatment). Open-pollinated seeds from each parental plant were collected in summer 1991 and individually weighed. In fall 1991, up to 40 weighed seeds from each parental genotype from each parental environment were planted into two mowed and two unmowed randomized blocks adjacent to the parental site, representing a total of 1422 seeds of known mass. We then repeatedly censused both generations to estimate cumulative fitness. Individual parental plants were censused for survival and inflorescence production from summer 1991 to 1995. The offspring were censused for germination in fall 1991 and spring 1992, and then for survival and inflorescence production from summer 1992 to 1995. We then estimated the fitness of each maternal genotype as the product of mean parental fitness and mean offspring fitness for each combination of parent and offspring environment.

Removal of canopy shade was clearly beneficial to individual plants, resulting in increased fitness for both parents and offspring grown in the mowed environment. However, maternal exposure to vegetation shade resulted in increased progeny fitness in both environments. Thus, from the perspective of offspring fitness, the observed maternal environmental effects could be considered adaptive in the unmowed environment but maladaptive in the mowed environment. It is important to note, however, that the relative magnitude of this effect was significantly greater for progeny in the unmowed environment (figure 9.1, top); the fitness of progeny from shaded parents relative to that of unshaded parents was 9.9 under the canopy shade of the unmowed plots, compared with 2.1 in mowed plots.

Consequently, if we look at the fitness consequences of maternal environmental effects from the fitness perspective of the maternal genotype, a different pattern emerges. Although offspring produced in the unmowed treatment were of higher quality, maternal parents in this environment produced far fewer inflorescences, approximately 20% as many as were produced in the mowed environment (figure 9.1, middle). However, those genotypes that produced larger seeds also produced more of them, so no trade-off between seed mass and number of seeds was apparent at the genotype level (see Mazer and Wolfe, this volume, for similar results). Since impact of the maternal effect on progeny fitness was greater in the unmowed environment, the product of offspring fitness and offspring number for the parents from two parental environments reversed in ranking between the two offspring environments, although the maternal environment and progeny environment interaction was only marginally significant with the available sample size. This trend suggests that maternal environmental effects result in increased fitness of the maternal genotype if offspring experience the same environment as the parent (figure 9.1, bottom). This result is consistent with the predictions of the adaptive plasticity hypothesis.

Can the observed adaptive maternal environmental effect be attributed to differences in direct selection on seed mass in the two offspring environments? Although we predicted that individual selection for larger seeds would be stronger in the unmowed progeny environment, this pattern was not observed. Overall, larger seeds had greater germination success and, consequently, higher fitness. However, contrary to prediction, the strongest selection for seed weight was observed for seeds from shaded (unmowed) parents in the mowed progeny environment. The mechanisms underlying this result are unclear, but the adaptive maternal effect we observed cannot be attributed to selection on seed mass per se. Since our measure of overall seed mass does not distinguish between allocation to seed coat and resource provisioning (e.g., Lacey 1996; Sultan 1996), patterns of selection on seed provisioning may have been masked.

There was significant genetic variation in reaction norms of seed mass to the maternal canopy environment, detected as significant maternal genotype by maternal environment variation in the analysis of variance. Although there was a significant overall increase in seed mass with parental exposure to canopy shade, the magnitude of this plasticity differed among

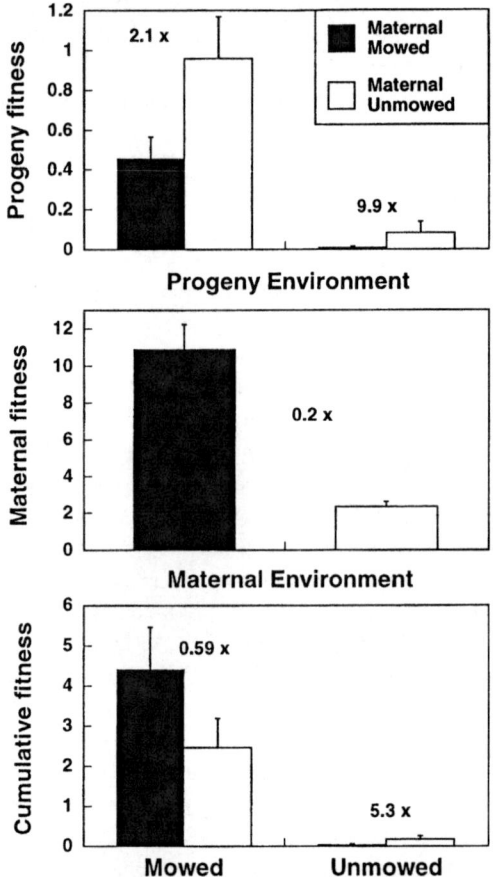

Figure 9.1 Progeny fitness (top), maternal fitness (middle), and cumulative fitness (bottom) in mowed and unmowed maternal and progeny (top and bottom only) environments. Progeny and maternal fitness were measured as the number of inflorescences. The numbers in the top panel indicate the degree of difference in fitness between progeny grown from seeds matured in the mowed as opposed to the unmowed maternal environment. The number in the middle panel is the difference in inflorescence production by maternal plants grown in the unmowed and mowed environment. Cumulative fitness is the product of maternal and progeny fitness calculated for each maternal and progeny environment. The numbers in the bottom panel indicate the degree of difference in cumulative fitness between progeny grown from seeds matured in the mowed as opposed to the unmowed maternal environment. From Schmitt and Donohue (in preparation).

maternal genotypes. Thus, the genetic potential exists for environmental maternal effects to evolve in response to heterogeneous selection on seed size in *P. lanceolata*.

This study illustrates several important points. Most important, the results indicate that the fitness consequences of maternal environmental effects may differ between the maternal and offspring generations. In this case, maternal environmental effects were not entirely adaptive from the fitness perspective of the progeny. However, they clearly increased the mean relative fitness of the maternal genotype when progeny experienced the same environment as the parent, a likely situation given the restricted seed dispersal of *P. lanceolata*. We hope that this experiment provides an example of how to examine such multilevel selection in the field. In addition, our results reinforce the emerging awareness that genetic variation for environmental maternal effects exists in natural populations, and thus the potential exists for selection at the maternal genotype level to result in evolutionary change.

Seed Dispersal in Cakile edentula var. lacustris

Previous studies of the adaptive value of dispersal have gone far in predicting the fitness consequences of dispersal to progeny in a range of environmental conditions. However, with much circumstantial evidence that dispersal may vary with the maternal environment, determining the total fitness consequences of dispersal requires more than measuring progeny fitness after dispersal. The following case study of seed dispersal in *C. edentula* var. *lacustris* examines seed dispersal within the context of maternal character evolution (Donohue 1993) and addresses the following questions: (1) How does the maternal environment influence seed dispersion patterns? (2) How predictable is the progeny environment from the maternal environment? (3) What are the fitness consequences of dispersal to the progeny? (4) What are the fitness consequences of dispersal to the maternal plant? By answering these questions, we can address the question of whether maternal responses that influence dispersal are adaptive.

   C. edentula var. *lacustris*, or the Great Lakes sea rocket, is an annual brassicaceous plant that grows along the shores of the Great Lakes. It has heteromorphic, single-seeded fruit segments that are dispersed by wind and water. The distal segment typically abscises from the proximal segment and is dispersed independently from it. The proximal segment can detach from the maternal plant by pedicel breakage but frequently remains attached to the dead maternal plant even through germination. Consequently, one sees extremely high-density clumps of seedlings at the site of the dead and buried maternal plant (hundreds of seedlings per square meter) as well as scattered individuals elsewhere on the beach whose nearest neighbor may be tens of meters away. Using this system in which progeny density varies tremendously in the field, field observational studies of seed dispersion patterns coupled with selection studies both on the progeny and on the maternal plants were employed to investigate the adaptive value of dispersal within the context of maternal character evolution.

   The maternal environment influenced seed dispersion patterns directly, but it did so even more strongly through its influence on plant morphological traits that influence dispersal (Donohue 1993). Plant morphological traits varied greatly as a function of the environment of the maternal plant. In general, low nutrient or water availability and high density resulted in shorter plants with fewer branches, fewer fruits, and a greater percentage of their fruits located on the primary stem. Most of these phenotypes enhanced dispersal.

   The direction of the phenotypic responses of morphological traits to density indicates that maternal and progeny environments are likely to vary greatly in density. Although environmental factors such as water and nutrient availability vary throughout *C. edentula's* habitat, these factors do not vary systematically as a function of dispersal distance (Donohue 1997). Sibling density declines with dispersal in this system, so plant morphological traits that respond to density are expected to vary as a function of dispersal itself. The nature of the response of maternal plant traits to density results in the prediction of positive phenotypic correlations between generations for some traits that influence dispersal, negative phenotypic

Table 9.1 Estimates of nongenetic correlations between generations for phenotype and crowding.

Trait	Phenotype → Density	Density → Phenotype	Between-generation phenotypic correlation
Total fruits	0.356*	−0.882***	−0.314
Height	0.373*	−0.746***	−0.278
Branches/cm	−0.144	−0.363**	0.052
% Primary fruits	0.076	0.783***	0.060
Correlation between generations for local crowding			−0.480

Estimates of between-generation correlations for phenotypic values in *Cakile edentula* are obtained by multiplying the standardized simple regression slopes of the effect of phenotype on crowding by the standardized simple regression slope of the effect of crowding on phenotype. The sum of these correlations gives the estimate of the between-generation correlation for crowding. This gives the total effect of crowding in one generation on crowding in the next generation.

From Donohue (in preparation).
* $p < 0.05$.
** $p < 0.01$.
*** $p < 0.001$.

correlations for others, and an overall negative correlation between generations for progeny density (table 9.1). Although these predicted phenotypic correlations between generations do not include any genetic component, the fact that some phenotypic correlations are negative suggests that the response to selection on dispersal may be greatly reduced, or may even be rendered negative (Kirkpatrick and Lande 1989). Even without considering the fitness effects of morphological traits on the maternal plants, viewing dispersal as a maternal character has already led to very different predictions concerning the evolution of dispersal than would be the case if dispersal were considered to be simply a trait of progeny under individual selection. This result also shows that density is expected to vary between generations.

So what are the fitness consequences of these responses, both to the progeny and to the maternal plant? Not surprisingly, seed dispersion patterns strongly influenced progeny fitness (Donohue 1997). Increased dispersal led to higher progeny fitness, as would be expected in a plant with long-distance water dispersal and limited opportunity for small-scale local adaptation. Again, environmental factors such as water and nutrient availability did not vary systematically as a function of dispersal distance even though they influenced fitness. Sibling density did vary as a function of dispersal, and density strongly influenced fitness. Therefore, selection on seed dispersion patterns at the level of the progeny was through the effects of sibling density rather than of dispersal distance per se.

Maternal traits that influence dispersal were also correlated with fitness of the maternal plant (Donohue in preparation). One trait, the total number of fruit segments, is in fact a measure of fitness itself for this highly selfing species, since almost all segments produce one seed. Increased height and more compact branch placement were associated with greater seed production at low density but not at high density. At low density, plants with a large proportion of their fruits located on the primary stem had greater seed production, whereas at high density the opposite was observed. Significant stabilizing selection was also observed on this trait in all density environments. Clearly, these maternal traits that influence dispersal are also associated with maternal plant fitness, sometimes allometrically (as in the case of fruit number), and sometimes not (as in the case of stabilizing selection on fruit placement). In

Table 9.2 Summary of effects of an increase in plant traits on maternal fitness and progeny fitness in *Cakile edentula* through the influence on dispersal.

Trait increase	Maternal fitness	Dispersal	Progeny fitness
Total fruits	Positive	Decreased	Negative
Height	Positive	Decreased	Negative
Branches/cm	Positive	Decreased	Negative
% Primary fruits	Negative	Neutral/Deceased	Neutral/Negative

From Donohue (in preparation).

short, maternal plant traits influence *progeny* fitness through their influence on dispersal, and dispersal influences *maternal* plant fitness through differing selection on maternal plant traits in different densities. Whether the traits themselves are under selection, or whether they merely reflect maternal vigor, the association between maternal traits that influence dispersal and maternal fitness will influence the total selective value of dispersal.

As in the case of seed mass in *Plantago*, the maternal response to its environment is predicted to enhance progeny fitness in one environment (i.e., when the maternal plant grows under high density, the progeny are more likely to be dispersed—good) but decrease progeny fitness in another environment (i.e., when the maternal plant is growing in low density, the progeny are less likely to be dispersed—bad). The total adaptive value of dispersal, however, depends both on the maternal plant's fitness in its density environment and on the progeny fitness as a function of its density environment. In this example, a response that increased progeny fitness through increasing dispersal was most often associated with reduced maternal fitness (table 9.2). By considering only the fitness of progeny under different dispersion patterns, one would have predicted a rapid evolution toward increasing dispersal ability. This would never explain the presence of the high-density clumps of seedlings so common on the beach. Nor does such an approach alone have promise for explaining the ubiquity of underdispersal in so many other diverse systems. By considering dispersal as a maternal character, with maternal plants under selection as well as progeny, we predict not only a dramatically reduced selective advantage to dispersal, but also a much slower or even negative response to such selection.

Conclusions

These case studies demonstrate possible promising methods for studying the adaptive value of maternal environmental effects. Selection studies on progeny traits should be conducted not in one environment, but in a range of environments experienced by maternal plants and their progeny. The magnitude of selection in the different environments should be compared to the observed maternal response to that environment in order to determine if such a response leads to advantageous progeny phenotypes in the different environments. Moreover, the fitness consequences to the maternal plant should be considered. In both case studies, the fitness of the maternal plant determined whether the maternal environmental effect was adaptive or not. Finally, one needs to know not only how maternal and progeny traits are selected in their respective environments, but also what these environments are, how they change between generations, and to what extent they are genotype specific or even evolving in themselves.

We raise these issues not to provide answers at this point, but to illustrate how we can test hypotheses concerning the evolution of maternally influenced traits such as seed mass, dormancy, and dispersal. Selection studies that investigate only fitness consequences to progeny of maternally determined characters may be greatly limited in their ability to predict the evolutionary trajectories of these characters. Moreover, studies that investigate the fitness consequences of maternal morphological traits may benefit from considering potential effects of these traits on progeny. Traits with intergenerational effects are probably much more common than is now appreciated. Dealing directly with the difficulties of intergenerational evolutionary dynamics by applying multilevel selection approaches promises to have great explanatory power for future research on the evolution of complex life history traits.

Acknowledgments   Renata D. Wulff has been a long-term collaborator on the *Plantago* maternal effects project, and we thank her for invaluable discussion and insight. Susan Gamble and Susan Dudley made major contributions to the *Plantago* experiment reported here. We are also grateful to all the research assistants and conscripted lab members who helped with the field experiment, especially Kyle Young, Colin Purrington, Lynn Adler, and Rachel Collin, and to Duncan Brown and Saara DeWalt for the figure. Susan Mazer, Michael Wade, and an anonymous reviewer gave many helpful comments on the manuscript. Special thanks are due to Timothy Mousseau for forceful encouragement to produce this review. This work was supported by NSF Grants BSR-8906291, BSR 9023389, and DEB 9306637.

References

Adler, L. S., K. Wikler, F. S. Wyndham, C. R. Linder, and J. Schmitt. 1993. Potential for persistence of genes escaped from canola: Germination cues in crop, wild, and crop-wild hybrid *Brassica rapa*. Funct. Ecol. 7:736–745.

Alexander, H. M., and R. Wulff. 1985. Experimental ecological genetics in *Plantago*. X. The effects of maternal temperature on seed and seedling characters in *Plantago lanceolata*. J. Ecol. 73:271–282.

Antonovics, J., K. Clay, and J. Schmitt. 1987. The measurement of small-scale environmental heterogeneity using clonal transplants of *Anthoxanthum odoratum and Danthonia spicata*. Oecologia 71:601–607.

Argyres, A. Z., and J. Schmitt. 1992. Neighborhood relatedness and competitive performance in *Impatiens capensis* (Balsaminaceae): A test of the resource partitioning hypothesis. Am. J. Bot. 79:181–185.

Augspurger, C. K. 1983. Seed dispersal of the tropical tree, *Platypodium elegans*, and the escape of its seedlings from fungal pathogens. J. Ecol. 71:759–771.

Augspurger, C. K., and K. P. Hogan. 1983. Wind dispersal of fruits with variable seed number in a tropical tree (*Lanchocarpus pentaphyllus*: Legumenosaea). Am. J. Bot. 70:1031–1037.

Augspurger, C. K., and K. Kitajima. 1992. Experimental studies of seedling recruitment from contrasting seed distributions. Ecology 73:1270–1284.

Baker, G. A., and D. J. O'Dowd. 1982. Effects of parent plant density on the production of achene types in the annual *Hypochoeris glabra*. J. Ecol. 70:201–215.

Baker, H. G. 1972. Seed weight in relation to environmental conditions in California. Ecology 53:997–1010.

Balkau, B. J., and M. W. Feldman. 1973. Selection of migration modification. Genetics 74:171–174.

Ballaré, C. L. 1994. Light gaps: Sensing the light opportunities in highly dynamic canopy environments. Pp. 73–110 in M. M. Caldwell and R. W. Pearcy (eds.), Exploitation of Environmental Heterogeneity by Plants: Ecophysiological Processes Above- and Below-ground. Academic Press, San Diego.

Beattie, A. J., and N. Lyons. 1975. Seed dispersal in *Viola* (Violaceae): Adaptation and strategies. Am. J. Bot. 62:714–722.

Bell, G., and M. J. Lechowicz. 1991. The ecology and genetics of fitness in forest plants. I. Environmental heterogeneity measured by explant trials. J. Ecol. 79:663–685.

Bell, G., and M. J. Lechowicz. 1994. Spatial heterogeneity at small scales and how plants respond to it. Pp. 391–414 in M. M. Caldwell and R. W. Pearcy (eds.), Exploitation of Environmental Heterogeneity by Plants: Ecophysiological Processes Above- and Below-ground. Academic Press, San Diego.

Bell, G., M. J. Lechowicz, and D. J. Schoen. 1991. The ecology and genetics of fitness in forest plants. III. Environmental variance in natural populations of *Impatiens pallida*. J. Ecol. 79:697–713.

Berntson, G. M., and J. Weiner. 1991. Size structure of populations within populations: Leaf number and size in crowded and uncrowded *Impatiens pallida* individuals. Oecologia 85:327–331.

Brown, W. V. 1952. The relation of soil moisture to cleistogamy in *Stipa leucotricha*. Bot. Gaz. 113:438–444.

Burdon, J. J., and G. A. Chilvers. 1975. A comparison between host density and innoculum density: Effects on the frequency of primary infection foci in *Pythium*-induced damping-off disease. Austral. J. Bot. 23:899–904.

Cheplick, G. P., and J. A. Quinn. 1982. *Amphicarpum purshii* and the "pessimistic strategy" in amphicarpic annuals with subterranean fruit. Oecologia 52:327–332.

Clay, K. 1982. Environmental and genetic determinants of cleistogamy in a natural population of the grass *Danthonia spicata*. Evolution 36:734–741.

Comins, H. N., W. D. Hamilton, and R. M. May. 1980. Evolutionary stable dispersal strategies. J. Theor. Biol. 82:205–230.

Cremer, K. W. 1977. Distance of dispersal in eucalypts estimated from seed weight. Austral. For. Res. 7:225–228.

Cresswell, E. G., and J. P. Grime. 1981. Induction of a light requirement during seed development and its ecological consequences. Nature 291:583–585.

Dolan, R. W. 1984. The effect of seed size and maternal source on individual size in a population of *Ludwigia leptocarpa* (Onagraceae). Am. J. Bot. 71:1320–1307.

Donohue, K. 1993. The evolution of seed dispersal in *Cakile edentula var. lacustris*. Ph.D. dissertation, University of Chicago.

Donohue, K. 1997. Seed dispersal in *Cakile edentula var. lacustris*: Decoupling the fitness effects of density and distance from the home site. Oecologia 110:520–527.

Dudley, S. A., and J. Schmitt. 1996. Testing the adaptive plasticity hypothesis: Density-dependent selection on manipulated stem length in *Impatiens capensis*. Am. Nat. 147:445–465.

Fenner, M. 1985. Seed Ecology. Chapman and Hall, New York.

Fenner, M. 1991. The effects of the parent environment on seed germinability. Seed Sci. Res. 1:75–84.

Garbutt, K., and J. R. Witcombe. 1986. The inheritance of seed dormancy in *Sinapis arvensis* L. Heredity 56:25–31.

Gerakis, P. A., F. P. Guerrero, and W. A. Williams. 1975. Growth, water relations, and nutrition of three grassland annuals as affected by drought. J. Appl. Ecol. 12:125–135.

Goodnight, C. J., J. M. Schwartz, and L. Stevens. 1992. Contextual analysis of models of group selection, soft selection, hard selection, and the evolution of altruism. Am. Nat. 140:743–761.

Gorski, T. 1975. Germination of seeds in the shadow of plants. Physiol. Plant. 34:342–346.

Green, D. S. 1980. The terminal velocity and dispersal of spinning samaras. Am. J. Bot. 67:1218–1224.

Gross, K. L., and A. D. Smith. 1991. Seed mass and emergence time effects on performance of *Panicum dichotomiflorum* Michx. across environments. Oecologia 87:270–278.

Gutterman, Y. 1980–81. Review: Influences on seed germinability: phenotypic maternal effects during seed maturation. Pp. 93–97 in A. M. Mayer (ed.), Control Mechanisms in Seed Germination. Vol. 29. Weizmann Science Press of Israel, Jerusalem.

Gutterman, Y. 1992. Maternal effects on seeds during development. Pp. 27–59 in M. Fenner (ed.), Seeds: The Ecology of Regeneration in Plant Communities. Redwood Press, Melksham.

Haig, D., and M. Westoby. 1988. On limits to seed production. Am. Nat. 131:757–759.

Hamilton, W. D., and R. M. May. 1977. Dispersal in stable habitats. Nature 269:578–581.

Hartgerink, A. P., and F. A. Bazzaz. 1984. Seedling scale environmental heterogeneity influences individual fitness and population structure. Ecology 65:198–206.

Hastings, A. 1983. Can spatial variation alone lead to selection for dispersal? Theor. Pop. Biol. 24:244–251.

Hayes, R. G., and W. H. Klein. 1974. Spectral quality influence of light during development of *Arabidopsis thaliana* plants in regulating seed germination. Plant and Cell Physiol. 15:643–653.

Holt, R. D. 1985. Population dynamics in two-patch environments: Some anomalous consequences of an optimal habitat distribution. Theor. Pop. Biol. 28:181–208.

Horvitz, C. C., and D. W. Schemske. 1986. Seed dispersal and environmental heterogeneity in a neotropical herb: A model of population and patch dynamics. Pp. 169–186 in A. Estrada and T. H. Flemming (eds.), Frugivores and Seed Dispersal. Dr. W. Junk Publishers, Dordrecht.

Howe, H. F., and J. Smallwood. 1982. Ecology of seed dispersal. Annu. Rev. Ecol. Syst. 7:469–495.

Janzen, D. H. 1971. Escape of *Cassia grandis* L. beans from predators in time and space. Ecology 52:964–979.

Janzen, D. H. 1972. Escape in space by *Sterculia apetala* from the bug *Dysdercus fasciatus* in a Cost Rican deciduous forest. Ecology 53:350–361.

Kalisz, S. 1986. Variable selection on the timing of germination in *Collinsia verna* (Scrophulariaceae). Evolution 40:479–491.

Kalisz, S. 1989. Fitness consequences of mating system, seed weight, and emergence date in a winter annual, *Collinsia verna*. Evolution 43:1263–1272.

Kirkpatrick, M., and R. Lande. 1989. The evolution of maternal characters. Evolution 43:485–503.

Kuiper, P. J. C., and M. Bos. 1992. Plantago: A Multidisciplinary Study. Ecological Studies, Vol. 89. Springer-Verlag, Berlin.

Lacey, E. P. 1991. Parental effects on life history traits in plants. Pp. 735–744 in E. C. Dudley (ed.), The Unity of Evolutionary Biology. Dioscorides Press, Portland, Ore.

Lacey, E. P. 1996. Parental effects in *Plantago lanceolata* L. I: A growth chamber experiment to examine pre- and postzygotic temperature effects. Evolution 50:865–878.

Lechowicz, M. J., and G. Bell. 1991. The ecology and genetics of fitness in forest plants II. Microspatial heterogeneity of the edaphic environment. J. Ecol. 79:687–696.

Levin, D. A. 1972. Plant density, cleistogamy, and self-fertilization in natural populations of *Lithospermum caroliniense*. Am. J. Bot. 59:71–77.

Levin, D. A., and H. W. Kerster. 1974. Gene flow in seed plants. Pp. 138–220 in T. Dobzhansky, M. K. Hecht, and W. C. Ateere (eds.), Evolutionary Biology. Vol. 7. Plenum Press, New York.

Levin, S. A., D. Cohen, and A. Hastings. 1984. Dispersal strategies in patchy environments. Theor. Pop. Biol. 26:165–191.

Liew, T. C., and F. O. Wong. 1973. Density, recruitment, mortality, and growth of dipterocarp seedlings in virgin and logged forests in Sabah. Malayan For. 36:3–15.

Lloyd, D. G. 1987. Selection of offspring size at independence and other size-versus-number strategies. Am. Nat. 129:800–817.

Marshall, D. 1986. Effects of seed size on seedling success in three species of *Sesbania* (Fabaceae). Am. J. Bot. 73:457–464.

Matlack, G. R. 1987. Diaspore size, shape, and fall behavior in wind-dispersed plant species. Am. J. Bot. 74:1150–1160.

Mazer, S. J. 1989. Ecological, taxonomic, and life history correlates of seed mass among Indiana dune angiosperms. Ecol. Monogr. 59:153–175.

Mazer, S. J., and L. M. Wolfe. 1992. Planting density influences the expression of genetic variation in seed mass in wild radish (*Raphanus sativus* L.: Brassicaceae). Am. J. Bot. 79:1185–1193.

McCanny, S. J., and P. B. Cavers. 1989. Parental effects on spatial patterns of plants: A contingency table approach. Ecology 70:368–378.

McCullough, J. M., and W. Shropshire Jr. 1970. Physiological predetermination of germination responses in *Arabidopsis thaliana* (L.). Plant and Cell Physiol. 11:139–148.

Menges, E. S. 1987. Biomass allocation and geometry of the clonal forest herb *Laportea canadensis*: Adaptive responses to the environment or allometric constraints? Am. J. Bot. 74:551–563.

Miao, S. L., F. A. Bazzaz, and R. B. Primack. 1991a. Effects of maternal nutrient pulse on reproduction of two colonizing *Plantago* species. Ecology 72:586–596.

Miao, S. L., F. A. Bazzaz, and R. B. Primack. 1991b. Persistence of maternal nutrient effects in *Plantago major*: The third generation. Ecology 72:1634–1642.

Mitchell, R. 1977. Bruchid beetles and seed packaging by Palo Verde. Ecology 58:644–651.

Morse, D. H., and J. Schmitt. 1985. Propagule size, dispersal ability, and seedling performance in *Asclepias syriaca*. Oecologia 67:372–379.

Nakamura, R. R., and M. L. Stanton. 1986. Embryo growth and seed size in *Raphanus sativus*: Maternal and paternal effects in vivo and in vitro. Evolution 43:1435–1443.

Pake, C. E., and D. L. Venable. 1996. Seed banks in desert annuals: Implications for persistence and coexistence in variable environments. Ecology 77:1427–1435.

Parrish, J. A. D., and F. A. Bazzaz. 1985. Nutrient content of *Abutilon theophrastii* seeds and the competitive ability of the resulting plants. Oecologia 65:247–251.

Peroni, P. A. 1994. Seed size and dispersal potential of *Acer rubrum* (Aceraceae) samaras produced by populations in early and late successional environments. Am. J. Bot. 81:1428–1448.

Philippi, T. 1993a. Bet-hedging germination of desert annuals: Beyond the first year. Am. Nat. 142:474–487.

Philippi, T. 1993b. Bet-hedging germination of desert annuals: Variation among populations and maternal effects in *Lepidium lasiocarpum*. Am. Nat. 142:488–507.

Platenkamp, G. A. J., and R. G. Shaw. 1993. Environmental and genetic maternal effects on seed characters in *Nemophila menziesii*. Evolution 47:540–555.

Platt, W. J. 1976. The natural history of a fugitive prairie plant (*Mirabilis hirsuta* (Pursh)). Oecologia 22:399–409.

Platt, W. J., and I. M. Weiss. 1977. Resource partitioning and competition among fugitive prairie plants. Am. Nat. 111:479–513.

Platt, W. J., and I. M. Weiss. 1987. Resource partitioning and competition among fugitive prairie plants. Ecology 66:708–720.

Potvin, C., and C. Charest. 1991. Maternal effects of temperature on metabolism in the C-4 weed *Echinochloa crus galli*. Ecology 72:1973–1979.

Rabinowitz, D. 1978. Dispersal properties of Mangrove propagules. Biotropica 10:47–57.

Rees, M., and V. K. Brown. 1991. The effect of established plants on recruitment in the annual forb *Sinapsis arvensis*. Oecologia 87:58–62.

Rees, M., and M. J. Long. 1992. Germination biology and the ecology of annual plants. Am. Nat. 139:484–508.

Rice, K. J. 1985. Responses of *Erodium* to varying microsites: The role of germination cuing. Ecology 66:1651–1657.

Roach, D. A., and R. D. Wulff. 1987. Maternal effects in plants. Annu. Rev. Ecol. Syst. 18:209–235.

Roff, D. A. 1975. Population stability and the evolution of dispersal in a heterogeneous environment. Oecologia 19:217–237.

Schmid, B., and C. Dolt. 1994. Effects of maternal and paternal environment and genotype on offspring phenotype in *Solidago altissima* L. Evolution 48:1525–1549.

Schmitt, J. 1995. Genotype-enviroment interaction, parental effects, and the evolution of plant reproductive traits. Pp. 199–214 in P. C. Hoch and A. G. ??? (eds.), Experimental and Molecular Approaches to Plant Biosystematics. Monographs in Systematic Botany. No. 53. Missouri Botanical Garden, St. Louis.

Schmitt, J., and J. Antonovics. 1986. Experimental studies of the evolutionary significance of sexual reproduction IV. Effect of neighbor relatedness and aphid infestation of seedling performance. Evolution 40:830–836.

Schmitt, J., and D. W. Ehrhardt. 1990. Enhancement of inbreeding depression by dominance and supression in *Impatiens capensis*. Evolution 44:269–278.

Schmitt, J., and S. E. Gamble. 1990. The effect of distance from the parental site on offspring performance and inbreeding depression in *Impatiens capensis*: A test of the local adaptation hypothesis. Evolution 44:2022–2030.

Schmitt, J., and R. D. Wulff. 1993. Light spectral quality, phytochrome, and plant competition. Trends Ecol. Evol. Biol. 8:47–51.

Schmitt, J., D. W. Ehrhardt, and D. Swartz. 1985. Differential dispersal of self-fertilized and outcrossed progeny in jewelweed (*Impatiens capensis*). Am. Nat. 126:570–575.

Schmitt, J., J. Niles, and R. D. Wulff. 1992. Norms of reaction of seed traits to maternal environments in *Plantago lanceolata*. Am. Nat. 139:451–466.

Sheldon, J. C., and P. M. Burrows. 1973. The dispersal effectiveness of the achene-pappus units of selected Compositae in steady winds with convection. New Phytol. 72:665–675.

Shipley, B., and J. Dion. 1992. The allometry of seed production in herbaceous angiosperms. Am. Nat. 139:467–483.

Silander, J. A. 1978. Density-dependent control of reproductive success in *Cassia biflora*. Biotropica 10:292–296.

Smith, B. 1983. Demography of *Floerkia proserpinacoides*, a forest floor annual II. Density-dependent reproduction. J. Ecol. 71:405–412.

Smith, C. C., and S. D. Fretwell. 1974. The optimal balance between size and number of offspring. Am. Nat. 108:499–506.

Smith, H. 1982. Light quality, photoreception, and plant strategy. Annu. Rev. Plant Physiol. 33:481–518.

Smith, H., J. J. Casal, and G. M. Jackson. 1990. Reflection signals and the perception by phytochrome of the proximity of neighboring vegetation. Plant Cell Envir. 13:73–78.

Stanton, M. L. 1984. Seed variation in wild radish: Effect of seed size on components of seedling and adult fitness. Ecology 65:1105–1112.

Stewart, S. C., and D. J. Schoen. 1987. Pattern of phenotypic viability and fecundity selection in a natural population of *Impatiens pallida*. Evolution 41:1290–1301.

Stratton, D. A. 1989. Competition prolongs expression of maternal effects in seedlings of *Erigeron annuus* (Asteraceae). Amer. J. Bot. 76:1646–1653.

Stratton, D. A. 1994. Genotype-by-environment interactions for fitness of *Erigeron annuus* show fine-scale selective heterogeneity. Evolution 48:1607–1618.

Stratton, D. A. 1995. Spatial scale of variation in fitness of *Erigeron annuus*. Am. Nat. 146:608–624.

Sultan, S. E. 1996. Phenotypic plasticity for offspring traits in *Polygonum persicaria*. Ecology 77:1791–1807.

Theide, D. A., and C. K. Augspurger. 1996. Intraspecific variation in seed dispersion of *Lepidium campestre* (Brassicaceae). Am. J. Bot. 83:856–866.

Thompson, D. A., and A. J. Beattie. 1981. Density-mediated seed and stolon production in *Viola* (Violaceae). Am. J. Bot. 68:383–388.

Van der Toorn, J., and P. H. Van Tienderen. 1992. Ecotypic differentiation in *Plantago lanceolata*. Pp. 269–288 in P. J. C. Kuiper and M. Bos (eds.), Plantago: A Mutidisciplinary Study. Springer-Verlag, Berlin.

Van Hinsberg, A. 1996. On phenotypic plasticity in *Plantago lanceolata*: Light quality and plant morphology. Ph.D. dissertation, Netherlands Institute of Ecology.

Van Tienderen, P. H. 1991. Evolution of generalists and specialists in spatially heterogeneous environments. Evolution 45:1317–1331.

Van Valen, L. 1971. Group selection and the evolution of dispersal. Evolution 24:591–598.

Vazquez-Yanes, C., and A. Orozco-Segovia. 1994. Signals for seeds to sense and respond to gaps. Pp. 209–236 in M. M. Caldwell and R. W. Pearcy (eds.), Exploitation of Environmental Heterogeneity by Plants: Ecophysiological Processes Above- and Belowground. Academic Press, San Diego.

Venable, D. L. 1985. The evolutionary ecology of seed heteromorphism. Am. Nat. 126:577–595.

Venable, D. L., and J. S. Brown. 1988. The selective interactions of dispersal, dormancy, and seed size as adaptations for reducing risk in variable environments. Am. Nat. 131:360–384.

Venable, D. L., and L. R. Lawlor. 1980. Delayed germination and dispersal in desert annuals: Escape in space and time. Oecologia 46:272–282.

Venable, D. L., and D. A. Levin. 1985. Ecology of the achene dimorphism in *Heterotheca latifolia*: I. Achene structure, germination, and dispersal. J. Ecol. 73:133–145.

Venable, D. L., A. Burquez, G. Corral, E. Morales, and F. Espinosa. 1987. The ecology of seed heteromorphism in *Heterosperma pinnatum* in central Mexico. Ecology 68:65–76.

Wade, M. J. 1985. Hard selection, soft selection, kin selection, and group selection. Am. Nat. 125:61–73.

Waller, D. M. 1980. Environmental determinants of outcrossing in *Impatiens capensis* (Balsaminaceae). Evolution 34:747–761.

Waller, D. M. 1985. The genesis of size hierarchies in seedling populations of *Impatiens capensis* Meerb. New Phytol. 100:243–260.

Weiner, J., and S. C. Thomas. 1992. Competition and allometry in three species of annual plants. Ecology 73:648–656.

Weiss, P. W. 1980. Germination, reproduction, and interference in the amphicarpic annual *Emex spinosa*. Oecologia 45:244–251.

Went, F. W. 1949. Ecology of desert plants. II. The effect of rain and temperature on germination and growth. Ecology 30:1–13.

Westoby, M., E. Jurado, and M. Leishman. 1992. Comparative ecology of seed size. Trends Ecol. Evol. 7:368–372.

Wilson, D. E., and D. J. Janzen. 1972. Predation on Scheelea palm seeds by bruchid beetles: Seed density and distance from the parent plant. Ecology 53:954–959.

Winn, A. A. 1988. Ecological and evolutionary consequences of seed size in *Prunella vulgaris*. Ecology 69:1537–1544.

Winn, A. A., and T. E. Miller. 1995. Effect of density on magnitude of directional selection on seed mass and emergence time in *Plantago wrightiana* Dcne. (Plantaginaceae). Oecologia 103:365–370.

Wulff, R. D. 1986. Seed size variation in *Desmodium paniculatum* II. Effects on seedling growth and physiological performance. J. Ecol. 74:99–114.

Wulff, R. D. 1995. Environmental maternal effects on seed quality and germination. Pp. 491–505 in J. Kigel and G. Galili (eds.), Seed Development and Germination. Dekker, New York.

Wulff, R., and F. A. Bazzaz. 1992. Effect of parental nutrient regime on growth of the progeny in *Abutilon theophrasti* (Malvaceae). Am. J. Bot. 79:1102–1107.

Wulff, R. D., C. Caceres, and J. Schmitt. 1994. Seed and seedling responses to maternal and offspring environments in *Plantago lanceolata*. Funct. Ecol. 8:763–769.

10

Maternal Effects as Adaptations for Transgenerational Phenotypic Plasticity in Insects

CHARLES W. FOX & TIMOTHY A. MOUSSEAU

In nature, organisms live in spatially and temporally variable environments, environments that vary in abiotic conditions, resource availability, resource quality, social interactions, and numerous other characteristics across both time and space. Variation in selection resulting from this environmental variation generally results in the maintenance of genetic variation within populations, the evolution of polymorphisms, and possibly even speciation. Flexibility of ecologically important behavioral, morphological, and life history traits (phenotypic plasticity), induced by environmental cues, also frequently evolves in response to a variable environment (Via 1993). Examples of phenotypic plasticity abound in nature, and there is general agreement that many of these examples are adaptive.

In many organisms, maternal environment may provide a reliable indicator of the environmental conditions that their progeny will encounter (either biotic or abiotic). In such cases, maternal effects may evolve as mechanisms for "transgenerational phenotypic plasticity" (Mousseau and Dingle 1991b), whereby, in response to a predictive environmental cue (e.g., high or low host density, short or long photoperiod), a mother can program a developmental switch in her offspring appropriate for the environmental conditions predicted by the cue. In other words, maternal environmental cues can stimulate phenotypic plasticity in progeny. For example, female insects that encounter rapidly cooling temperatures or decreasing day lengths may produce offspring that immediately enter diapause (a quiescent state), and females that must oviposit on low-quality larval substrates may provide more resources to their progeny than if they had encountered higher-quality oviposition substrates.

This chapter provides a brief overview of the types of adaptive phenotypic plasticity that are commonly under maternal control in insects. However, while the emphasis here is on maternal effects as adaptations, it should be acknowledged that distinguishing between adaptive and nonadaptive transgenerational phenotypic plasticity is not a trivial endeavor. Organisms of identical genotype may produce different phenotypes in response to environmental variation due to constraints on development (such as food shortage), rather than as a result of adaptive evolution (Scheiner 1993; Gotthard and Nylin 1995). Substantial discussion has centered on the general problem of defining criteria necessary to demonstrate that a character is an adaptation (Williams 1966; Lewontin 1978; Coddington 1994). While a review of this intellectual exchange on adaptation is not the objective of this chapter, it should be noted that few examples of phenotypic plasticity are understood well enough to meet any strict set of criteria for demonstrating adaptation (Gotthard and Nylin 1995). Nonetheless, it is now widely accepted that many types of phenotypic plasticity can evolve, are frequently under selection, and have often evolved in response to selection favoring different phenotypes in different environments. In this review we discuss examples of transgenerational phenotypic plasticity that have been demonstrated to be adaptive as well as examples that are likely to be adaptive.

10.1 Diapause in Insects

One of the best-studied and most clearly adaptive maternal effects is the influence of maternal environment on the probability of offspring diapause. Diapause, a state of centrally controlled suppressed development, provides a mechanism by which insects can survive cyclic (generally seasonal) periods of adverse conditions, such as drought, extreme temperatures, or food shortage (Beck 1980; Danks 1987). An individual that fails to diapause at the appropriate time risks mortality from cold, desiccation, or starvation. However, in environments that allow multiple generations per year, selection favors diapause when progeny are produced late in the season but favors the inhibition of diapause when near-future environmental conditions are suitable for the development of another generation.

Thus, some species of insects are obligate diapausers (Tauber et al. 1986; Danks 1987), while many are facultative, relying on environmental cues such as photoperiod or temperature to trigger the induction or avoidance of diapause. For these facultative diapausers, either the developmental stage that diapauses or the stage immediately prior to the diapausing stage is generally the most sensitive to environmental triggers (Mousseau and Dingle 1991b). However, in many insects diapause is triggered by the environmental conditions experienced by the parental generation (reviewed in Pinger and Eldridge 1977; Tauber et al. 1986; Danks 1987; Mousseau and Dingle 1991a,b), and, in some aphids, may even be triggered by the grandparental environment (in which photoperiod stimulates the production of sexual morphs, which mate and lay diapausing eggs; Blackman 1975). However, in most cases in which the parental or grandparental environment stimulates diapause, offspring environmental conditions can modify the probability of diapause (Pinger and Eldridge 1977; Olvido et al. in press).

Maternal exposure to short or decreasing photoperiod and/or cool temperatures, indicators of approaching cold (winter) seasons, generally result in an increased incidence of diapause in progeny (reviewed in Tauber et al. 1986; Mousseau and Dingle 1991a,b). For example, in the cricket *Allonomobius fasciatus*, short day lengths and cool temperatures stimulate the production of diapausing eggs, although the egg's environment mediates the magnitude of the maternal effect (Olvido et al. in press). Similarly, maternal exposure to long or increasing photoperiods and/or hot temperatures, indicators that winter is still far away, can inhibit progeny diapause. In the flesh fly, *Sarcophaga bullata*, females that have experienced pupal diapause produce progeny that cannot enter diapause, even when reared in a strongly diapause-inducing environment (Henrich and Denlinger 1982; Denlinger, this volume). This maternal effect prevents individuals developing in early spring from responding to short-day conditions by unnecessarily entering diapause (because their parents underwent diapause to overwinter) but allows individuals developing in late fall to enter diapause in response to short day length.

Maternal age also affects the incidence of diapause in many insects (Danks 1987). In seasonal environments, the amount of time available for a second generation to develop before the advent of adverse conditions is shorter for progeny produced by older mothers. Thus, selection should favor an increasing proportion of progeny that diapause with increasing maternal age because the probability that the progeny produced by a first generation female will have time to complete development decreases as the female gets older. Mousseau (1991) tested this hypothesis in a comparative study of 10 populations of the striped ground cricket, *Allonemobius fasciatus*, that varied in voltinism and season length. Maternal age (clutch order) did not affect the incidence of progeny diapause in univoltine populations, in which season length is too short to produce a second generation. However, within bivoltine and transition zone (mixed univoltine and bivoltine) populations the incidence of diapause in offspring increased with increasing maternal age. These results support the hypothesis that maternal age effects on diapause have evolved as an adaptive mechanism for diapause regulation in populations where the growing season is long enough to produce a second generation if eggs are laid early enough in the summer, but not if eggs are laid much later.

Both maternal host-plant condition and maternal crowding can also affect whether progeny enter diapause (e.g., in aphids; reviewed in Tauber et al. 1986). More thorough reviews of maternal effects on progeny diapause can be found in Tauber et al. (1986), Danks (1987), and Mousseau and Dingle (1991a,b).

10.2 Flight Polymorphisms

Another well-studied maternal effect that is generally acknowledged to be adaptive is the production of flying and/or dispersing phenotypes. In general, for insects using seasonal/ephemeral resources, environmental cues such as crowding, temperature, or photoperiod may be predictable indicators of future deterioration of habitat quality and impending food shortage. In many insects, the environment that progeny experience during development is the primary determinant of whether they develop into flying or flightless morphs. However, in some insects maternal (and sometimes paternal) environmental conditions are major determinants of progeny flight phenotype.

For example, selection often favors the production of sedentary (flightless) progeny at low population densities but favors increased production of dispersing progeny as population density increases (reviewed in Dingle 1996). In many aphids, crowding stimulates the production of alate (winged) progeny, which disperse to a new host plant, often of a different species (Dixon 1977). These alate females then generally produce wingless progeny, which, while still at low density, continue to produce wingless progeny (Hardie and Lees 1985). The degree to which maternal crowding affects the production of dispersing progeny varies among aphid species. In *Acyrthosiphon pisum*, a species that uses ephemeral host plants, relatively low maternal densities stimulate the production of alate progeny (Lamb and MacKay 1979), while in *Aphis nerii*, an aphid infesting long-lived perennial trees, intense maternal crowding is necessary to stimulate the production of alates (Hall and Ehler 1980). Also, there is often genetic variation among aphid clones in their sensitivity to crowding and the subsequent production of alates (Lamb and MacKay 1979; Groeters and Dingle 1989). In some aphids, maternal crowding itself may not stimulate the production of alates but may mediate how sensitive progeny are to subsequent maternal crowding (Awram 1968).

Although the most complicated density-dependent flight polymorphisms appear to occur in aphids, the best-known example of a maternally induced flight polymorphism occurs in grasshoppers (Uvarov 1921). In many grasshopper species, crowding of nymphs results in agregarization, the development of a gregarious (more tolerant of conspecifics) and often swarming phenotype, whereas rearing nymphs at low density results in the development of a more solitary (less tolerant of conspecifics) nonswarming phenotype, generally with a continuous range of intermediate phenotypes as well. These two extreme morphs, gregarious and solitary, differ in their morphology, coloration, reproductive biology, development, physiology, and many aspects of their behavior, including, as noted above, their tendency to swarm (review in Pener 1991). Phase transformations, from a solitary morph to a more gregarious morph, generally take multiple generations (Albrecht et al. 1959), clearly involving maternal effects (Hunter-Jones 1958), and possibly environmentally based paternal effects (Islam et al. 1994b). For example, in the grasshopper *Schistocerca gregaria*, both maternal and paternal crowding independently have a gregarizing effect on progeny (Islam et al. 1994b). Maternal crowding of *S. gregaria* during oviposition also causes progeny to behave gregariously, while females that oviposit in isolation produce progeny that are less tolerant of crowding, independent of maternal preoviposition environment (Islam et al. 1994a). Thus, even a brief stimulus, such as density during oviposition, is enough to cue mothers to produce more or less gregarious progeny.

Like crowding, temperature and photoperiod can also be predictable indicators of future environmental quality because environments typically exhibit seasonal variation in many variables (e.g., nutritional quality of a host plant) that determine their suitability for insect development. Thus, maternal, and even grandmaternal, temperature and/or photoperiod often stimulate the production of diapausing (discussed in section 10.1) or flying progeny. For example, in the red linen bug, *Pyrrhocoris apterus*, mothers reared under long-day conditions produce progeny that are almost all long-winged, whereas mothers reared under short-day conditions produce progeny that become long-winged when developing in long-day conditions and short-winged when developing in short-day conditions (Honek 1980). In many aphids, such as *Myzus persicae* and *Megoura viciae*, parental and grandparental photoperiods affect the production of alate (and sexual) progeny (Lees 1959; Blackman 1975), with more alates produced at shorter photoperiods.

The production of dispersing progeny often increases with maternal age, as well. In the pea aphid (*Acyrthosiphon pisum*), older mothers are more likely to produce alate progeny, presumably because the quality of their host decreases throughout the season, which corresponds to increasing maternal age (because mothers are aging along with the host plants; Sutherland 1969; MacKay and Lamb 1979). For the seed beetle *Callosobruchus maculatus*, eggs laid by older females are more likely to develop into a dispersal morph than eggs laid by younger females (Sano-Fujii 1979).

In addition to environmental characters predictive of future host quality, host-plant condition itself can affect the production of alate offspring in aphids, often mediated through maternal effects (Sutherland 1969). In some aphids, particularly those that alternate between herbaceous and woody hosts, the proportion of alates produced increases as host-plant quality decreases (e.g., Forrest 1970). However, in many monophagous aphids alate production increases when the host plants are most nutritious, possibly increasing the chances that dispersing progeny will successfully colonize a new host (reviewed in Harrison 1980). Maternal host-plant condition may also mediate how sensitive parents are to the effects of crowding. For example, in *Acyrthosiphon pisum*, females do not produce many alate progeny in response to crowding when reared exclusively on seedlings of *Vicia faba*, but do produce alates in response to crowding when reared on mature *V. faba* leaves (Sutherland 1969).

10.3 Maternal Effects and Patterns of Host-Plant Suitability

For herbivorous insects, host plants represent heterogeneous resources due to temporal and spatial variability in both host suitability and availability. Thus, these herbivores experience coarse-grained spatial variation in natural selection on characters influencing adaptation to host plants. Parental host plant can affect progeny life histories in two ways. (1) Maternal oviposition decisions, and thus progeny rearing environment, may be influenced by a female's larval feeding environment (generally mediated through early adult experiences). Substantial attention has been dedicated to the ecology and evolution of learning of oviposition preferences in herbivorous insect (e.g., Papaj and Lewis 1993), so this topic is not discussed further here. (2) The type of eggs a female lays can also be affected by her larval experiences. Because less attention has been devoted to how larval host-plant experiences affect the kind of progeny produced by mothers, independent of single-generation responses to selection (i.e., independent of progeny genotype), it is the topic of this section.

Maternal Host Plant Affects Progeny Phenotype

Based on the results of Schmidt (1953) that the gypsy moth, *Lymantria dispar*, undergoes a progressive deterioration in survival and fertility as successive generations are reared on hosts other than oaks, and McMorran (1965) that the lack of essential nutrients in an insect artificial diet often becomes evident after only the second or third successive generation, Morris (1967), speculated that nutritional deficiencies in parents are transmissible to their progeny, presumably mediated through the quality and/or quantity of their egg yolks. Morris demonstrated this type of maternal effect for *Hyphantria cunea* by showing that parental foliage quality (age of apple leaves) affects the viability of a female's eggs and the ability of her progeny to become established on their food source.

Since Morris's speculation and observation, some researchers have begun to recognize that the plant species (or plant part) on which a mother develops can have dramatic consequences for the phenotype of her offspring (Rossiter 1991, 1993), especially affecting observed patterns of host suitability (Gould 1988; Fox et al. 1995b, 1997) and subsequent population dynamics (Rossiter 1994; Ginzburg, this volume), mediated via maternal or paternal effects. In *L. dispar*, the plant species that a mother develops on can affect the composition of her eggs (Rossiter 1993, 1994), which subsequently affects the growth and development of her progeny. For example, daughters of black oak–reared mothers develop substantially faster when reared on chestnut oak than do daughters of red oak–reared mothers (Rossiter 1991), and progeny of quaking aspen–reared parents survived periods of neonate starvation better than progeny of red oak-reared parents (Rossiter 1994). In the

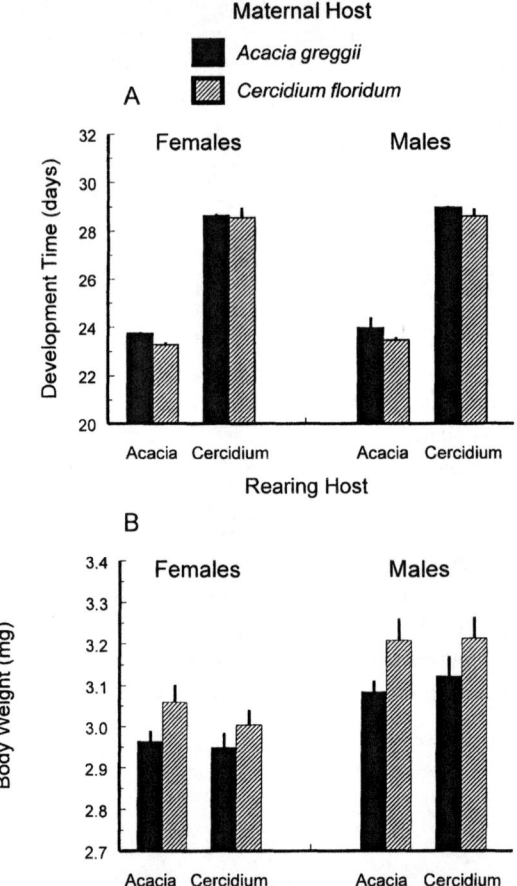

Figure 10.1 The effect of maternal host plant on progeny development time (A) and body size (B) at adult emergence for *Stator limbatus* larvae reared on *Acacia greggii* and *Cercidium floridum*. Note that parents reared on *C. floridum* (hatched bars) produce progeny that develop faster and attain larger adult size regardless of the progeny-rearing host, due to a nongenetic maternal effect (see also Fox et al. 1995b).

greenbug *Schizaphis graminum*, offspring of corn-reared parents attain larger adult size than offspring of sorghum-reared parents (McCauley et al. 1990). For the moth *Epirrata autumnata* and chrysomelid beetle *Agelastica alni*, maternal foliage quality affects subsequent progeny survivorship on poor-quality foliage (Jerker 1981, cited in Rossiter 1994; Haukioja and Neuvonen 1987).

In the seed beetle *Stator limbatus*, mothers reared on seeds of the blue palo verde, *Cercidium floridum*, produce progeny that develop faster and attain larger body size than progeny of mothers reared on seeds of catclaw acacia, *Acacia greggii*, regardless of progeny rearing host (figure 10.1; Fox et al. 1995b). This maternal effect has numerous ecological consequences for *S. limbatus*, including facilitating the expansion of the beetle onto an introduced tree (Texas ebony, *Chloroleucon ebano*) in the southwestern United States. Parents

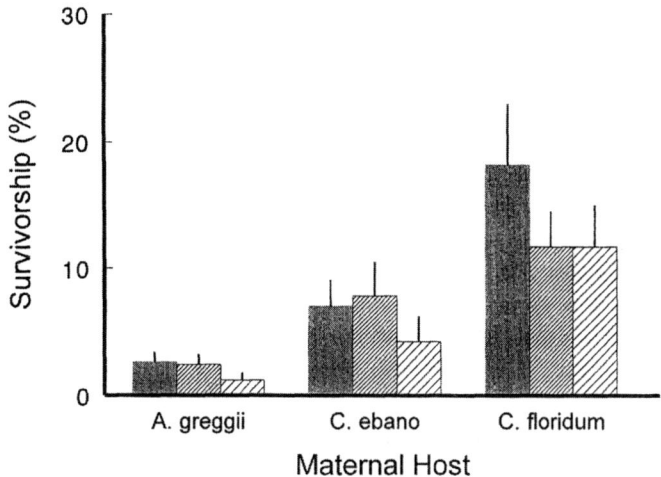

Figure 10.2 Effect of maternal rearing host (*Acacia greggii, Chloroleucon ebano,* or *Cercidium floridum*) on survivorship of *Stator limbatus* larvae reared on Texas ebony (*Chloroleucon ebano*). The differently shaded bars represent three populations of *S. limbatus*. Note that the maternal host differences represent a maternal effect and not a response to selection during the maternal generation (see also Fox et al. 1997).

reared on *C. floridum* produce progeny that can use the introduced host (survivorship ≈ 15%), while progeny of parents reared on *A. greggii* cannot (survivorship ≈ 3%), and these patterns are mediated entirely via a maternal effect (figure 10.2; Fox et al. 1997b).

While documented parental host-plant effects on progeny are generally maternally mediated, paternal host plant can also affect progeny phenotypes. For example, nongenetic paternal host effects were also suggested for *S. limbatus* development time in the experiment described above; relative to fathers reared on *A. greggii* fathers reared on *C. floridum* produced offspring that developed *slower* on *C. floridum*, and *faster* on *A. greggii* (Fox et al. 1995b), although the significance of this paternal effect is unknown. In the moth *Utetheisa ornatrix*, alkaloids sequestered from the host plant of the male as well as the female parent are transferred to eggs of their progeny, where they appear to serve a defensive function (Droussard et al. 1988). Considering that paternally derived substances are frequently incorporated into eggs during oogenesis (Boggs 1990), such paternal host-plant effects are likely to be common in herbivorous insects.

Aside from the sequestration of allelochemicals in eggs, the mechanisms by which parental rearing host affects progeny phenotype are largely unknown. Here we speculate on three potential mechanisms that need further study. First, and likely most commonly, maternal rearing host may affect the size and/or composition of the eggs that she lays. For example, rearing host has been shown to affect egg vitellogen levels in *L. dispar* (Rossiter 1993). Likewise, paternal rearing host may affect the composition and/or size of his ejaculate, which may in turn affect the size and/or composition of the eggs laid by his mate. Subsequent effects of among-egg variation on progeny growth and development would be detected as a maternal (or paternal) effect, and may provide a mechanism by which females can adaptively change the type of progeny they produce. Alternatively, parents reared on poor-quality host plants may produce small, low-quality eggs, resulting in poor performance of their progeny relative to parents reared on more suitable host plants.

Second, symbiotic microorganisms transmitted from a mother to her progeny may provide progeny the ability to detoxify allelochemicals or may provide some essential nutrient not present in a host plant. Almost all insects examined to date harbor both extracellular and intracellular microbial symbionts that provide essential nutrients and/or detoxify plant allelochemicals (Jones 1984; Dowd 1991). These symbionts are generally transmitted from parent to offspring through eggs, either by entering the oocytes directly or by being smeared on the outside of the egg as it is laid and subsequently being ingested by the hatching insect (Brooks 1963). Parents successfully surviving to adult on one host may develop a dense microbial flora that assists in development on this host. This flora may then be transmitted to progeny, providing a developmental advantage over progeny not inoculated with the appropriate symbionts or progeny inoculated with an inadequate number of symbionts.

Third, enzyme induction by environmental cues may improve the ability of larvae to detoxify xenobiotic compounds (Nebert 1979; Harshman et al. 1991). Induction of detoxification enzymes after exposure to host plants has been reported for many herbivorous insects (e.g., Yu et al. 1979; Farnsworth et al. 1981). Interactions with highly toxic hosts may result in the induction of detoxification enzymes, which may be metabolically expensive to produce when not needed (Dykhuizen 1978). However, for enzyme induction to explain the maternal effects, enzymes induced in parents must be expressed in progeny. Vertical transfer of proteins and other organic compounds is common in animals (see review in Heath and Blouw, this volume). Also, for a short time after fertilization, embryonic mRNA is primarily maternally derived, such that mRNA transcribed from maternally induced genes is translated to protein in progeny, providing a mechanism for jump-starting protein synthesis in the progeny (reviewed in Matzke and Matzke 1993; Jablonka and Lamb 1995; Heath and Blouw, this volume). This mechanism of transgenerational expression of induced genes, as well as others, may be common in herbivorous insects and is certainly worthy of further examination.

Maternal Effects as Adaptations to Predictable Host-Plant Suitability or Availability

If a female's (or male's) rearing host is indicative of local or future host-plant availability, then it is advantageous to produce offspring that are "acclimated" to the host on which a parent has been reared. However, the only available tests of the hypothesis that mothers, due to nongenetic effects, produce offspring that are acclimated to the host on which the maternal parent has been reared found no evidence of acclimation of larval performance. In aphids, maternal effects on offspring quality are particularly likely because embryos of granddaughters are often developing within embryos of daughters while a mother is feeding on a host. Thus, embryos of daughters and granddaughters are exposed to maternal host-plant compounds. Using replicate lineages of two specialist pea aphid clones (*Acyrthosiphon pisum*), Via (1991) tested the hypothesis that progeny performance on alfalfa and red clover was improved by maternal feeding on the same host. However, she was unable to demonstrate that maternal host-plant species had any effect on the relative performance of either clone on either host. Similarly, Fox et al. (1995b) found no evidence that environmentally based maternal effects caused larvae of the seed beetle *S. limbatus* to develop better when reared on the same host as their mother.

However, generalizations cannot be made from only two studies. Other systems need to be examined. Maternal conditioning of host suitability, if demonstrated for any insect, could have profound implications for our understanding of host-use evolution of herbivores, host-race formation, and sympatric speciation. For example, host experience often influences oviposition preference of females (Papaj and Lewis 1993), and if host experience also influences larval performance on these hosts, then nongenetic correlations between oviposition preference and larval performance may be maintained in a randomly mating population through the effect of maternal host experience (see Wade, this volume).

10.4 Egg-Size Plasticity

Propagule size is a particularly important life history trait involved in maternal effects expression because it is simultaneously a maternal and offspring character (Sinervo 1991; Bernardo 1996); eggs are produced by mothers but also determine initial offspring resources and size. Thus, the amount and quality of resources allocated to eggs by mothers often profoundly influence the growth and survival of their progeny (Roach and Wulff 1987; Platenkamp and Shaw 1993). In animals, progeny developing from large eggs generally grow faster, attain larger size, and have higher survivorship than progeny developing from small eggs (reviewed in Fleming and Gross 1990; Kaplan 1991; Reznick 1991; Fox 1994a,b). However, mothers laying large eggs must lay fewer eggs due to the trade-off between size and number of progeny (Smith and Fretwell 1974; Fleming and Gross 1990; Berrigan 1991; Fox et al. 1997b), resulting in an egg size that is a balance between selection for large eggs and selection for many eggs.

In natural populations, the consequences of egg-size variation depend on environmental conditions (Semlitsch and Gibbons 1990), with fitness differences between large and small eggs generally greatest in adverse environments (Janzen 1977; Braby 1994; Fox and Mousseau 1996), and selection favors eggs of different sizes in different environments (Sibly and Calow 1983; Parker and Begon 1986), resulting in substantial geographic variation in egg size in many organisms (Fleming and Gross 1990; Rowe 1994; Heath, this volume). In temporally and spatially heterogeneous environments, selection variably favors large eggs and small eggs, depending on the time and location of oviposition. Alternatively, selection may favor egg-size plasticity (Kawecki 1995; but see McGinley et al. 1987), whereby a mother that experiences a predictive environmental cue variably allocates resources to her offspring appropriate for the conditions predicted by the cue (Mousseau and Dingle 1991a).

In many animals, egg size varies within and among females, depending on the environmental conditions experienced by the female. However, in most of these examples egg size varies simply with maternal nutrient status or age (e.g., references in Fox 1993) and probably reflects a constraint (stressed females lay smaller and/or poorer quality eggs) rather than an adaptation. Yet in some animals, egg-size plasticity appears to be adaptive. For example, in some cladocerans, females respond to decreased food concentration, and possibly to increased concentration of waste products, by producing larger eggs (Smith 1963; Perrin 1989) that are higher in protein and lipid content (Guisande and Gliwicz 1992). This increased size and quality of eggs results in larger neonates (Guisande and Gliwicz 1992) that have higher survivorship under food stress (Gliwicz and Guisande 1992). This increased egg size does come at a cost, however; females laying larger eggs lay fewer eggs, due to a trade-off between egg size and number.

The well-studied examples of egg-size plasticity in marine arthropods demonstrate that maternal rearing environment can affect the type of progeny she produces, generally mediated through maternal physiological state. However, females can sometimes respond rapidly to their immediate oviposition environment by modifying the size and/or composition of the eggs they produce, independently of their physiological state. In the next section, we examine in more detail this latter type of egg-size plasticity in insects in which females respond to their oviposition environment by adjusting egg size.

Adaptive Egg-Size Plasticity in Response to Oviposition Site Quality

To our knowledge, the most thoroughly studied example of egg-size plasticity in response to oviposition site quality is our research on egg size in the seed beetle *Stator limbatus*. *S. limbatus* is a generalist seed parasite distributed from northern South America to the southwestern United States (Johnson and Kingsolver 1976; Johnson et al. 1989; Nilsson and Johnson 1993). Throughout its large geographic range, *S. limbatus* has been reared from seeds of >50 plant species in at least nine genera. In central Arizona, *S. limbatus* uses primarily three host plants: *Acacia greggii*, *Cercidium floridum*, and *C. microphyllum* (Johnson and Kingsolver 1976; Johnson et al. 1989). These hosts differ substantially in their quality as substrates for larval development: egg-to-adult survivorship of larvae developing on *C. floridum* is very low (generally <50%) while survivorship of larvae developing on *A. greggii* or *C. microphyllum* is comparatively very high (generally >90% and >85%, respectively; Siemens and Johnson 1990; Siemens et al. 1992; Fox et al. 1994, 1995b, 1996). Likewise, egg-to-adult development time on *C. floridum* is quite long relative to development time on either *A. greggii* or *C. microphyllum* ($\approx$4–6 days longer on *C. floridum* at 28–29°C). However, despite being a relatively poor host for *S. limbatus*, *C. floridum* is abundant throughout most of the lower elevation desert regions of Arizona and parts of California, whereas *A. greggii* is largely restricted to higher elevations. Also, mature fruits of *C. floridum* are available on the trees for many months in the summer and fall, well after seeds of *C. microphyllum* have dispersed (and largely been harvested by rodents). Thus, despite being a comparatively poor–quality host for *S. limbatus*, *C. floridum* is readily available at more locations and times than *A. greggii* or *C. microphyllum*.

The fitness consequences of egg size also vary substantially among host plants. On *C. floridum*, on which egg-to-adult survivorship is very poor, larvae from large eggs survive substantially better than larvae from small eggs (figure 10.3A; Fox and Mousseau 1996). However, on *A. greggii* and *C. microphyllum*, on which survivorship is comparatively very high, there is no effect of egg size on egg-to-adult survivorship, at least for the range of egg sizes normally expressed by beetles (figure 10.3B,C; Fox and Mousseau 1996; C. W. Fox, unpublished data). Thus, there is intense selection for females to lay large eggs when their larvae will develop on *C. floridum* (because small eggs almost all die), but selection to lay small eggs when larvae will develop on *A. greggii* or *C. microphyllum* (because small eggs survive as well as large eggs on these hosts but females laying small eggs can lay more eggs due to an egg size/number trade-off; Fox et al. 1997).

Population comparisons support these predictions about differing selection on *C. floridum* and *A. greggii*. Populations collected from *C. floridum* lay fewer and larger eggs than popu-

A - Cercidium floridum

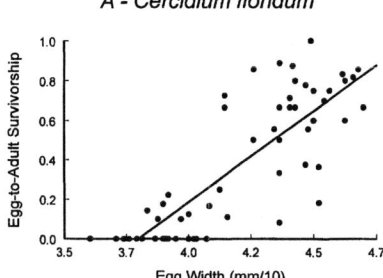

B - Cercidium microphyllum

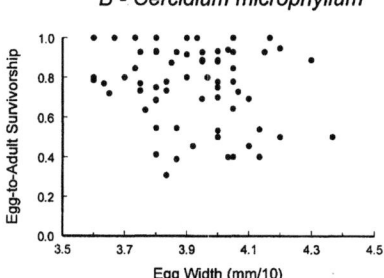

C - Acacia greggii

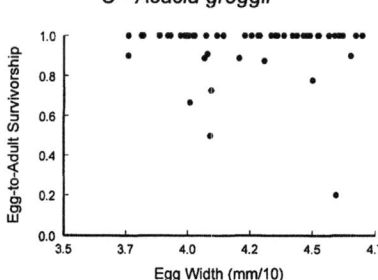

Figure 10.3 The effect of egg size on egg-to-adult survivorship of *Stator limbatus* larvae reared on *Cercidium floridum* (A), *Cercidium microphyllum* (B), and *Acacia greggii* (C). Note that egg size has a large effect on survivorship of larvae reared on *C. floridum* but not on survivorship of larvae reared on *C. microphyllum* or *A. greggii*. See also Fox and Mousseau (in press).

lations collected from *A. greggii* (Fox et al. 1995a; Fox and Mousseau 1996). Interestingly, eggs laid on *C. floridum* in nature are larger than eggs laid on *A. greggii* even when host plants are sympatric (i.e., within beetle populations; C. W. Fox, unpublished data). This latter observation represents egg-size plasticity (Fox et al. 1997b). Because female *S. limbatus* delay oviposition for at least 24 hours after emergence, during which time they finish maturing eggs (C. W. Fox, unpublished data), they are often in contact with their oviposition substrate (host plant) during egg maturation, providing the opportunity for facultative responses to host species. Female *S. limbatus* take advantage of this to variably allocate resources to offspring (e.g., figure 10.4A; Fox et al. 1997b; C. W. Fox, unpublished data). Laying large eggs on *C. floridum*, however, comes at a cost to females, who lay substantially fewer eggs on *C. floridum* than on *A. greggii* or *C. microphyllum* (figure 10.4B), due to a trade-off between egg size and egg number. A trade-off between egg size and egg number is also evident from the negative correlation between egg size and egg number, after controlling for body size, on each host (Fox et al. 1997b).

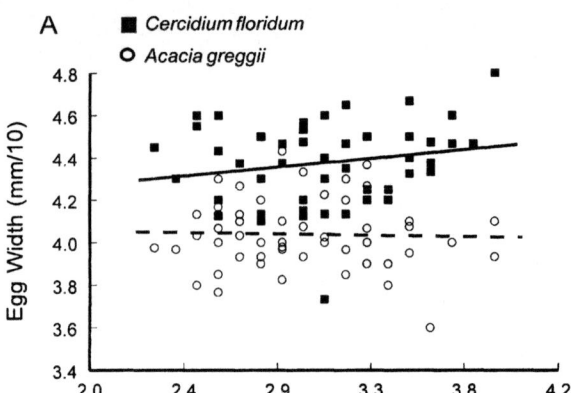

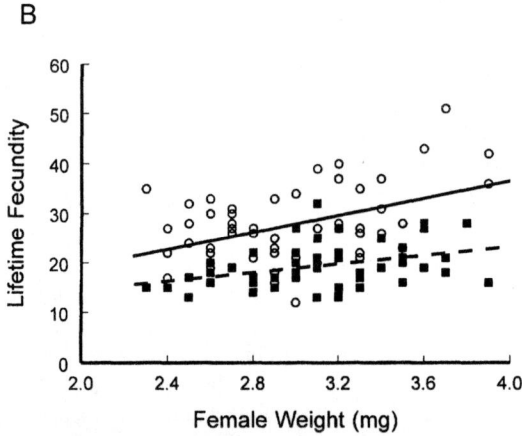

Figure 10.4 Phenotypically plastic egg size in *Stator limbatus*. Note that females oviposit-ing on *Cercidium floridum* lay both larger eggs (A) and fewer eggs (B) than females oviposit-ing on *Acacia greggii*. See also Fox et al. (1997b).

Phenotypically plastic egg size in *S. limbatus* is not the result of female reluctance to lay eggs on *C. floridum*. Because larval survivorship is very low on *C. floridum*, we might expect females to evolve avoidance of this host in preference for *A. greggii*. Such reluctance to oviposit on *C. floridum* would result in prolonged egg retention when encountering *C. floridum*, and thus possibly increased resources allocated to eggs. This is supported by the observation that females laying on *C. floridum* defer egg laying approximately 6–12 hours longer than females on *A. greggii*. However, paired preference tests indicate that, despite high larval mortality, females either prefer to oviposit on *C. floridum* (when compared with *A. greggii*) or show no preference for either host (Fox et al. 1994), indicating that female *S. limbatus* are not reluctant to lay on *C. floridum*. Also, when deprived of hosts during egg maturation, and thus forced to retain eggs longer, first-laid eggs are small (i.e., not signifi-cantly different from the sizes of eggs normally laid on *A. greggii* but significantly smaller than eggs normally laid on *C. floridum* by non-host-deprived females) regardless of the host

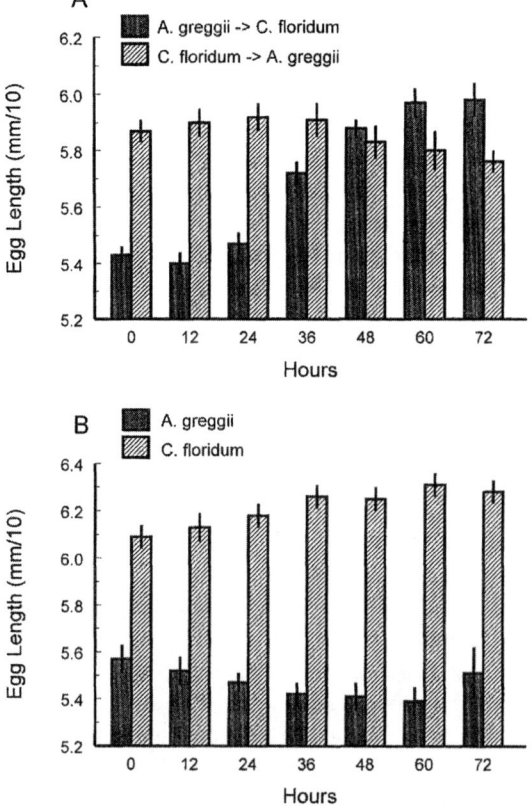

Figure 10.5 Female *Stator limbatus* that initially lay eggs on *Acacia greggii* and are then switched to *Cercidium floridum* readjust egg size and begin laying large eggs after 24–48 hours. Females initially laying on *C. floridum* also respond to a host switch but less dramatically. Control data (no host switch) are given in B. The difference in mean egg size between A and B reflects the fact that the experiments were not performed simultaneously. See also Fox et al. (in press-a).

that eggs are laid on. When females are forced to lay on other unpreferred hosts, on which egg laying is delayed, they lay small, *Acacia*-size eggs (C. W. Fox, unpublished data).

To test whether egg-size plasticity in *S. limbatus* is adaptive, we took advantage of the following convenient observation: when females are switched to a new host (from *A. greggii* to *C. floridum* or from *C. floridum* to *A. greggii*), egg size stays relatively constant for the next 24–36 hours before they begin to readjust egg size, producing progressively larger eggs on *C. floridum* and smaller eggs on *A. greggii* (Fox et al. 1997; figure 10.5). Thus, we conditioned females to lay large eggs (conditioned on *C. floridum*) or small eggs (conditioned on *A. greggii*), and then forced all females to lay for 24 hours on *C. floridum*. Eggs from the *A. greggii*–conditioned females subsequently had very low egg-to-adult survivorship on *C. floridum* (0.3 ± 0.2%), while eggs laid by females that were conditioned on *C. floridum* had substantially higher egg-to-adult survivorship on this host (23.7 ± 2.1%).

These experiments demonstrate that within- and among-female variation in egg size in *S.*

limbatus in part represents an adaptive response to a changing optimal egg size; females variably allocate resources to offspring in a manner likely to increase individual fitness by laying large eggs on *C. floridum* (on which large eggs have higher larval survivorship than small eggs), and laying small eggs on *A. greggii* (on which egg size does not effect offspring survivorship). This plastic egg size is adaptive in *S. limbatus*—females conditioned to lay large eggs produce progeny that have substantially higher fitness on *C. floridum* than females conditioned to lay small eggs, while females laying small eggs on *A. greggii* lay substantially more eggs than females laying large eggs.

Similar examples of egg-size plasticity have also been reported for two other insects. Data from Kawecki (1995) suggest that females of *Callosobruchus maculatus*, another seed beetle that develops on discrete resource patches (seeds), lay larger eggs in response to increasing adult density—an indicator that their larvae will encounter severe competition for food. Similarly, females of *Nasonia vitripennis*, a parasitoid wasp that also develops on discrete resources patches (muscoid fly larvae), facultatively increase the size of their eggs when ovipositing on previously parasitized hosts (Niehaus and Lalonde submitted).

10.5 Negative Maternal Effects

For some organisms, particularly those with bivoltine life cycles, the environmental conditions likely to be experienced by progeny are predictable, independent of information about the maternal environment. In such cases, selection may favor parents that produce progeny that are different from themselves regardless of the environmental conditions that the parents experience. In many cases, progeny that express a phenotype opposite from that of the parent may be produced (a negative maternal effect). For example, in the milkweed bug, *Oncopeltus fasciatus*, females that undergo a reproductive diapause produce progeny that do not enter reproductive diapause, or delay entering diapause (Groeters and Dingle 1987). Also in *O. fasciatus*, selection for adult diapause, and thus delayed age at first reproduction, results in earlier reproducing progeny (Dingle et al. 1977) due to a negative maternal effect (H. Dingle, personal communication). In some aphids, clones derived from eggs that have overwintered in diapause produce only virginoparae, even when exposed to short day length (Lees 1960).

One intriguing example of a negative maternal effect was reported by Janssen et al. (1988). In the collembolan *Ochesella cincta*, females produce progeny with a phenotype opposite from themselves—early-maturing mothers produce late-maturing progeny (and vice versa) and small mothers produce large progeny (and vice versa). The two successive generations produced each year (summer and winter) encounter differences in temperature, humidity, and probably food abundance and rates of predation. Thus, selection likely favors the production of different kinds of progeny in each generation, and, although this negative maternal effect has not been demonstrated to be adaptive, as we have seen throughout this chapter, maternal effects are excellent mechanisms for adapting to such predictably variable environments.

Conclusions

Maternal effects are common in nature and often have dramatic consequences for progeny growth and development. This and other articles in this volume clearly demonstrate that many maternal effects are under selection and likely adaptive. In particular, evolutionary responses to selection may be manifest through the evolution of maternal effects, particularly in heterogeneous environments—when parental environment is an accurate predictor of progeny environment, parents can adjust the kind of progeny they make, either by modifying propagule size and quality or by initiating specific developmental pathways in their progeny that are most appropriate for the environments predicted by the environmental cues. The evidence reported here, in previous reviews (e.g., Mousseau and Dingle 1991a,b), and in other chapters in this volume strongly suggest that many maternal effects have evolved as mechanisms for transgenerational phenotypic plasticity that increases offspring fitness in heterogeneous environments. Thus, understanding the evolution of maternal effects is often essential for understanding the evolution of life history and behavior patterns in nature.

Acknowledgments   We thank D. Byers, H. Dingle, S. Mazer, and U. Savalli for helpful comments on the manuscript. K. Stafford graciously provided access to the entomology library at the Connecticut Agricultural Experiment Station in New Haven. This essay is contribution 144 from the Louis Calder Center of Fordham University.

References

Albrecht, F. O., M. Verdier, and R. E. Blackith. 1959. Maternal control of ovariole number in the progeny of the migratory locust. Nature 184:103–104.

Awram, W. J. 1968. Effects of crowding on wing morphogenesis in *Myzus persicae* Sulz (Aphididae: Homoptera). Q. Entomol. 4:3–29.

Beck, S. D. 1980. Insect Photoperiodism. 2nd ed. Academic Press, New York.

Bernardo, J. 1996. The particular maternal effect of propagule size, especially egg size: Patterns, models, quality of evidence and interpretations. Am. Zool. 36:216–236.

Berrigan, D. 1991. The allometry of egg size and number in insects. Oikos 60:313–321.

Blackman, R. L. 1975. Photoperiodic determination of the male and female sexual morphs of *Myzus persicae*. J. Insect Physiol. 21:435–453.

Boggs, C. L. 1990. A general model of the role of male-donated nutrients in female insects—reproduction. Am. Nat. 136:598–617.

Braby, M. F. 1994. The significance of egg size variation in butterflies in relation to host plant quality. Oikos 71:119–129.

Brooks, M. A. 1963. Symbiosis and aposymbiosis in arthropods. Symp. Soc. Gen. Microbiol. 13:200–231.

Coddington, J. A. 1994. Homology and convergence in the study of adaptation. Pp. 53–78 in P. Eggleton, and R. Vane-Wright (eds.), Phylogenetics and Ecology. Academic Press, London.

Danks, H. V. 1987. Insect Dormancy: An Ecological Perspective. Biological Survey of Canada Monograph Series. No. 1. Ottawa, Canada.

Dingle, H. 1996. Migration: The Biology of Life on the Move. Oxford University Press, New York.

Dingle, H., C. K. Brown, and J. P. Hegmann. 1977. The nature of genetic variance influencing photoperiodic diapause in a migrant insect, *Oncopeltus fasciatus*. Am. Nat. 111:1047–1059.

Dixon, A. F. G. 1977. Aphid ecology: Life cycles, polymorphism, and population regulation. Annu. Rev. Ecol. Syst. 8:329–353.

Dowd, P. F. 1991. Symbiont-mediated detoxification in insect herbivores. Pp. 411–440 in P. Barbosa, V. A. Krischik, and C. G. Jones (eds.), Microbial Mediation of Plant-Herbivore Interactions. Wiley, New York.

Droussard, D. E., K. Ubik, C. Harvis, J. Resch, J. Meinwald, and T. Eisner. 1988. Biparental defensive endowment of eggs with acquired plant alkaloid in the moth *Utetheisa ornatrix*. Proc. Natl. Acad. Sci. USA 85:5992–5996.

Dykhuizen, D. (1978) Selection for tryptophan auxotrophs of *Escherichia coli* in glucose-limited chemostats as a test of the energy conservation hypothesis of evolution. Evolution 32:125–150.

Farnsworth, D. E., R. E. Berry, S. J. Yu, and L. C. Terriere. 1981. Aldrin epoxidase activity and cytochrome P-450 content of some microsomes prepared from alfalfa and cabbage looper larvae fed various plant diets. Pest. Biochem. Physiol. 15:158–165.

Fleming, I. A., and M. T. Gross. 1990. Latitudinal clines: A trade-off between egg number and size in pacific salmon. Ecology 71:1–11.

Forrest, J. M. S. 1970. The effect of maternal and larval experience on morph determination in *Dysaphis devecta*. J. Insect Physiol. 16:2281–2292.

Fox, C. W. 1993. The influence of maternal age and mating frequency on egg size and offspring performance in *Callosobruchus maculatus* (Coleoptera: Bruchidae). Oecologia 96:139–146.

Fox, C. W. 1994a. The influence of egg size on offspring performance in the seed beetle, *Callosobruchus maculatus*. Oikos 71:321–325.

Fox, C. W. 1994b. Maternal and genetic effects on egg size and larval performance in the seed beetle, *Callosobruchus maculatus*. Heredity 73:509–517.

Fox, C. W., and T. A. Mousseau. 1996. Larval host plant affects the fitness consequences of egg size in the seed beetle *Stator limbatus*. Oecologia 107:541–548.

Fox, C. W, K. J. Waddell, and T. A. Mousseau. 1994. Host-associated fitness variation in a seed beetle (Coleoptera: Bruchidae): Evidence for local adaptation to a poor quality host. Oecologia 99:329–336.

Fox, C. W., L. A. McLennan, and T. A. Mousseau. 1995a. Male body size affects female lifetime reproductive success in a seed beetle. Anim. Behav. 50:281–284.

Fox, C. W., K. J. Waddell, and T. A. Mousseau. 1995b. Parental host plant affects offspring life histories in a seed beetle. Ecology 76:402–411.

Fox, C. W., J. A. Nilsson, and T. A. Mousseau. 1997a. The ecology of diet expansion in a seed-feeding beetle—pre-existing variation, rapid adaptation and maternal effects? Evol. Ecol. 11:183–195.

Fox, C. W., M. S. Thakar, and T. A. Mousseau. 1997b. Egg size plasticity in a seed beetle: An adaptive maternal effect. Am. Nat. 149:149–163.

Gliwicz, Z. M., and C. Guisande. 1992. Family planning in *Daphnia*: Resistance to starvation in offspring born to mothers at different food levels. Oecologia 91:463–467.

Gotthard, K., and S. Nylin. 1995. Adaptive plasticity and plasticity as an adaptation: A selective review of plasticity in animal morphology and life history. Oikos 74:3–17.

Gould, F. 1988. Stress specificity of maternal effects in *Heliothis virescens* (Boddie) (Lepidoptera: Noctuidae) larvae. Mem. Entomol. Soc. Can. 146:191–197.

Groeters, F. R., and H. Dingle. 1989. Genetic and maternal influences on life history plasticity in response to photoperiod by milkweed bugs (*Oncopeltus fasciatus*). Am. Nat. 129:332–346.

Guisande, C., and Z. M. Gliwicz. 1992. Egg size and clutch size in two *Daphnia* species grown at different food levels. J. Plankt. Res. 14:997–1007.

Hall, R. W., and L. E. Ehler. 1980. Population ecology of *Aphis nerii* on oleander. Envir. Entomol. 9:338–344.

Hardie, J., and A. D. Lees. 1985. The induction of normal and teratoid viviparae by a juvenile hormone and kinoprene in two species of aphids. Physiol. Entomol. 10:65–74.

Harrison, R. G. 1980. Dispersal polymorphisms in insects. Annu. Rev. Ecol. Syst. 11:95–118.

Harshman, L. G., J. A. Ottea, and B. D. Hammock. 1991. Evolved environment-dependent expression of detoxification enzyme activity in *Drosophila melanogaster*. Evolution 45:791–795.

Haukioja, E., and S. Neuvonen. 1987. Insect population dynamics and induction of plant resistance: The test of hypotheses. Pp. 411–432 in P. Barbosa and J. C. Schulz (eds.) Insect Outbreaks. Academic Press, New York.

Henrich, V. C., and D. L. Denlinger. 1982. A maternal effect that eliminates pupal diapause in progeny of the flesh fly, *Sarcophaga bullata*. J. Insect Physiol. 28:881–884.

Honek, A. 1980. Maternal regulation of wing polymorphism in *Pyrrchoris apterus*: Effect of cold activation. Experientia 36:418–419.

Hunter-Jones, P. 1958. Laboratory studies on the inheritance of phase characters in locusts. Ant-Locust Bull. 29:1–32.

Islam, M. S., P. Roessingh, S. J. Simpson, and A. R. McCaffery. 1994a. Effects of population density experienced by parents during mating and oviposition on the phase of hatchling desert locusts, *Schistocerca gregaria*. Proc. R. Soc. Lond. (ser. B) 257:93–98.

Islam, M. S., P. Roessingh, S. J. Simpson, and A. R. McCaffery. 1994b. Parental effects on the behaviour and coloration of nymphs of the desert locust *Schistocerca gregaria*. J. Insect Physiol. 40:173–181.

Jablonka, E., and M. J. Lamb. 1995. Epigenetic Inheritance and Evolution: The Lamarkian Dimension. Oxford University Press, Oxford.

Janssen, G. M., G. De Jong, E. N. G. Josse, and W. Scharloo. 1988. A negative maternal effect in springtails. Evolution 42: 828–834.

Janzen, D. H. 1977. Variation in seed size within a crop of a Costa Rican *Micuna andreana* (Leguminosae). Am. J. Bot. 64:347–349.

Johnson, C. D., and J. M. Kingsolver. 1976. Systematics of *Stator* of North and Central America (Coleoptera: Bruchidae). USDA Tech. Bull. 1537:1–101.

Johnson, C. D., J. M. Kingsolver, and A. L. Teran. 1989. Sistematica del genero *Stator* (Insecta: Coleoptera: Bruchidae) en Sudamerica. Opera Lilloana 37:1–105 (in Spanish).

Jones, C. G. 1984. Microorganisms as mediators of plant resource exploitation by insect herbivores. Pp. 53–99 in P. W. Price, C. N. Slobodchikoff, and W. S. Gaud (eds.), A New Ecology: Novel Approaches to Interactive Systems. Wiley, New York.

Kaplan, R. H. 1991. Developmental plasticity and maternal effects in amphibian life histories. Pp. 794–799 in E. C. Dudley (ed.), The Unity of Evolutionary Biology. Vol. 2. Dioscorides Press, Portland, Ore.

Kawecki, T. J. 1995. Adaptive plasticity of egg size in response to competition in the cowpea weevil, *Callosobruchus maculatus* (Coleoptera: Bruchidae). Oecologia 102:81–85.

Lamb, R. J., and P. A. MacKay. 1979. Variability in migratory tendency within and among natural populations of the pea aphid, *Acyrthosiphon pisum*. Oecologia 39:289–299.

Lees, A. D. 1959. The role of photoperiod and temperature in the determination of parthenogenetic and sexual forms in the aphid *Megoura viciae* Bucton—I. The influence of these factors on apterous virginoparae and their progeny. J. Insect Physiol. 3:92–117.

Lees, A. D. 1960. The role of photoperiod and temperature in the determination of parthenogenetic and sexual forms in the aphid *Megoura viciae* Bucton—II. The operation of the "interval timer" in young clones. J. Insect Physiol. 4:154–175.

Lewontin, R. C. 1978. Adaptation. Sci. Am. 239:212–230.

MacKay, P. A., and R. J. Lamb. 1979. Migratory tendency in aging populations of the pea aphid *Acyrthosiphon pisum*. Oecologia 39:301–308.

Matzke, M., and A. J. M. Matzke. 1993. Genomic imprinting in plants: Parental effects and trans-inactivation phenomena. Annu. Rev. Plant Physiol. Plant Mol. Biol. 44:53–76.

McCauley, G. W., Jr., D. C. Margolies, and J. C. Resse. 1990. Feeding behavior, fecundity and weight of sorghum- and corn-reared greenbugs on corn. Entomol. Exp. Appl. 55:183–190.

McGinley, M. A., D. H. Temme, and M. A. Geber. 1987. Parental investment in offspring in variable environments: Theoretical and empirical considerations. Am. Nat. 130:370–398.

McMorran, A. 1965. A synthetic diet for the spruce budworm, *Choristoneura fumiferana* (Clem.) (Lepidoptera: Tortricidae). Can. Entomol. 97:58–62.

Morris, R. F. 1967. Influence of parental food quality on the survival of *Hyphantria cunea*. Can. Entomol. 99:24–33.

Mousseau, T. A. 1991. Geographic variation in maternal-age effects on diapause in a cricket. Evolution 45:1053–1059.

Mousseau, T. A., and H. Dingle. 1991a. Maternal effects in insect life histories. Annu. Rev. Entomol. 36:511–534.

Mousseau, T. A., and H. Dingle. 1991b. Maternal effects in insects: examples, constraints, and geographic variation. Pp. 745–761 in E. C. Dudley (ed.), The Unity of Evolutionary Biology. Vol. 2. Dioscorides Press, Portland, Ore.

Nebert, D. W. 1979. Multiple forms of inducible drug-metabolizing enzymes: A reasonable mechanism by which any organism can cope with adversity. Mol. Cell. Biol. 27:27–45.

Niehaus, J. A., and R. G. Lalonde. Submitted. Unusual plasticity in a maternal trait: Parasitoids increase egg size if offspring must compete.

Nilsson, J. A., and C. D. Johnson. 1993. Laboratory hybridization of *Stator beali* and *S. limbatus*, with new host records for *S. limbatus* and *Mimosestes amicus* (Coleoptera: Bruchidae). Southwest. Natur. 38:385–387.

Olvido, A. E., S. Busby, and T. A. Mousseau. In press. Relative effect of maternal programming behaviour on offspring developmental plasticity in a cricket. Anim. Behav.

Papaj, D. R., and A. C. Lewis (eds.). 1993. Insect Learning: Ecological and Evolutionary Perspectives. Chapman and Hall, New York.

Parker, G. A., and M. Begon. 1986. Optimal egg size and clutch size: Effects of environment and maternal phenotype. Am. Nat. 128:573–592.

Pener, M. P. 1991. Locust phase polymorphism and its endocrine relations. Adv. Insect Physiol. 23:1–79.

Perrin, N. 1989. Population density and offspring size in the cladoceran *Simocephalus vetulus* (Muller.). Funct. Ecol. 3:29–36.

Pinger, R. R., and B. F. Eldridge. 1977. The effect of photoperiod on diapause induction in *Aedes canadensis* and *Psorophora ferox* (Diptera: Culicidae). Ann. Entomol. Soc. Am. 70:437–441.

Platenkamp, G. A. J., and R. G. Shaw. 1993. Environmental and genetic maternal effects on seed characters in *Nemophila menziesii*. Evolution 47:540–555.

Reznick, D. N. 1991. Maternal effects in fish life histories. Pp. 78–93 in E. C. Dudley (ed.), The Unity of Evolutionary Biology. Vol. 2. Dioscorides Press, Portland, Ore.

Roach, D. A., and R. Wulff. 1987. Maternal effects in plants: evidence and ecological and evolutionary significance. Annu. Rev. Ecol. Syst. 18:209–235.

Rossiter, M. C. 1991. Environmentally-based maternal effects: A hidden force in insect population dynamics. Oecologia 87: 288–294.

Rossiter, M. C. 1993. Initiation of maternal effects in *Lymantria dispar*: Genetic and ecological components of egg provisioning. J. Evol. Biol. 6:577–589.

Rossiter, M. C. 1994. Maternal effects hypothesis of herbivore outbreak. Bioscience 44:752–763.

Rowe, J. W. 1994. Reproductive variation and the egg size-clutch size trade-off within and among populations of painted turtles (*Chrysemys picta bellii*). Oecologia 99:35–44.

Sano-Fujii, I. 1979. The effect of parental age and development rate on the production of active form of *Callosobruchus maculatus* (F.) (Coleoptera: Bruchidae). J. Stored Prod. Res. 2:187–195.

Scheiner, S. M. 1993. Genetics and the evolution of phenotypic plasticity. Annu. Rev. Ecol. Syst. 24:35–68.

Schmidt, L. 1953. The influence of food on the development of gypsy moth (*Lymantria dispar* L.). Yugos. Acad. Sci. Zagreb. 105–166 (as cited in Morris 1967).

Semlitsch, R. D., and J. W. Gibbons. 1990. Effects of egg size on success of larval salamanders in complex aquatic environments. Ecology 71:1789–1795.

Sibly, R. M., and P. Calow. 1983. An integrated approach to life-cycle evolution using selective landscapes. J. Theor. Biol. 102:527–547.

Siemens, D. H., and C. D. Johnson. 1990. Host-associated differences in fitness within and between populations of a seed beetle (Bruchidae): Effects of plant variability. Oecologia 82:408–423.

Siemens, D. H., C. D. Johnson, and K. V. Ribardo. 1992. Alternative seed defense mechanisms in congeneric plants. Ecology 73:2152–2166.

Sinervo, B. 1991. Experimental and comparative analyses of egg size in lizards: Constraints on the adaptive evolution of maternal investment per offspring. Pp. 725–734 in E. C. Dudley (ed.), The Unity of Evolutionary Biology. Vol. 2. Dioscorides Press, Portland, Ore.

Smith, C. C., and S. D. Fretwell. 1974. The optimal balance between size and number of offspring. Am. Nat. 108:499–506.

Smith, F. E. 1963. Population dynamics in *Daphnia magna* and a new model for population growth. Ecology 44:651–663.

Sutherland, O. R. W. 1969. The role of the host plant in the production of winged forms by two strains of the pea aphid, *Acyrthosiphon pisum*. J. Insect Physiol. 15:2179–2201.

Tauber, M. J., C. A. Tauber, and S. Masaki. 1986. Seasonal Adaptations of Insects. Oxford University Press, New York.

Uvarov, B. P. 1921. A revision of the genus *Locusta*, L. (=*Pachytylus*, Fieb.) with a new theory as to the periodicity and migrations of locusts. Bull. Entomol. Res. 12:135–163.

Via, S. 1991. Specialized host plant performance of pea aphid clones is not altered by experience. Ecology 72:1420–1427.

Via, S. 1993. Adaptive phenotypic plasticity: Target or by-product of selection in a variable environment? Am. Nat. 142:352–365.

Williams, G. C. 1966. Adaptation and Natural Selection. Princeton University Press, Princeton, N.J.

Yu, S. J., Berry, R. E., and Terriere, L. C. 1979. Host plant stimulation of detoxification enzymes in a phytophagous insect. Pest. Biochem. Physiol. 12:280–284.

11

Are Maternal Effects in Fish Adaptive
or Merely Physiological Side Effects?

DANIEL D. HEATH AND D. MAX BLOUW

In general, maternal effects in fishes have received little attention except as confounding factors in quantitative genetic experiments. However, due to commercial interest in certain fish species, considerable research provides evidence for large and ubiquitous maternal effects in fishes. Here, we review that evidence with a special emphasis on the potential, or demonstrated, adaptive nature of those effects. There is virtually no direct evidence for adaptive maternal effects in fishes; however, strong inferential evidence exists. The bulk of the published research deals with maternal effects in oviparous species, although important work has also been done on viviparous species. Maternal effects on offspring size are best documented; however, evidence for maternal effects on disease resistance, survival and mortality, embryonic levels of hormones, mRNA, environmental contaminants, and behavior is presented as well. We conclude that the study of adaptive maternal effects in fishes is still rudimentary and predict that further research in this area will provide useful and important insights into the evolution of fishes.

11.1 Introduction

Our general objective in this chapter is to review literature dealing with maternal effects in fishes. Within this context we specifically try to identify adaptive components of maternal effects. For the purpose of this review we define maternal effects as nongenetic influences derived from parental phenotypes or environments that have an impact on offspring phenotypes. We focus principally on morphological and physiological maternal effects, but some of the fascinating research opportunities associated with behavioral phenotypes we address in the closing section.

As noted by other authors in this volume, when maternal effects have been identified, they have generally been either ignored or trivialized. Perhaps this is not surprising for fishes. Much of the genetic research on fish life history and physiological traits has been motivated by commercial brood stock improvement goals. From this perspective significant maternal effects have been interpreted, understandably, as " troublesome source(s) of environmental resemblance" (Falconer 1981, p. 145). However, within the last decade more interest in the theoretical aspects of maternal effect evolution has emerged (Reznick 1991; Mousseau and Dingle 1991a,b; Fox 1993). This developing theory supports the interpretation of maternal effects as potentially powerful influences in the evolution of life history strategies (Kirkpatrick and Lande 1989; Reznick 1991; Cheverud and Moore 1994).

Although many researchers now acknowledge the *potential* adaptive role of maternal effects, unambiguous identification and quantification of their adaptive value are very difficult. This difficulty is no different for fishes than for other organisms. Nevertheless, we show that a combination of (1) evidence that specific traits are influenced by maternal effects and (2) research that shows the impact of those traits on individual fitness, can lead to strong inference of an adaptive role for maternal effects in fishes.

Fishes are arguably the most diverse group of vertebrates, certainly in morphology and physiology, but also in life history. The dramatic differences that occur in life histories among fishes can profoundly affect the nature and evolutionary influence of maternal effects. A primary life history distinction among fishes is oviparity (egg laying) versus viviparity (live birth). We discuss maternal effects in viviparous fishes, but our primary focus is on oviparous species. Within oviparous species we discuss the effects of further life history distinctions, especially semelparity (single reproduction) versus iteroparity (multiple reproduction). The characterization of a species as semelparous is usually straightforward (death following reproduction; see Roff 1992). Iteroparity, however, may apply to species that breed more than once, or to those that oviposit more than once. For the purposes of our discussion we define iteroparity as multiple breeding events, not multiple ovoposing from one mating event. Iteroparous species have the option to alter their reproductive strategy over time, which admits the possibility that these adjustments may be adaptive to immediate environmental context. Semelparous fish, on the other hand, have only a single opportunity for successful reproduction, with no opportunity to "optimize" their reproductive allocation over a number of reproductive episodes. Clearly, this fundamental life history difference between semelparous and iteroparous fishes has the potential to influence strongly the evolutionary consequences of maternal effects.

Finally, we note differences between commercially valuable and noncommercial species. This difference influences the quantity and nature of quantitative genetic information that has been reported in the literature. For example, a great volume of published research exists for salmonids relative to all other fish families. Accordingly, the bias of information toward salmonid fish species presented in this review reflects a true research bias and not a personal bias of the authors.

11.2 Viviparous (Live-Bearing) Fishes

Ovoviviparity is considered primitive to viviparity (Thibault and Schultz 1978; Wourms 1981). This and the observation that viviparity has evolved independently in widely divergent orders and families of fish (Wourms 1981) suggest that it has adaptive value (see Roff 1992), at least in certain environments. Also, greater opportunity for maternal effects occurs among live-bearing fish than among egg-bearing species. This is because of the longer developmental period during which viviparous offspring are physiologically dependent on their mothers (Lombardi 1996). Analogies to mammalian maternal influences are obvious, but they cannot be taken very far because profound differences exist among viviparous fishes in the nature of the mother-offspring interaction (Wourms 1981). For example, some viviparous fish gestate virtually complete eggs, providing little more than physical protection ("*ovo*viviparity"). Other viviparous species are dependent on the mother for nourishment throughout development (Wourms 1981; Lombardi 1996). Furthermore, classification of fishes as ovoviviparous or viviparous may be an oversimplification because some species exhibit reproductive strategies between the two extremes. Nevertheless, the distinction is important; the pattern of maternal investment in offspring has important consequences for the interpretation of maternal effects.

With a few exceptions (i.e., guppies and mosquito fish), little research has been done on maternal effects in viviparous fish despite the expectation that maternal impact on developing offspring may be large (Lombardi 1996). There is virtually no research reported on the possible adaptive role of maternal effects in viviparous fish. However, by analogy to the extensive literature on maternal effects in other viviparous vertebrates, we expect that adaptive maternal effects in viviparous fish will become apparent on investigation (Lombardi 1996). Below we discuss two general forms of maternal effects in viviparous fish species (maternal-offspring size correlations and the occurrence of multibrooding, or superfetation) and the "grandfather effect" experiments of Reznick (1981, 1982).

By far the most commonly reported maternal effect, in general, is positive correlation between maternal size and offspring size. Many researchers working with viviparous fish species acknowledge the adaptive importance of offspring size as well as the ubiquitous nature of maternal-offspring size correlations. Surprisingly, they generally report little or no correlation between female and offspring size in viviparous fish (e.g., *Heterandria formosa*: Travis et al. 1987; *Gambusia affinis*: Reznick 1981; Meffe 1987; *G. holbrooki*: Meffe 1990; *Poecilia latipinna* and *P. reticulata*: Reznick 1991). These results contradict much of the reported research in oviparous fish, where maternal-egg size correlations are the rule (Reznick 1991; Roff 1992). There are, of course, some examples of viviparous fish species exhibiting pronounced mother-offspring size correlations (see Reznick 1991). Nevertheless, the mechanism of the interaction between mothers and offspring that influences offspring size appears to differ between viviparous and oviparous fishes, at least as indicated by the strength of mother-offspring correlations.

Superfetation, or the presence of multiple broods in the reproductive system of a single female, is a characteristic of viviparous fish that is not found in oviparous species (Thibault and Schultz 1978; Wourms 1981; Travis et al. 1987). Many viviparous fishes are facultatively capable of superfetation, depending on food supply (Thibault and Schultz 1978). The simultaneous production of multiple clutches may influence the level of maternal investment per offspring. Unfortunately, the impact of superfetation on individual offspring phenotypes or fitness has not been extensively studied. Thibault and Schultz (1978) showed that the level of superfetation among five species of poeciliids corresponds to that expected based on habitat stability and food availability. Travis et al. (1987) showed that superfetation in *Heterandria formosa* increases with female age and that neither superfetation levels nor female size has significant effect on offspring size (see also Reznick et al. 1996). The occurrence of superfetation has been variously interpreted as a method for (1) allocating the energetic requirements of the female more uniformly over time (Thibault and Schultz 1978; Meffe 1990) and

(2) increasing total female reproductive output, given the physical limitations of the brood-rearing capability of viviparous fish species (Travis et al. 1987). Both of these strategies are clearly adaptive given specific environmental conditions (e.g., low but constant food supply, or a nonlimiting food supply with female size limitations).

The single best-documented example of maternal genetic effects in fish involves the "grandfather effect" in the mosquito fish *Gambusia affinis* and in guppies, *Poecilia reticulata* (Reznick 1981, 1982; see Reznick 1991). In controlled breeding experiments Reznick (1981, 1982) found offspring size to have a significant dam component of additive genetic variation but no sire component. Interestingly, the male effect was found to be significant in the subsequent (F_2) generation (i.e., a "grandfather" effect). Such a pattern of inheritance can be interpreted as a cross-generational maternal genetic effect. That is, the genotype of the mother influences offspring size prior to expression of the offspring's genotype, but the paternal contribution to the offspring's genome has little impact until the offspring itself reproduces. Thus, in the F_2 generation, maternal (but not paternal) additive genetic effects are observed, and a significant additive genetic contribution is also observed from the male maternal grandparent (maternal grandfather). Whether such a maternal effect is adaptive is not clear. However, the general body of evidence indicating that offspring size influences fitness (table 11.1; see Roff 1992) implies potential adaptive implications of the grandfather effect.

11.3 Oviparous (Egg-Bearing) Fishes

By far the majority of fish species, particularly teleosts, are egg bearing. Unfortunately, despite a large body of evidence for significant maternal effects, there is virtually no published research on the adaptive value of those maternal effects. Below we discuss the *potential* adaptive value of the various documented maternal effects from three perspectives: (1) quantitative genetic analysis of maternal effects, (2) maternal-offspring size correlations, and (3) direct maternal effects. The section on quantitative genetic analysis deals with maternal effects identified in breeding experiments designed to estimate narrow-sense heritabilities. Maternal-offspring size correlations refer almost exclusively to female-egg size correlations, along with evidence concerning the adaptive value of such effects. Direct maternal effects involve specific contributions by the mother to the cytoplasmic composition of eggs and embryos (e.g., hormones, mRNA, and environmental contaminants).

Quantitative Genetic Analyses

Much of the literature dealing with fish-breeding experiments focuses on cultured species, probably due to the availability of rearing facilities and strong economic motivation (see Gjedrem 1983; Kinghorn 1983). Breeding experiments are basically of two designs: hierarchical nested and diallel crossed. Nested designs do not allow unambiguous identification of maternal effects (indeed, such is not the intent of their application). Diallel cross designs enable maternal effects to be estimated specifically.

Hierarchical nested breeding designs usually consist of replicates of a single male mated to two or more females (though multiple males mated to a single female is also a nested design, it is rarely implemented; Heath et al. 1993). Most studies involving hierarchical

Table 11.1. Reported correlations between fish egg size or mother size and survival.

Species	Common Name	Survival	Starvation Resistance	Reference
Clupea harengus	Atlantic herring		+	Blaxter & Hempel 1963
Cyprinus carpio	Carp	NS; +; −		Zonova 1973; Tomita et al. 1980; Kirpichnikov 1966
Engraulis anchoita	Argentine anchovy		NS	Ciechomski 1966
Etheostoma spectabile	Orangethroat darter		+	Marsh 1986
Gadus morhua	Atlantic cod		+	Knutsen & Tilseth 1985
Ictalurus punctatus	Channel catfish	NS		Broussard & Stickney 1981
Leuciscus leuciscus	Dace		+	Mann & Mills 1985
Mallotus villosus	Capelin		NS	Chambers et al. 1989
Morone saxatilis	Striped bass	NS		Monteleone & Houde 1990
Oncorhynchus keta	Chum salmon	−		Beacham & Murray 1985
O. kisutch	Coho salmon	−		Allen 1957
O. mykiss	Rainbow trout	+(3); NS(2)		Gall 1974; Springate & Bromage 1985; Pitman 1979
O. tshawytscha	Chinook salmon	+; −(2); NS		Fowler 1972; Heath (unpub. Data)
Oreochromis mossambicus	Tilapia		+	Rana 1985
Poecilia latipinna	Sailfin molly	+		Travis et al. 1992
Salmo salar	Atlantic salmon	−; NS		Glebe et al. 1979
S. trutta	Brown trout		+	Bagenal 1969
Salvelinus alpinus	Arctic charr	+		Wallace & Aasjord 1984
S. fontinalis	Brook trout	+; NS		Robison & Luempert 1984; Hutchings 1991

Survival refers to survival under cultured conditions, with food supplied. Starvation resistance refers to time to starvation, with no food supplied. +, significant positive relation; −, negative relation; NS, significant correlation at the $P < 0.05$ level. A number in parentheses indicates more than one study with the same result for the same species.

nested designs show dam-component heritabilities to be higher than sire-component heritabilities, at least in early life. Unfortunately, maternal, nonadditive genetic, and "tank" (rearing environment) effects confound almost all cases. These studies cover a wide range of traits, for example, disease resistance (Gjedrem and Gjøen 1995), length and weight at age (Refstie and Steine 1978; Refstie 1980; Robison and Leumpert 1984; Bailey and Loudenslager 1986; Withler et al. 1987; Heath et al. 1993), survival and mortality (Kanis et al. 1976; Robison and Leumpert 1984; Withler et al. 1987), and incidence of precocious maturation (Silverstein and Hershberger 1992). We mention these studies because they specifically discuss maternal effects. However, there are probably hundreds of other published hierarchical fish breeding experiments, particularly involving salmonid species, that reflect the prevalence and duration of maternal effects in fishes.

Diallel, or factorial, breeding designs involve mating each male to all of the females in the experiment, and vice versa. Clearly, factorial designs can only be applied to certain species; however, many fish species do lend themselves to factorial breeding design. Despite logistical drawbacks, factorial breeding designs are extremely powerful in estimating genetic and environmental determinants of phenotypic variation. Accordingly, factorial breeding designs have been used to investigate maternal effects in a variety of salmonid species: rainbow trout,

Oncorhynchus mykiss (McKay et al. 1986a,b; Gjerde 1988; DeMarch 1991a); pink salmon, *O. gorbuscha* (Beacham 1988, 1989); chum salmon, *O. keta* (Beacham 1988), Arctic charr, *Salvelinus alpinus* (DeMarch 1991a,b; Nilsson 1992); brook trout, *S. fontinalis* (Robison and Luempert 1984); Atlantic salmon, *Salmo salar* (Gjerde and Refstie 1984); and interspecific hybridizations (Ayles 1974; Blanc and Poisson 1983). Size-related traits are most commonly studied, and it has been repeatedly shown that strong maternal effects exist for length and weight during the egg, larval (alevin), and early juvenile stages (Gjerde and Refstie 1984; Robison and Luempert 1984; McKay et al. 1986b; Beacham 1988, 1989; Gjerde 1988; DeMarch 1991a,b; Wangila and Dick 1996). However, maternal effects for these traits are small, or nonexistent, in older offspring (Gjerde and Refstie 1984; Robison and Luempert 1984; McKay et al. 1986a; Beacham 1989; Nilsson 1992; Wangila and Dick 1996). Two exceptions to this pattern have been reported: Gjerde (1988) and Beacham (1989) found significant maternal effects on body size past 1 year of age in juvenile salmonids. However, Gjerde speculated that his results may have been due to some form of sampling error.

Survival (or mortality) has also been widely investigated using diallel crosses. The simplest measurement of survival for early ontogenetic stages is hatching success. All of the reviewed reports agree that hatching success is strongly influenced by maternal effects, with little or no evidence of an additive genetic component (Ayles 1974; Blanc and Poisson 1983; Robison and Luempert 1984; Beacham 1988; DeMarch 1991a,b). Survival rates well into the juvenile stage were also found to have significant maternal component (Blanc and Poisson 1983; Gjerde and Refstie 1984; Robison and Luempert 1984; DeMarch 1991a; but see Gjerde 1988).

The quantitative genetic analyses reported above make it clear that maternal effects have a measurable impact on traits that are likely to influence fitness in fishes. Although we cannot rule out the possibility that the identified maternal effects are due to simple physiological constraints, it seems probable from their obvious relationship to fitness that they are also subject to adaptive evolutionary mechanisms.

Kinghorn (1983) showed that heritabilities for production traits (size, growth rate, survival, food conversion efficiency, etc.) in farmed fish tend to increase with age. He suggested that this was due to a reduction in the magnitude of maternal effects over time. In fact, this pattern recurs in the published record: maternal effects in fishes are usually negligible beyond the early juvenile life stages regardless of the method used to identify them.

Maternal-Offspring Size Correlations

The two most often reported correlations are positive relationships between fecundity and female body size, and egg size and female body size. The high frequency with which these relationships are observed is probably a result of the ease of measurement of the three traits and because they are critical variables for life history theorists, fisheries scientists, and fisheries managers.

Significant correlations between female body size and egg size have been found in a wide variety of fish species (table 11.2; Sargent et al. 1987; Chambers and Leggett 1996). In fact, the maternal-offspring size correlation extends beyond fishes to include not only other animals but also plants (see Roff 1992; Visman et al. 1996). These correlations (and variations) have been used to generate and test a number of life history models (see Jonsson and Hindar 1982; Reznick 1991; Roff 1992); however, the unstated assumption is that egg size has a her-

Table 11.2 Correlation coefficients (r) generated from regressions of female body size with egg size (fem-egg) and egg size with juvenile body size (egg-juv) in various fish species.

Species	Common name	r (fem-egg)	r (egg-juv)	Reference
Allocyttus niger	Black oreo	NS		Conroy and Pankhurst 1989
Arius c.f. leptaspis	Fork-tailed catfish	0.66*		Coates 1988
A. solidus		0.74*		
A. sp.3		0.85*		
Brustiarius nox	Fork-tailed catfish	0.87*		Coates 1988
Cichlasoma citrinellum	Midas cichlid	NS	0.08	Lagomarsino et al. 1988
Clupea harengus	Atlantic herring	0.34		Blaxter and Hempel 1963
		NS		
		0.46		
		0.70		
		NS		
		NS		
		0.45*		Bradford and Stephenson 1992
		0.71		
		NS		
Cyprinus carpio	Carp	NS		Zonova 1973
Engraulis anchoita	Argentine anchovy	0.40*		Ciechomski 1966
Etheostoma spectabile	Orangethroat darter		0.36	Marsh 1986
		NS		Marsh 1984
Gadus morhua	Atlantic cod		0.30	Miller et al. 1995
			0.87	Knutsen and Tilseth 1985
Gasterosteus aculeatus	Three-spined stickleback	0.51		Wootton 1973
Hemipimelodus velutinus	Fork-tailed catfish	0.71*		Coates 1988
Hoplostethus atlanticus	Orange roughy	NS		Koslow et al. 1995
		NS		Pankhurst and Conroy 1987
Ictalurus punctatus	Channel catfish	−0.57		Broussard and Stickney 1981
		−0.44		
			0.68	Reagan and Conley 1977
		NS		Walser and Phelps 1993
Leuciscus leuciscus	Dace	0.88*	0.82	Mann and Mills 1985
		0.71		
		0.83		
		0.81		
		0.82		
		0.84		
		0.85		
		0.67		
Mallotus villosus	Capelin		0.31	Chambers et al. 1989
Melanogrammus aeglefinus	Haddock	0.45*		Hislop 1988
Morone saxatilis	Striped bass	NS*	0.92	Monteleone and Houde 1990
Oncorhynchus keta	Chum salmon	0.65*	0.65	Beacham and Murray 1985
			0.64	
		NS		
O. kisutch	Coho salmon		0.48	Silverstein and Hershberger 1992
		0.65		Sargent et al. 1987
		.40*		Fleming and Gross 1990

Table 11.2 (continued)

Species	Common name	r (fem-egg)	r (egg-juv)	Reference
O. kisutch	Coho salmon	NS*		
		0.65*		
		NS*		
		NS*		
		NS*		
		0.67*		
		0.52*		
		NS*		
		NS*		
		0.66*		
		0.57*		
		NS*		
		0.77*		
		0.49*		
		0.41*		
		NS*		
		0.47*		
		0.64*		
		0.79*		
		NS*		
		0.57*		
		0.53*		
		0.82*		
		0.88*		
		0.81*		
		0.61*		
		0.75*		
		NS*		
		0.81*		
O. mykiss	Rainbow trout		0.43	Gall 1974
			0.88*	Springate and Bromage 1985
		0.48		Galkina 1970
		NS		Scott 1962
		0.73		Bromage et al. 1990
O. nerka	Sockeye salmon	0.79	0.71	Bilton 1970
		0.79*		Bilton 1971
O. nerka	Sockeye salmon (22 populations)	0.21–0.76		Quinn et al. 1995
O. tshawytscha	Chinook salmon	0.49*	0.84	Fowler 1972
		0.69*	0.35	Heath, unpublished data
Oreochromis mossambicus	Tilapia		0.89	Rana 1985
Pseudocyttus maculatus	Smooth oreo	NS		Conroy and Pankhurst 1989
Salmo salar	Atlantic salmon	0.55*		Jonsson et al. 1996
		0.73*	0.94	Thorpe et al. 1984
		0.91	0.88	Kazakov 1981
		0.46		Galkina 1970
		0.56		

Table 11.2 (continued)

Species	Common name	r (fem-egg)	r (egg-juv)	Reference
S. trutta	Brown trout		0.94	Elliott 1984
		0.74*		L'Abee-Lund and Hindar 1990
		NS*		
		0.76*		
		NS*		
		0.68*		
		0.59*		
		0.74*		
		NS*		
Salvelinus alpinus	Arctic charr	0.43*		Jonsson and Hindar 1982
S. fontinalis	Brook trout	0.39		Hutchings 1996
		0.77		
		0.57		
		0.65*		Wydoski and Cooper 1966
		0.63*		Liskauskas and Ferguson 1990
			0.55	Hutchings 1991
			0.76	
Theragra chalcogramma	Walleye pollock	0.13	0.38	Hinckley 1990

All reported r values are significant at the $P < 0.05$ level or higher, NS indicates a nonsignificant regression. *Data used for figure 11.3.

itable component and can respond to natural selection in a predictable manner. If egg-size variation is mostly explained by maternal effects, then its response to selection could be largely unpredictable (Kirkpatrick and Lande 1989). One of our goals is to evaluate whether the correlation between female body size and egg size is merely a physiological side effect (and hence of little evolutionary consequence) or a consequence of adaptive evolution.

There is little argument that propagule size has important fitness consequences (Bell 1980; Reznick 1991; Roff 1992; Quinn et al. 1995; Bernardo 1996; among others). Among fishes there is a considerable body of evidence that larger eggs hatch into larger juveniles/larvae (Reznick 1991; Roff 1992; Chambers and Leggett 1996; table 11.2). Large egg size is also often associated with higher survival, although some have failed to find such a relationship (table 11.1; see Ware 1975). Large egg size and/or juvenile body size is associated with faster growth (Gall 1974; Pitman 1979; Wallace and Aasjord 1984; Reznick 1991), shorter development time (Ware 1975; Bradford and Peterman 1987), higher incidence of precocious sexual maturation (Silverstein and Hershberger 1992), and better swimming performance (Bams 1967). Unfortunately, only a handful of studies follow maternal groups past the larval period. In such longer term studies the correlation of survival (and/or growth) with egg size drops off rapidly with increasing offspring age (Reznick 1991; Silverstein and Hershberger 1992; D.D. Heath, unpublished data).

From the foregoing it is clear that egg size in fish is correlated with physiological traits that potentially have important fitness consequences. Two questions present themselves: (1)

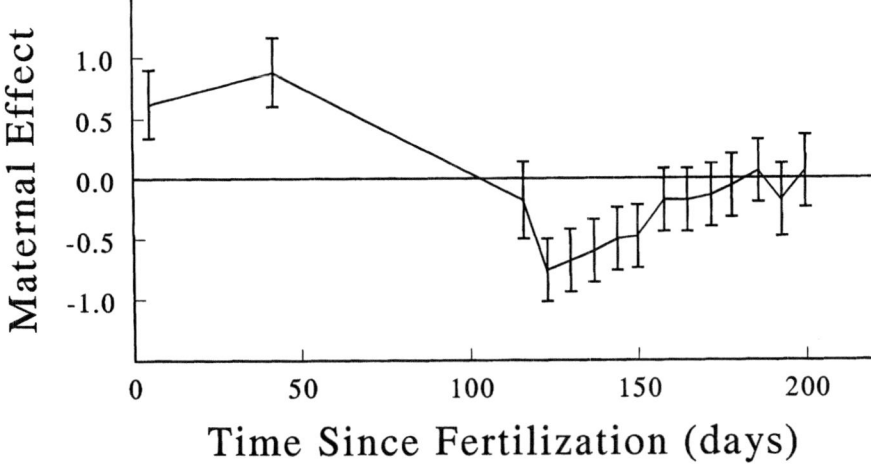

Figure 11.1 Calculated maternal effects (±2 SE) for 60 full-sibling chinook salmon (*Oncorhynchus tshawytscha*) during egg, larval, and juvenile stages. Maternal effect was calculated by subtracting the sire-offspring h^2 from the dam-offspring h^2.

Does the correlation between female body size and egg size reflect only maternal effects, or are there also genetic effects? (2) If the correlations are indeed maternal effects, do they represent adaptations? We deal with those two questions at three scales of analysis: within a population, among populations, and among species. We present three studies that address the adaptive value of maternal effects, one at each of these scales of analysis.

Heath (unpublished data) investigated the nature of the correlation between female body size and offspring size in a study conducted within a cultured chinook salmon (*Oncorhynchus tshawytscha*) population. Sixty males and females of different sizes and ages taken from a commercial brood stock were mated in a one-to-one design. The offspring from the resulting 60 full-sibling families were weighed twice during the egg stage, and 13 times during the fry stage. The mean offspring weight for each family was regressed against female and male weight for each sampling period (i.e., parent-offspring regressions). Mean family weights were standardized (by dividing by the overall mean), as was the parental weight. This design allowed the explicit calculation of maternal effects by subtracting the heritability calculated by the sire-offspring regression from the dam-offspring heritability. Figure 11.1 shows the results of the calculations for the family groups for 200 days of postfertilization development and growth.

A number of important observations can be made from figure 11.1. First, the female-egg size correlation is clearly a true maternal effect (the sire-offspring regression, not shown, gave a zero heritability for the first three sampling dates). Second, the maternal effect in this population drops to zero soon after the fry begin feeding (i.e., they become dependent on exogenous energy sources). Finally, the significant *negative* maternal effect seen from the fourth to the eighth sampling periods (figure 11.1) has, to our knowledge, not been reported previously. The cause of this negative maternal effect is a negative slope for the mother-offspring regression, coupled with a positive slope for the father-offspring regression (data not shown). The effect is best explained as a cost associated with the initial size advantage of

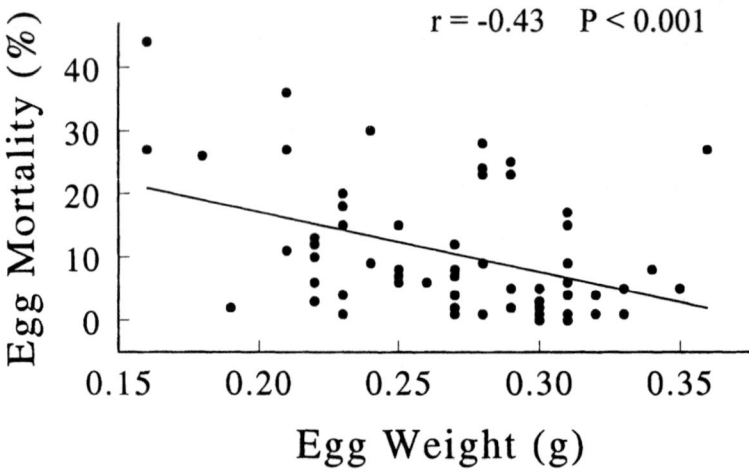

Figure 11.2 Regression of egg mortality (percent) on mean egg weight (g) for 71 full-sibling chinook salmon (*Oncorhynchus tshawytscha*) families. The regression was little changed by the transformation of egg mortality (arcsin-square root) and egg weight (ln); thus, we present untransformed data for ease of interpretation. The correlation coefficient and significance of the regression line are given.

families with larger eggs. Although a number of physiological mechanisms for this are possible, perhaps the most likely is that fry derived from the larger eggs have a lower metabolic rate. Such depression of metabolic rate may result from metabolic down-regulation during egg incubation as a consequence of the lower surface area:volume ratio of larger eggs. Experiments designed to test this possibility are ongoing.

Finally, the mean egg weight for each family was regressed against mortality during the egg and alevin, or prefeeding, stages. A significant negative correlation was found (figure 11.2). Note the range in mortality that was observed (0-45%, figure 11.2). Such mortality rates facilitate a potentially large selection opportunity and are certainly consistent with opportunity for the evolution of adaptive maternal effects.

The above experiment suggests the potential for evolution of the female-egg size correlation. More evidence exists at the next higher level of biological organization: among populations. Fleming and Gross (1990) presented data on the number and average size of eggs taken from populations of coho salmon (*Oncorhynchus kisutch*) over a wide latitudinal range on the west coast of North America. They found a significant latitudinal increase in egg number and a significant latitudinal decrease in egg size after correcting for the effect of female size (Fleming and Gross 1990). The authors concluded that the simplest explanation of their findings was that eggs incubated at higher latitudes experience lower incubation temperatures, with a concomitant increase in yolk conversion efficiency (Beacham and Murray 1985, 1990). That is, if a small egg (with a lower yolk energetic content; Quattro and Weeks 1991) is incubated at a sufficiently reduced temperature, it will develop into the same-sized juvenile as an embryo from a large egg reared at a higher temperature. The implication is that salmon populations have characteristic eggs sizes that are adaptive to local environmental conditions, and the variation in egg size is adaptive across populations.

Although Fleming and Gross's (1990) geographical pattern in maternal allocation is suggestive of adaptation for egg size in salmon, it cannot be used to conclude that the female-egg size correlations are adaptive. Manipulative experiments are needed to further explore the adaptive value of egg size variation.

Our final line of inference concerning the adaptive value of maternal-egg size correlations in fish is based on a cross-species comparison. We found 58 reports in which either the regression equation for female body-egg size was explicitly given or the raw data were provided. For comparison among species we standardized egg size and female body size measurements. When both egg and parent measurements were either mass based (i.e., weight or volume) or length based (i.e., length or diameter), the slope was multiplied by the parental:offspring mean ratios. Where parental size was a length measurement while egg size was a weight or volume measurement, we transformed the egg weight/volume measurements to approximate length measurements by taking the cube root. The slope was then standardized as above. Finally, for publications without regression equations or primary data, the figures were digitized and the primary data so derived were reanalyzed. Once standardized slopes were calculated, we estimated heritability as twice the slope of the maternal-offspring regression (Falconer 1981). We assumed that additive genetic contribution to this heritability was negligible, and thus we assumed that the h^2 estimate was equivalent to an estimate of maternal effect. Maternal effects so calculated may be inflated by confounding additive genetic components; however, it is reasonable to assume that little or no component of the observed variation would be due to expression of genes in the offspring's genome (see below), and thus our estimate of maternal effect should not be greatly affected. Calculated maternal effects are shown in figure 11.3; iteroparous and semelparous species are shown separately.

Semelparity is a "big bang" reproductive strategy; all reproductive effort goes into a single, very costly, mating event (Bell 1980; Roff 1992). Since female semelparous fish have no opportunity for variation in reproductive effort or allocation, they are expected to experience greater selection pressure on reproduction than iteroparous fish who have a "second chance." Accordingly, we predict that, if maternal effects on egg size have adaptive value then, all other things being equal, semelparous species should show greater maternal effects than iteroparous species. Alternatively, if maternal effects on egg size are merely due to physiological side effects (i.e., nonadaptive) then we expect no difference between the two life history categories.

Based on the 58 estimates of maternal effects, semelparous species exhibit a significantly larger maternal effect on egg size than iteroparous species ($F = 7.44$, df = 1, $P < 0.01$). Thus, interspecific comparisons support the hypothesis that, among fishes, the female-egg size correlation does have an adaptive component. However, it is also possible that the observed difference in the magnitude of maternal effects is due to differences in species composition of the iteroparous and semelparous groups.

Determination of the adaptive value of a trait is never straightforward, and female-egg size correlations in fishes are no exception. We have attempted to make an inferential case for the adaptive nature of the female-egg size correlation that is so prevalent within and among different fishes. However, to our knowledge, no experiment has been performed that definitively demonstrates an adaptive value of the correlation. Roff (1992) suggested that such an experiment would ideally involve experimental manipulation of egg size within a population (or within a clutch). Another approach would be to follow the performance of

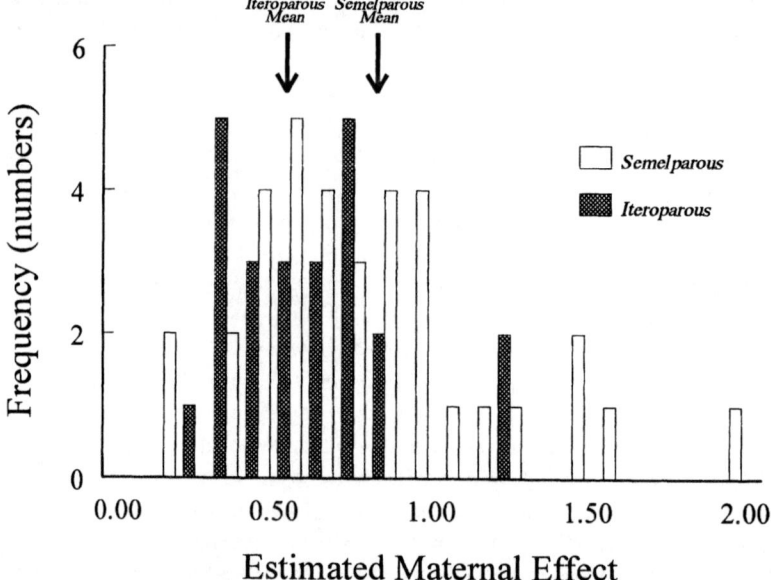

Figure 11.3 Frequency distribution of 58 estimations of maternal effects on egg size calculated from published reports of egg-mother size correlations (see section 11.3). Data for iteroparous and semelparous species are plotted separately; the means for each are indicated by the labeled arrows. The two distributions differ significantly ($F = 7.44$, $P < 0.01$). Data sources are indicated by asterisks in table 11.2.

offspring from eggs of different sizes within a population. Genetic factors could be accounted for by examining the within-clutch variation in egg size that occurs in many fishes (Elgar 1990; Chambers and Leggett 1996). Although direct evidence is currently unavailable we anticipate that it will soon be forthcoming and that it will demonstrate an adaptive component.

Direct Maternal Effects

Direct transfer of specific substances from mother to offspring has been documented in developing fish eggs and larvae. In some cases, a significant maternal effect has resulted (see Lombardi 1996). Probably the best-understood direct maternal effects are those involving transfer of proteins (hormones), organic compounds, and mRNA from mother to the offspring.

The transfer of maternal hormones to offspring is demonstrable in two ways: (1) time series data showing high concentrations at fertilization, declining during incubation, followed by an increase, generally associated with the depletion of the yolk sac; and (2) injection of females with supernormal amounts of the hormone followed by demonstration of elevated concentrations in the offspring (relative to the concentrations in offspring of noninjected females). Such studies have shown that the initial hormone content in eggs and developing larvae is derived from the mother, and that the maternal contribution can be extremely important for normal embryo development (Lam 1994). For example, concentrations of thyroid hormones thyroxine (T4) and triiodothyronine (T3) have been shown to be elevated in fertilized and unfertilized salmonid eggs (Tagawa and Hirano 1987, 1990; Greenblatt et al. 1989) and to decrease through embryonic development until near complete yolk-sac absorption, when the circulating levels begin to increase, suggesting offspring de novo synthesis (Brown et al. 1987; Sullivan et al. 1987; Tagawa and Hirano 1987, 1990; Greenblatt et al. 1989). Ayson and Lam (1993) found that injection of female rabbitfish (*Siganus guttatus*) with T4 was correlated with increased levels of T4 (and T3) in the offspring, indicating a transfer of the elevated hormonal concentrations to the offspring. The elevated offspring hormonal concentrations were also found to be associated with increased larval size and survival (Ayson and Lam 1993). In a similar experiment with brown trout (*Salmo trutta*), offspring hormonal levels were also found to be elevated following maternal injection; however, no effect on the size of the resulting juveniles was observed (Mylonas et al. 1994). Thus, transfer of maternal hormones is well documented, but evidence for an effect on offspring phenotype (indicative of a maternal effect) is limited. Experiments involving experimentally elevated cortisol and immunoglobin M-like protein concentrations in tilapia (*Oreochromis mossambicus*) have shown elevated concentrations in the eggs (Hwang et al. 1992; Takemura 1993). The adaptive value, if any, of maternally derived cortisol is unclear, but the transfer of the immune factors such as immunoglobin M-like proteins obviously may have survival implications.

"Vertical" transfer (i.e., from mother to offspring) of both infectious agents (Brown et al. 1990, 1996; Pascho and Elliott 1993; Elliott et al. 1995) and disease resistance (Mor and Avtalion 1990; Takemura 1993; Brown et al. 1996) occurs among fishes. Given the strong evidence for such an effect in mammals, it is perhaps not surprising that female fish can influence the immunological competency of their offspring. Although research on immunological maternal effects in fishes is still in its infancy, the fitness implications are obvious for both egg and juvenile offspring.

It is not difficult to imagine how the maternally produced proteins described above could be of benefit to the offspring. However, not all substances transferred from the mother are benign. A number of studies have documented maternal transfer of various environmental organic compounds, and deleterious effects of these transferred chemicals on the growth and survival of the offspring were also documented (Spies and Rice 1988; Ankley et al. 1991; Tilghman Hall and Oris 1991; Miller 1993). Clearly, this specific form of maternal effect is maladaptive. Such transfers are probably side effects of the maternal transfer of essential proteins and hormones necessary for the development of the embryo. It is thus difficult to imagine how this maladaptive form of substance transfer could evolve to become less deleterious except perhaps through the evolution of biochemical mechanisms with the capacity to decouple the transfer of deleterious compounds from the transfer of essential compounds.

Transfer of mRNA from mother to offspring has been well documented (Mechali et al. 1990; Moreau et al. 1995) and is speculated to be a method of "jump-starting" protein synthesis in the developing embryo. Little of the research on the transfer of maternal mRNA to offspring has been done with fish (but see Kanaya et al. 1996). There is, however, no reason to expect that the role played by maternal mRNA in the development of eggs and larvae differs between fish and other oviparous species, and clearly, it may generate maternal effects.

Documentation of the transfer of mRNA and proteins from mother to egg underscores an important point in the study of maternal effects. The observation that maternal effects are most highly expressed during the early stages of development is not simply a physiological curiosity. Rather, it reflects the key role of noninherited maternal influence transmitted cytoplasmically and by other proximate mechanisms associated with close physical association during early ontogeny of mother and offspring. As organisms develop, many genes must be activated and expressed in closely orchestrated coordination. It seems unlikely that the (initially) single-celled zygote is capable of activation of all of its genetic systems immediately at fertilization, nor is it clear that the full array of genetic activity can be orchestrated autonomously at that early stage. Embryonic dependency on the cytoplasmic legacy of enzymes, hormones, mRNA, and genetic modifying agents from the mother seems very likely. As the embryo develops, more of its own genome will be expressed and the relative effect of cytoplasmically inherited maternal factors will decrease. In this context it is interesting to note that Reznick's "grandfather effect" (see Reznick 1991) is similar to what would be expected from mitochondrial-nuclear DNA epistasis, in essence, a cytoplasmically inherited effect coupled with traditional quantitative genetic inheritance.

11.4 Maternal Effects on Behavioral Traits

It is well known that the behavior of individuals of one generation may have large influence over the expression of phenotypes in subsequent generations. In addition to this general consideration, two other observations lead us to think that maternal effects in behavior will be common, diverse and, often, adaptive in fishes. First, selection acts on phenotypes, and because behavior is perhaps the most immediate and malleable source of interaction between the phenotype of an organism and its environment, we expect adaptive evolution to be common at this interface. Second, the kinds and the extent of interaction between offspring and their kin (whether parents or siblings) are enormously variable among fishes. For example, *all* forms of parental care occur (none, male only, female only, biparental; Gross and Sargent 1985; Clutton-Brock 1991), and maternal investment in gametes and offspring is hugely variable, as discussed in sections 11.2 and 11.3. Accordingly, it is appropriate to think of the influence of relatives in terms of "kin effect" models (Cheverud and Moore 1994) rather than simply as "maternal" effects models.

Gender in fishes may be influenced by maternal effects. For example, Kallman (1975) reviews evidence for maternal effects in sex determination in the viviparous swordtail *Xiphophorous maculatus*. Reproductive behaviors in this internally fertilizing species are clearly sex specific and central to fitness. But it remains unclear how the observed maternal effects on sex determination (in other words, the deviations from segregation female:male ratios expected on the basis of sex chromosome segregation) may be adaptive. The reviews of Price (1984) and Francis (1992) reveal the extent to which sex determination in fishes is subject to both genetic and environmental influences. This lability in a trait that is basic to a wide range of behaviors, and is fundamentally relevant to fitness, perhaps best demonstrates the likelihood that adaptive evolution may occur in behavioral maternal effects in fishes.

The behaviors of organisms are often related to their size (see Travis 1994). To the extent that size is also influenced by maternal effects in fishes (see section 11.3), it seems probable that a cascade of secondary behavioral consequences may follow from size variation that has its origin in maternal effects. Travis (1994) points out the statistical and experimental design complexities in isolating the various contributions to behavioral variance that are size dependent.

Finally, maternal effects have been demonstrated in many of the studies in which quantitative genetic estimates of behavioral attributes have been made (e.g., Ferguson and Noakes 1982; Farr 1983; Fleming and Gross 1992; Rodd 1994; see section 11.3), and we expect, for the reasons given above, that this will be a general finding among fishes. Like the morphological and physiological traits discussed above, however, maternal effects in behavior have not generally been a subject of interest or study in their own right.

Unusual opportunities exist among fishes for the examination of the evolutionary consequences and adaptive nature of behavioral maternal effects that we feel would richly repay study. The opportunities associated with labile sex determination have been mentioned earlier. Another opportunity is associated with the recent discovery of an evolutionary reversal in parental care (from male care to no care) in a stickleback fish (Blouw 1996). Genetic crosses between fish showing parental care and emancipation are viable, and numerous ecological, physiological, and morphological differences occur among them (Blouw and Hagen 1990; MacDonald et al. 1995a,b). With carefully designed experiments along the lines suggested by Cheverud and Moore (1994), it should be possible to estimate maternal effects, and it may be possible in the context of the reversal in parental care to discover adaptive components.

Conclusions

An important conclusion to be drawn from the research reviewed here is that maternal effects in fishes should no longer be approached simply as a confounding factor in breeding experiments. The adaptive nature of a few of the maternal effects described here is obscure; nevertheless, for most of the maternal effects we have reviewed, a potentially high component is possible. Therefore, we predict that natural selection has driven the evolution of many of the maternal effects reported here toward an adaptive role.

One limitation of the few examples of direct evidence for adaptive maternal effects (such as Fox and Mousseau, this volume) is that it usually involves laboratory populations subjected to experimental perturbations. Ideally, tests of the adaptive component in maternal effects should be performed in the field. The logistical problems with such an approach would have been insurmountable as recently as 5–10 years ago. Recent developments in DNA fingerprinting and associated techniques provide a set of powerful new tools for evolutionary biologists. A pilot study to test the efficacy of DNA fingerprint analysis in determining maternal and paternal effects on growth and survival in a population of farmed rainbow trout, *Oncorhynchus mykiss*, showed remarkable success (Herbinger et al. 1995). Molecular markers have also been used to determine heritabilities and genetic correlations (see Ritland 1996) in a captive population of chinook salmon, *Oncorhynchus tshawytscha* (Mousseau et al. in press), and in the monkeyflower (Ritland and Ritland 1996). Molecular markers provide an exciting new opportunity to study the adaptive nature and magnitude of maternal effects in the field.

Although we have presented inferential evidence that we believe to be compelling that adaptive maternal effects occur in fishes, direct evidence does not yet exist. We predict that the study of maternal effects in fish will yield interesting and significant results within the next few years, both because a huge volume of quantitative genetic data exists for many fish species, particularly those that are commercially exploited, and because of a general change in attitude toward the evolutionary significance of maternal effects as exemplified in this volume.

Acknowledgments We thank Yellow Island Aquaculture Ltd. for their support of the experimental work described in this chapter. Grace Cho, Davidson Heath, Allison Street, and Stevan Springer all assisted with the literature search and receive our thanks. We also thank Julie Smit and Charles Fox for their critical review of the manuscript. This work was supported in part by a grant from the Natural Science and Engineering Research Council, Canada, to D.D.H.

References

Allen, G. H. 1957. Survival through hatching of eggs from silver salmon (*Oncorhynchus kisutch*). Trans. Am. Fish. Soc. 87:207–219.

Ankley, G. T., D. E. Tillitt, J. P. Giesy, P. D. Jones, and D. A. Verbrugge. 1991. Bioassay-derived 2,3,7,8-tetrachlorodibenzo-*p*-dioxin equivalents in PCB-containing extracts from the flesh and eggs of Lake Michigan chinook salmon (*Oncorhynchus tshawytscha*) and possible implications for reproduction. Can. J. Fish. Aquat. Sci. 48:1685–1690.

Ayles, G. B. 1974. Relative importance of additive genetic and maternal sources of variation in early survival of young splake hybrids (*Salvelinus fontinalis* X *S. namaycush*). J. Fish. Res. Board Can. 31:1499–1502.

Ayson, F. G., and T. J. Lam. 1993. Thyroxine injection of female rabbitfish (*Siganus guttatus*) broodstock: Changes in thyroid hormone levels in plasma, eggs, and yolk-sac larvae, and its effect on larval growth and survival. Aquaculture 109:83–93.

Bagenal, T. B. 1969. Relationship between egg size and fry survival in brown trout *Salmo trutta* L. J. Fish. Biol. 1:349–353.

Bailey, J. K., and E. J. Loudenslager. 1986. Genetic and environmental components of variation for growth of juvenile Atlantic salmon (*Salmo salar*). Aquaculture 57:125–132.

Bams, R. A. 1967. Differences in performance of naturally and artificially propagated sockeye salmon migrant fry, as measured with swimming and predation tests. J. Fish. Res. Board Can. 24:1117–1153.

Beacham, T. D. 1988. A genetic analysis of early development in pink salmon (*Oncorhynchus gorbuscha*) and chum salmon (*Oncorhynchus keta*) at three different temperatures. Genome 30:89–96.

Beacham, T. D. 1989. Genetic variation in body weight of pink salmon (*Oncorhynchus gorbuscha*). Genome 32:227–231.

Beacham, T. D., and C. B. Murray. 1985. Effect of female size, egg size, and water temperature on developmental biology of chum salmon (*Oncorhynchus keta*) from the Ninat River, British Columbia. Can. J. Fish. Aquat. Sci. 42:1755–1765.

Beacham, T. D., and C. B. Murray. 1990. Temperature, egg size, and development of embryos and alevins of five species of Pacific salmon: A comparative analysis. Trans. Am. Fish. Soc. 119:927–941.

Bell, G. 1980. The costs of reproduction and their consequences. Am. Nat. 116:45–76.

Bernardo, J. 1996. The particular maternal effect of propagule size, especially egg size: Patterns, models, quality of evidence and interpretations. Am. Zool. 36:216–236.

Bilton, H. T. 1970. Maternal influences on the age at maturity of Skeena River sockeye salmon (*Oncorhynchus nerka*). Fisheries Research Board of Canada, Technical Report No. 167. Ottawa, Canada.

Bilton, H. T. 1971. A hypothesis of alternation of age of return in successive generations of Skeena River sockeye salmon (*Oncorhynchus nerka*). J. Fish. Res. Board Can. 28:513–516.

Blanc, J. M., and H. Poisson. 1983. Parental sources of variation in hatching and early survival rates of *Salmo trutta* female × *Salvelinus fontinalis* male hybrid. Aquaculture
32:115–122.

Blaxter, J. H. S., and G. Hempel. 1963. The influence of egg size on herring larvae (*Clupea
harengus* L.). J. Conserv. Perm. Int. Expl. Mer. 28:211–240.

Blouw, D. M. 1996. Evolution of offspring desertion in a stickleback fish. Ecoscience
3:18–24.

Blouw, D. M., and D. W. Hagen. 1990. Breeding ecology and evidence of reproductive isolation of a widespread stickleback fish (Gasterosteidae) in Nova Scotia, Canada. Biol. J.
Linn. Soc. 39:195–217.

Bradford, M. J., and R. M. Peterman. 1987. Maternal size effects may explain positive correlations between age at maturity of parent and offspring sockeye salmon
(*Oncorhynchus nerka*). Can. Spec. Publ. Fish. Aquat. Sci. 96:90–100.

Bradford, R. G., and R. L. Stephenson. 1992. Egg weight, fecundity, and gonad weight variability among northwest Atlantic herring (*Clupea harengus*) populations. Can. J. Fish.
Aquat. Sci. 49:2045–2054.

Bromage, N., P. Hardiman, J. Jones, J. Springate, and V. Bye. 1990. Fecundity, egg size and
total egg volume differences in 12 stocks of rainbow trout, *Oncorhynchus mykiss*
Richardson. Aquacult. Fish. Man. 21:269–284.

Broussard, M. C., and R. R. Stickney. 1981. Evaluation of reproductive characters for four
strains of channel catfish. Trans. Am. Fish. Soc. 110:502–506.

Brown, C. L., C. V. Sullivan, W. W. Dickhoff, and H. A. Bern. 1987. Occurrence of thyroid
hormones in early developmental stages of teleost fish. Trans. Am. Fish. Soc. Symp.
2:144–150.

Brown, L. L., L. J. Albright, and T. P. T. Evelyn. 1990. Control of vertical transmission of
Renibacterium salmoninarum by injection of antibiotics into maturing female coho
salmon *Oncorhynchus kisutch*. Dis. Aquat. Org. 9:127–131.

Brown, L. L., G. K. Iwama, and T. P. T. Evelyn. 1996. The effect of early exposure of coho
salmon (*Oncorhynchus kisutch*) eggs to the p57 protein of the *Renibacterium salmoninarum* on the development of immunity to the pathogen. Fish Shellfish Immun. 6:149–
165.

Chambers, R. C., and W. C. Leggett. 1996. Maternal influences on variation in egg sizes in
temperate marine fishes. Am. Zool. 36:180–196.

Chambers, R. C., W. C. Leggett, and J. A. Brown. 1989. Egg size, female effects, and the correlations between early life history traits of capelin, *Mallotus villosus*: An appraisal at
the individual level. Fish. Bull. US 87:515–523.

Cheverud, J. M., and A. J. Moore. 1994. Quantitative genetics and the role of the environ-
ment provided by relatives in behavioral evolution. Pp. 67–100 in C. R. B. Boake (ed.),
Quantitative Genetic Studies of Behavioral Evolution. University of Chicago Press,
Chicago.

Ciechomski, J. D. 1966. Development of the larvae and variations in the size of the eggs of
the Argentine anchovy, *Engraulis anchoita* Hubbs and Marini. J. Conserv. Perm. Int.
Explor. Mer. 30:281–290.

Clutton-Brock, T. H. 1991. The Evolution of Parental Care. Princeton University Press,
Princeton, N. J.

Coates, D. 1988. Length-dependent changes in egg size and fecundity in females, and
brooded embryo size in males, of fork-tailed catfishes (Pisces: Ariidae) from the Sepik
River, Papua New Guinea, with some implications for stock assessments. J. Fish. Biol.
33:455–464.

Conroy, A. M., and N. W. Pankhurst. 1989. Size-fecundity relationships in the smooth oreo,

Pseudocyttus maculatus, and the black oreo, *Allocyttus niger* (Pisces: Oreosomatidae). N. Zeal. J. Mar. Freshw. Res. 23:525–528.

DeMarch, B. G. E. 1991a. Genetic, maternal, and tank determinants of growth in hatchery-reared juvenile Arctic charr (*Salvelinus alpinus*). Can. J. Zool. 69:655–660.

DeMarch, B. G. E. 1991b. Hatchery growth of pure strains and intraspecific hybrids of juvenile Arctic charr, *Salvelinus alpinus* (Canadian × Norwegian charr). Can. J. Fish. Aquat. Sci. 48:1109–1116.

Elgar, M. A. 1990. Evolutionary compromise between a few large and many small eggs: Comparitive evidence in teleost fish. Oikos 59:283–287.

Elliott, J. M. 1984. Growth, size, biomass and production of young migratory trout *Salmo trutta* in a lake district stream, 1966–83. J. Anim. Ecol. 53:979–994.

Elliott, D. G., R. J. Pascho, and A. N. Palmisano. 1995. Brood stock segregation for the control of bacterial kidney disease can affect mortality of progeny chinook salmon (*Oncorhynchus tshawytscha*) in seawater. Aquaculture 132:133–144.

Falconer, D. S. 1981. Introduction to Quantitative Genetics. 2nd ed. Longman, New York.

Farr, J. A. 1983. The inheritance of quantitative fitness traits in guppies, *Poecilia reticulata* (Pisces: Poeciliidae). Evolution 37:1193–1209.

Ferguson, M. M., and D. L. G. Noakes. 1982. Genetics of social behaviour in charrs (*Salvelinus* species). Anim. Behav. 30:128–134.

Fleming, I. A., and M. R. Gross. 1990. Latitudinal clines: A trade-off between egg number and size in Pacific salmon. Ecology 71:1–11.

Fowler, L. G. 1972. Growth and mortality of fingerling chinook salmon as affected by egg size. Prog. Fish-Cult. 34(2):66–69.

Fox, C. W. 1993. Maternal and genetic influences on egg size and larval performance in a seed beetle (*Callosobruchus maculatus*): Multigenerational transmission of a maternal effect? Heredity 73:509–517.

Francis, R. C. 1992. Sexual lability in teleosts: Developmental factors. Q. Rev. Biol. 67:1–18.

Galkina, Z. I. 1970. Dependence of egg size on the size and age of female salmon (*Salmo salar* L.) and rainbow trout (*Salmo irideus* Gib.). J. Ichthyol. 10:625–633.

Gall, G. A. E. 1974. Influence of size of eggs and age of female on hatchability and growth in rainbow trout. Calif. Fish Game 60:26–35.

Gjedrem, T. 1983. Genetic variation in quantitative traits and selective breeding in fish and shellfish. Aquaculture 33:51–72.

Gjedrem, T., and H. M. Gjøen. 1995. Genetic variation in susceptibility of Atlantic salmon, *Salmo salar* L., to furunculosis, BKD and cold water vibriosis. Aquacult. Res. 26:129–134.

Gjerde, B. 1988. Complete diallele cross between six inbred groups of rainbow trout, *Salmo gairdneri*. Aquaculture 75:71–87.

Gjerde, B., and T. Refstie. 1984. Complete diallel cross between five strains of Atlantic salmon. Aquaculture 11:207–226.

Glebe, B. D., T. D. Appy, and R. L. Saunders. 1979. Variation in Atlantic salmon (*Salmo salar*). Intl. Coun. Expl. Sea J. Mar. Sci. 23:1–11.

Greenblatt, M., C. L. Brown, M. Lee, S. Dauder, and H. A. Bern. 1989. Changes in thyroid hormone levels in eggs and larvae and in iodide uptake by eggs of coho and chinook salmon, *Oncorhynchus kisutch* and *O. tschawytscha*. Fish Physiol. Biochem. 6:261–278.

Gross, M. R., and R. C. Sargent. 1985. The evolution of male and female parental care in fishes. Am. Zool. 25:807–822.

Heath, D. D., N. J. Bernier, J. W. Heath, and G. K. Iwama. 1993. Genetic, environmental, and interaction effects on growth and stress response of chinook salmon fry. Can. J. Fish. Aquat. Sci. 50:435–442.

Herbinger, C. M., R. W. Doyle, E. R. Pitman, D. Paquet, K. A. Mesa, D. B. Morris, J. M. Wright, and D. Cook. 1995. DNA fingerprint based analysis of paternal and maternal effects on offspring growth and survival in communally reared rainbow trout. Aquaculture 137:245–256.

Hinckley, S. 1990. Variation of egg size of walleye pollock *Theragra chalcogramma* with a preliminary examination of the effect of egg size on larval size. Fish. Bull. US 88:471–483.

Hislop, J. R. G. 1988. The influence of maternal length and age on the size and weight of the eggs and the relative fecundity of the haddock, *Melanogrammus aeglefinus*, in British waters. J. Fish. Biol. 32:923–930.

Hutchings, J. A. 1991. Fitness consequences of variation in egg size and food abundance in brook trout *Salvelinus fontinalis*. Evolution 45:1162–1168.

Hutchings, J. A.. 1996. adaptive phenotypic plasticity in brook trout, *Salelinus fontinalis*, life histories. Ecoscience 3:25–32.

Hwang, P.-P., S.-M. Wu, J.-H. Lin, and L.-S. Wu. 1992. Cortisol content of eggs and larvae of teleosts. Gen. Comp. Endocrinol. 86:189–196.

Jonsson, B., and K. Hindar. 1982. Reproductive strategy of dwarf and normal Arctic charr (*Salvelinus alpinus*) from Vangsvatnet Lake, western Norway. Can. J. Fish. Aquat. Sci. 39:1404–1413.

Jonsson, N., B. Jonsson, and I. A. Fleming. 1996. Does early growth cause a phenotypically plastic response in egg production of Atlantic salmon? Funct. Ecol. 10:89–96.

Kallman, K. D. 1975. The platyfish, *Xiphophorous maculatus*. Pp. 81–132 in R. C. King (ed.), Handbook of Genetics: Vertebrates of Genetic Interest. Vol. 4. Plenum Press, New York.

Kanaya, S., Y. Kudo, S. Mokitoda, K. Katsura, and C. Delcarpio. 1996. Synchronous gene expressions during embryogenesis of *Oncorhynchus masou* (Yamame). Biochem. Mol. Biol. Intl. 39:261–266.

Kanis, E., T. Refstie, and T. Gjedrem. 1976. A genetic analysis of egg, alevin and fry mortality in salmon (*Salmo salar*), sea trout (*Salmo trutta*) and rainbow trout (*Salmo gairdneri*). Aquaculture 8:259–268.

Kazakov, R. V. 1981. The effect of the size of Atlantic salmon, *Salmo salar* L., eggs on embryos and alevins. J. Fish. Biol. 19:353–360.

Kinghorn, B. P. 1983. A review of quantitative genetics in fish breeding. Aquaculture 31:283–304.

Kirkpatrick, M., and R. Lande. 1989. The evolution of maternal characters. Evolution 43:485–503.

Kirpichnikov, V. S. 1966. Goals and methods in carp selection. Bull. St. Sci. Res. Inst. Lake River Fish. 62:41–42. (translated from Russian).

Knutsen, G. M., and S. Tilseth. 1985. Growth, development, and feeding success of Atlantic cod larvae *Gadus morhua* related to egg size. Trans. Am. Fish. Soc. 114:507–511.

Koslow, J. A., J. Bell, P. Virtue, and D. C. Smith. 1995. Fecundity and its variability in orange roughy: Effects of population density, condition, egg size, and senescence. J. Fish. Biol. 47:1063–1080.

L'Abee-Lund, J. H., and K. Hindar. 1990. Interpopulational variation in reproductive traits of anadromous female brown trout, *Salmo trutta* L. J. Fish. Biol. 37:755–763.

Lagomarsino, I. V., R. C. Francis, and G. W. Barlow. 1988. The lack of correlation between size of egg and size of hatchling in the midas cichlid, *Cichlasoma citrinellum*. Copeia 4:1086–1089.

Lam, T. J. 1994. Hormones and egg/larval quality in fish. J. World Aquat. Soc. 25:2–12.

Liskauskas, A. P., and M. M. Ferguson. 1990. Enzyme heterozygosity and fecundity in a naturalized population of brook trout (*Salvelinus fontinalis*). Can. J. Fish. Aquat. Sci. 47:2010–2015.

Lombardi, J. 1996. Postzygotic maternal influences and the maternal embryonic relationship of viviparous fishes. Am. Zool. 36:106–115.

MacDonald, J. F., J. Bekkers, S. M. MacIsaac, and D. M. Blouw. 1995a. Intertidal breeding and aerial development of embryos of a stickleback fish (*Gasterosteus*). Behaviour 132:1183–1206.

MacDonald, J. F., S. M. MacIsaac, J. Bekkers, and D. M. Blouw. 1995b. Experiments on embryo survivorship, habitat selection, and competitive ability of a stickleback fish (*Gasterosteus*) which nests in the rocky intertidal zone. Behaviour 132:1207–1221.

Mann, R. H. K., and C. A. Mills. 1985. Variation in the sizes of gonads, eggs and larvae of the dace, *Leuciscus leuciscus*. Environ. Biol. Fish. 13:277–287.

Marsh, E. 1984. Egg size variation in central Texas populations of *Etheostoma spectabile* (Pisces: Percidae). Copeia 2:291–301.

Marsh, E. 1986. Effects of egg size on offspring fitness and maternal fecundity in the orangethroat darter, *Etheostoma spectabile* (Pisces: Percidae). Copeia 1:18–30.

McKay, L. R., P. E. Ihssen, and G. W. Friars. 1986a. Genetic parameters of growth in rainbow trout, *Salmo gairdneri*, as a function of age and maturity. Aquaculture 58:241–254.

McKay, L. R., P. E. Ihssen, and G. W. Friars. 1986b. Genetic parameters of growth in rainbow trout, *Salmo gairdneri*, prior to maturation. Can. J. Genet. Cytol. 28:306–312.

Mechali, M., G. Almouzni, J. Moreau, S. Vriz, M. Leibovici, J. Hourdry, J. Geraudie, T. Soussi, and M. Gusse. 1990. Genes and mechanisms involved in early embryonic development in *Xenopus laevis*. Intl. J. Dev. Biol. 34:51–59.

Meffe, G. K. 1987. Embryo size variation in mosquitofish: Optimality vs plasticity in propagule size. Copeia 3:762–768.

Meffe, G. M. 1990. Offspring size variations in eastern mosquitofish (*Gambusia holbrooki*: Poeciliidae) from contrasting thermal environments. Copeia 1990:10–18.

Miller, M. A. 1993. Maternal transfer of organochlorine compounds in salmonines to their eggs. Can. J. Fish. Aquat. Sci. 50:1405–1413.

Miller, T. J., T. Herra, and W. C. Leggett. 1995. An individual-based analysis of the variability of eggs and their newly hatched larvae of Atlantic cod (*Gadus morhua*) on the Scotian shelf. Can. J. Fish. Aquat. Sci. 52:1083–1093.

Monteleone, D. M., and E. D. Houde. 1990. Influence of maternal size on survival and growth of striped bass *Morone saxatilis* Walbaum eggs and larvae. J. Exp. Mar. Biol. Ecol. 140:1–11.

Mor, A., and R. R. Avtalion. 1990. Transfer of antibody activity from immunized mother to embryo in tilapias. J. Fish. Biol. 37:249–255.

Moreau, J., N. Iouzalen, and M. Mechali. 1995. Isolation of cDNAs from maternal mRNAs specifically present during early development. Mol. Reprod. Dev. 41:1–7.

Mousseau, T. A., and H. Dingle. 1991a. Maternal effects in insect life histories. Annu. Rev. Entomol. 36:511–534.

Mousseau, T. A., and H. Dingle. 1991b. Maternal effects in insects: examples, constraints, and geographic variation. Pp. 745–761 in E. C. Dudley (ed.), The Unity of Evolutionary Biology, vol. 2. Dioscorides Press, Portland, Ore.

Mousseau, T. A., K. Ritland, and D. D. Heath. 1997. Estimating the heritability of life history traits using molecular markers. Heredity, in press.

Mylonas, C. C., C. V. Sullivan, and J. M. Hinshaw. 1994. Thyroid hormones in brown trout (*Salmo trutta*) reproduction and early development. Fish Physiol. Biochem. 13:485–493.

Nilsson, J. 1992. Genetic parameters of growth and sexual maturity in Arctic char (*Salvelinus alpinus*). Aquaculture 106:9–19.

Pankhurst, N. W., and A. M. Conroy. 1987. Size fecundity in the orange roughy, *Hoplostethus atlanticus*. N. Zeal. J. Mar. Freshw. Res. 21:295–300.

Pascho, R. J., and D. G. Elliott. 1993. Monitoring of the in-river migration of smolts from two groups of spring chinook salmon, *Oncorhynchus tshawytscha* (Walbaum), with different profiles of *Renibacterium salmoninarum* infection. Aquacult. Fish. Man. 24:163–169.

Pitman, R. W. 1979. Effects of female age and egg size on growth and mortality in rainbow trout. Prog. Fish-Cult. 41:202–204.

Price, D. J. (1984). Genetics of sex determination—a brief review. Pp. 77–89 in G. W. Potts and R. J. Wootton (eds.), Fish Reproduction: Strategies and Tactics. Academic Press, London.

Quattro, J. M., and S. C. Weeks. 1991. Correlations between egg size and egg energetic content within and among biotypes of the genus *Poeciliopsis*. J. Fish. Biol. 38:331–334.

Quinn, T. P., A. P. Hendry, and L. A. Wetzel. 1995. The influence of life history trade-offs and the size of incubation gravels on egg size variation in sockeye salmon (*Oncorhynchus nerka*). Oikos 74:425–438.

Rana, K. J. 1985. Influence of egg size on the growth, onset of feeding, point-of-no-return, and survival of unfed *Oreochromis mossambicus* fry. Aquaculture 46:119–131.

Reagan, R. E. J., and C. M. Conley. 1977. Effect of egg diameter on growth of channel catfish. Prog. Fish-Cult. 39:133–134.

Refstie, T. 1980. Genetic and environmental sources of variation in body weight and length of rainbow trout fingerlings. Aquaculture 19:351–357.

Refstie, T., and T. A. Steine. 1978. Selection experiments with salmon III. Genetic and environmental sources of variation in length and weight of Atlantic salmon in the freshwater phase. Aquaculture 14:221–234.

Reznick, D. 1981. "Grandfather effects": The genetics of interpopulation differences in offspring size in the mosquito fish. Evolution 35:941–953.

Reznick, D. 1982. Genetic determination of offspring size in the guppy (*Poecilia reticulata*). Am. Nat. 120:181–187.

Reznick, D. 1990. Maternal effects in fish life histories. Pp. 780–793 in E. C. Dudley (ed.), The Unity of Evolutionary Biology. Vol. 2. Dioscorides Press, Portland, Ore.

Reznick, D., H. Callahan, and R. Lauredo. 1996. Maternal effects on offspring quality in poeciliid fishes. Am. Zool. 36:147–156.

Ritland, K. 1996. A marker-based method for inferences about quantitative inheritance in natural populations. Evolution 50:1062–1073.

Ritland, K., and C. Ritland. 1996. Inferences about quantitative inheritance based on natural population structure in the yellow monkeyflower, *Mimulus guttatus*. Evolution 50:1074–1082.

Robison, O. W., and L. G. Luempert. 1984. Genetic variation in weight and survival of brook trout (*Salvelinus fontinalis*). Aquaculture 38:155–170.

Rodd, F. H. 1994. Phenotypic plasticity in the life history and sexual behaviour of the Trinidadian guppies (*Poecilia reticulata*) in response to their social environment. Ph.D. dissertation, York University, Toronto.

Roff, D. A. 1992. The evolution of life histories; theory and analysis. Chapman and Hall, New York.

Sargent, R. C., P. D. Taylor, and M. R. Gross. 1987. Parental care and the evolution of egg sizes in fishes. Am. Nat. 129:32–46.

Scott, D. P. 1962. Effect of food quantity on fecundity of rainbow trout, *Salmo gairdneri*. J. Fish. Res. Board Can. 19:715–731.

Silverstein, J. T., and W. K. Hershberger. 1992. Precocious maturation in coho salmon (*Oncorhynchus kisutch*): Estimation of the additive genetic variance. J. Hered. 83:282–286.

Spies, R. B., and D. W. Rice. 1988. Effects of organic contaminants on reproduction of the starry flounder *Platichthys stellatus* in San Francisco Bay. Mar. Biol. 98:191–200.

Springate, J. R. C., and N. R. Bromage. 1985. Effects of egg size on early growth and survival in rainbow trout (*Salmo gairdneri* Richardson). Aquaculture 47:163–172.

Sullivan, C. V., R. N. Iwamoto, and W. W. Dickhoff. 1987. Thyroid hormones in blood plasma of developing salmon embryos. Gen. Comp. Endocrinol. 65:337–345.

Tagawa, M., and T. Hirano. 1987. Presence of thyroxine in eggs and changes in its content during early development of chum salmon, *Oncorhynchus keta*. Gen. Comp. Endocrinol. 68:129–135.

Tagawa, M., and T. Hirano. 1990. Changes in tissue and blood concentrations of thyroid hormones in developing chum salmon. Gen. Comp. Endocrinol. 76:437–443.

Takemura, A. 1993. Changes in an immunoglobulin M (IgM)-like protein during larval stages in tilapia, *Oreochromis mossambicus*. Aquaculture 115:233–241.

Thibault, R. E., and R. J. Schultz. 1978. Reproductive adaptations among viviparous fishes (Cyprinodontiformes: Poeciliidae). Evolution 32:320–333.

Thorpe, J. E., M. S. Miles, and D. S. Keay. 1984. Developmental rate, fecundity and egg size in Atlantic salmon, *Salmo salar* L. Aquaculture 43:289–305.

Tilghman Hall, A., and J. T. Oris. 1991. Anthracene reduces reproductive potential and is maternally transferred during long-term exposure in fathead minnows. Aquat. Toxicol. 19:249–264.

Tomita, M., M. Iwahashi, and R. Suzuki. 1980. Number of spawned eggs and ovarian eggs and egg diameter and percent eyed eggs with reference to the size of female carp. Bull. Jap. Soc. Fish 96:1077–1081.

Travis, J. 1994. Size dependent behavioral variation and its genetic control within and among populations. Pp. 165–187 in C. R. B. Boake (ed.), Quantitative Genetic Studies of Behavioral Evolution. University of Chicago Press, Chicago.

Travis, J., J. A. Farr, S. Henrich, and R. T. Cheong. 1987. Testing theories of clutch overlap with the reproductive ecology of *Heterandria formosa*. Ecology 68:611–623.

Travis, J., J. C. Trexler, and M. Mcmanus. 1992. Effects of habitat and body size on mortality rates of *Poecilia latipinna*. Ecology 73:2224–2236.

Visman, V., S. Pesant, J. Dion, B. Shipley, and R. H. Peters. 1996. Joint effects of maternal and offspring sizes on clutch mass and fecundity in plants and animals. Ecoscience 3:173–182.

Wallace, J. C., and D. Aasjord. 1984. An investigation of the consequences of egg size for the culture of Arctic charr, *Salvelinus alpinus* (L.). J. Fish. Biol. 24:427–435.

Walser, C. A., and R. P. Phelps. 1993. Factors influencing the enumeration of channel catfish eggs. Prog. Fish-Cult. 55:195–198.

Wangila, B. C. C., and T. A. Dick. 1996. Genetic effects and the growth performance in pure and hybrid strains of rainbow trout, *Oncorhynchus mykiss* (Walbaum) (Order: Salmoniformes, family: Salmonidae). Aquacult. Res. 27:35–41.

Ware, D. M. 1975. Relation between egg size, growth, and natural mortality of larval fish. J. Fish. Res. Board Can. 32:2503–2512.

Withler, R. E., W. C. Clarke, B. E. Riddell, and H. Kreiberg. 1987. Genetic variation in freshwater survival and growth of chinook salmon (*Oncorhynchus tshawytscha*). Aquaculture 64:85–96.

Wootton, R. J. 1973. The effect of size of food ration on egg production in the female threespined stickleback, *Gasterosteus aculeatus* L. J. Fish. Biol. 5:89–96.

Wourms, J. P. 1981. Viviparity: The maternal-fetal relationship in fishes. Am. Zool. 21:473–515.

Wydoski, R. S., and E. L. Cooper. 1966. Maturation and fecundity of brook trout from infertile streams. J. Fish. Res. Board Can. 23:623–649.

Zonova, A. S. 1973. The connection between egg size and some of the characters of female carp (*Cyprinus carpio* L.). J. Ichthyol. 13:679–689.

12

Maternal and Paternal Effects in Birds

Effects on Offspring Fitness

TREVOR PRICE

There are many direct influences of parent birds on the fitness of their offspring. Females vary in the amount of nutrients, hormones, and antibodies they pass into the egg, and all have been shown to influence offspring fitness. Provisioning of food by both sexes has been shown to affect immunocompetence, body mass, male secondary sexual characteristics, and clutch size of offspring. Males affect offspring through food and other care they provide, as well as indirectly through effects on their mate. In addition, song is paternally transmitted in some species, and this can affect subsequent mating success. This review concentrates on maternal effects on offspring fitness. At the time of laying a clutch, females can invest in (1) the quality of their offspring, (2) the number of eggs they lay, and (3) their own future survival and reproduction prospects. Trade-offs among offspring quality, offspring number, and parental future reproductive success may result in many strategies with more or less equivalent fitnesses for the parents but very different fitnesses for offspring. In addition, some parents invest more in reproduction than others because they are in better condition and hence have more resources to invest. Usually parents in high condition invest more in both offspring number and offspring quality than parents in low condition. In the collared flycatcher, *Ficedula albicollis*, >25% of the variation in the size of clutches laid by 1-year-old offspring has been attributed to direct influences of their mother. Parental effects on offspring fitness in many species may well be higher than this because they also affect offspring survival.

12.1 Introduction

All birds lay eggs, and the majority care for their young during a period in which they mature and acquire the skills needed for independence. Consequently, the potential for maternal and paternal effects is large. Either the male, the female, or both control where to nest, what time to nest, clutch size, how much to invest in each egg, how much to incubate the eggs and brood the young, and how much to feed the young. All of these characteristics have been shown to affect offspring fitness and/or growth rate. In addition, there is some cultural transmission of behaviors, notably song. Here I review the evidence for maternal and paternal effects in birds, concentrating on effects of parents on their offspring's fitness. While I consider both maternal and paternal effects, I use the term "maternal" rather than "parental" in general discussion, since this has been the way parental effects have been analyzed in most studies.

The most precise definition of a maternal effect is the partial regression of the offspring's phenotype on the mother's phenotype, holding genetic sources of variation constant (Kirkpatrick and Lande 1989); by "genetic variation" is meant variation due to gene expression in the offspring. Maternal effects include both maternal inheritance and maternal selection (Kirkpatrick and Lande 1989). Maternal inheritance is defined by the resemblance between the mother and her offspring for measured traits. Maternal selection is defined as a direct influence on offspring fitness as a result of the mother's actions (Kirkpatrick and Lande 1989); thus, maternal selection is emphasized in this review. However, it is usually possible to redefine maternal selection in terms of maternal inheritance. Thus, if the amount of care a mother provides affects the survival of her chick, this could perhaps be recast in terms of a mother's care affecting traits in the chick (such as body size), and these traits affecting chick survival. I adopt this approach here. It has the advantage of focusing on those traits of an offspring that are affected by the mother and that also affect its own fitness. A conclusion from this review is that much remains to be done to understand what these traits are.

Lack (1947, 1948) was the first to consider how parents might influence the quality of their offspring. He viewed optimal clutch size as arising from a trade-off between the number of offspring and the quality of each one and presented some evidence that the optimal clutch size was indeed that which maximized the product of offspring survival and number. There have been two refinements of this basic proposition. First, Lack (1947, 1948, 1954) recognized that such a trade-off might not be observed if parents of different quality were combined. He separated parents into groups according to their date of breeding in the season, based on the argument that latebreeding parents may be in generally lower condition and hence have both smaller clutches and poorer quality offspring. This resulted in evidence for a trade-off between offspring number and quality in some species. The very largest broods produced fewer surviving young than slightly smaller broods, after breeding date was controlled (e.g., the starling *Sturnus vulgaris*, figure 12.1, left). However, in the starling, broods larger than the average still tend to be more productive than broods smaller than the average, and evidence for a trade-off is less compelling in other species (e.g., the great tit, *Parus major*, figure 12.1, right). Lack (1954, p. 29) suggested that the tendency for larger broods to be more productive than smaller broods could be attributed to other aspects of quality differences among individuals, such as food supply on a territory, which are not controlled by a crude division into early and late breeding.

A second modification of Lack's original proposal was that greater parental exhaustion might allow more young to be raised at one time, while reducing the chances of the parents surviving to breed again (paraphrased from Lack 1954, p. 49). Thus, females might balance prospects of future reproductive success against investment in offspring from the current brood, and this could account for the observation that larger broods tend to produce more surviving young than smaller broods. Lack did not develop this further (Charnov and Krebs 1974), perhaps because data on parental survival were unavailable. Manipulations have now shown that although more offspring are recruited from experimentally enlarged broods (Van-

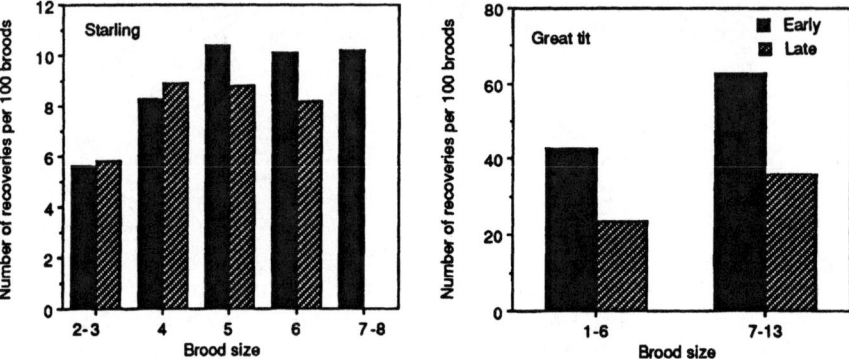

Figure 12.1 The number of young more than 3 months old recovered from broods of different sizes, separated according to whether they hatched early or late in the season (from Lack 1954, pp. 27–29). The number of young recovered is considered representative of survival to age 3 months. Left, starling, *Sturnus vulgaris*, in Switzerland. Right, great tit *Parus major*, in Holland. There is slight evidence for trade-off between offspring quality and offspring number in the starling in that the largest broods produce slightly fewer offspring than intermediate brood sizes.

derWerf 1992), brood enlargement does indeed have deleterious effects on parental survival and/or future reproductive success (Dijkstra et al. 1990a).

Much research over the past 20 years has been aimed at understanding how the various factors affecting parental fitness are resolved in nature, and hence how variation among parents in the quality of their offspring arises and is maintained. Many correlative and experimental studies have measured the costs and benefits of alternative strategies. In addition, theoretical models have been refined. The most elegant study is that on the European kestrel, *Falco tinnunculus*, in Holland. Daan and co-workers (e.g., Daan et al. 1990; Meijer et al. 1990) have measured reproductive costs and benefits in nature and combined them with models for the joint optimization of clutch size, offspring reproductive success, and parental future reproductive success (as described in section 12.4). From measurements of food supply on a territory they were able to predict the optimal clutch size and lay date (which is a main determinant of offspring survival). These predictions agreed reasonably well with their observations (Daan et al. 1990).

In this chapter I first describe the framework used by Daan et al. (1990) and others to investigate how offspring quality varies as a result of parents being selected to maximize their own fitness (section 12.2). Then I review empirical tests of this framework. I consider what offspring traits are maternally affected, and how these traits might affect offspring fitness (sections 12.3–12.5). The review implies that maternal effects are large. In section 12.6 I show how the methods of quantitative genetics can be adapted to provide a quantitative measure of the importance of maternal effects and summarize estimates from studies where all the needed data are available.

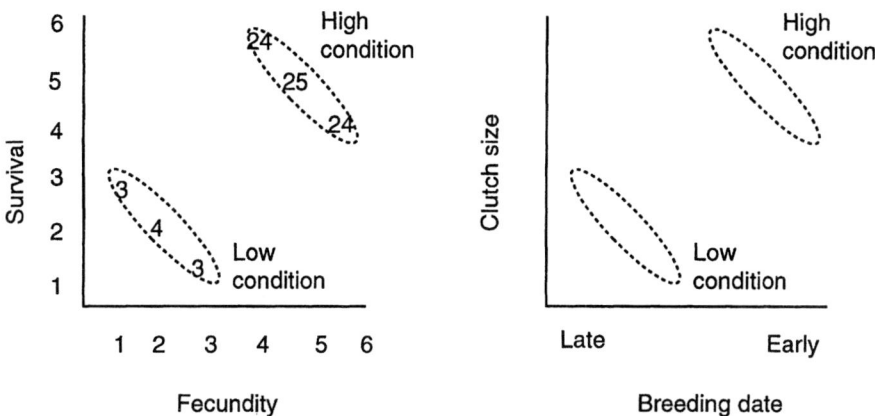

Figure 12.2 Left, associations between probability of survival and fecundity in a population can be partitioned into positive associations, due to the positive effect of condition on both life history components and negative associations due to trade-offs within condition classes (modified from van Noordwijk and de Jong 1986). Numbers along the axes are arbitrary fitness values; ellipses describe the range of individuals for a given condition class, and their fitnesses are obtained as the product of longevity × fecundity. Right, similar principles can be applied to subcomponents of fitness, such as clutch size and laying date (which has a strong influence on offspring quality).

12.2 Background

I assume that the value of an individual's life history trait—such as her fecundity—depends on values of the many underlying traits that make up her phenotype (Price and Schluter 1991; Schluter et al. 1991), as well as the environment. These influences on fecundity can be broadly partitioned into two kinds. First, some traits are rather inflexible, such as the depth of the beak or genetic resistance to disease. These traits lead to trade-offs between life history stages, when the value of the trait that is optimal at one stage differs from that at another stage (Schluter et al. 1991). For example, in a population of Darwin's finches, females with small beaks had higher reproductive success than individuals with large beaks, but lower survival prospects (Price 1984). Second, some traits are phenotypically plastic and respond to changes in the quality of the environment. For example, more food on a territory is reflected in higher body condition and/or increased harvesting rates (Daan et al. 1990). The joint influence of these two kinds of traits can lead to a pattern of positive correlations between life history traits (if the influence of condition is high), or negative correlations between life history traits (if the influence of condition is relatively low). Van Noordwijk and de Jong (1986) discuss this view of life histories in more detail. They partition the two kinds of traits into *acquisition* and *allocation*, with variation in acquisition leading to positive correlations and variation in allocation leading to negative correlations among life history traits. Such a partitioning is depicted for fecundity and longevity in the left panel of figure 12.2 and can be applied to other components of fitness (such as clutch size and offspring quality; figure 12.2, right).

The framework of figure 12.2 has been widely used to study the joint optimization of

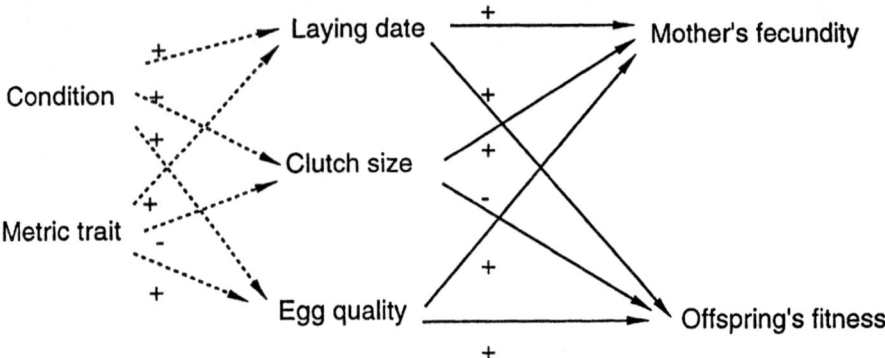

Figure 12.3 Path model showing some influences on female fecundity and offspring fitness at the time the female lays a clutch of eggs. An arbitrary metric trait (Price and Schluter 1993; e.g., beak depth) and condition are depicted as influencing the three reproductive traits, which in turn influence mother fecundity and the offspring's fitness. For clarity, paths from the three reproductive traits to mother survival and mother future reproductive success are omitted, but these paths have also been studied in recent experiments (table 12.1).

parental influence on offspring quality with other parental traits, such as clutch size. One immediate conclusion can be drawn. If trade-offs work more or less additively (because a female has a given amount of resources to invest among different components of fitness), but benefits are more nearly multiplicative (e.g., total fitness is the product of fecundity and survival), then both life history traits should receive relatively similar levels of investment. Thus, a low-condition bird with four units of resources to invest maximizes fitness by placing two in one component and two in the other, whereas a high-condition bird with 10 units to invest maximizes fitness by placing five in one component and five in the other (figure 12.2). This is essentially the argument put forward to explain why clutch size (female fecundity) is negatively correlated with lay date (which has a strong influence on offspring survival): females in high condition will maximize fitness by investing their additional resources partly in offspring quality (through early breeding) and partly in their own fecundity (through clutch size), rather than just one of these components (Drent and Daan 1980). The argument is developed in more depth by Drent and Daan (1980) and Rowe et al. (1994), and in section 12.4.

For simplicity, many models study the trade-off between just two components of fitness, assuming total investment in these components is constrained (e.g., Drent and Daan 1980; Rowe et al. 1994). More detailed models consider simultaneous trade-offs between several components so that, for example, longevity, future reproductive success, clutch size, and off-spring quality are jointly optimized (Daan et al. 1990; Rowe et al. 1994). One example of a more complex model (but that still ignores parental survival or future reproductive success) is shown in figure 12.3. It has been chosen because it summarizes the three main influences on offspring fitness as determined by the mother at time of laying: (1) when to lay in the season, (2) what clutch size to lay, and (3) how much to invest in each egg.

The traits females invest in have important implications for the offspring. From the point of the view of the mother, increases in reproductive investment either in clutch size, in the quality of the egg, or in breeding early increases her reproductive fitness, but from the point of view of the individual offspring this is not the case. The offspring would prefer to come from a large, nutritious egg, and hatch early, but from as small a brood as possible (figure 12.3). The signs of some path coefficients are clear: increasing a mother's clutch will increase her fecundity, all other things being held constant. Yet the magnitude of the paths, and the signs of some of them, are empirically open questions. For example, the effect of a particular metric trait on clutch size will depend on the trait and may be either positive or negative. The paths from female condition to each of clutch size, egg quality, and breeding date are of particular interest, because they represent an evolved strategy. Females of different condition may apportion their investment differently among each of the three traits (e.g., Nilsson 1994). The path diagram includes only effects on immediate fecundity of the mother, and the strength of paths from clutch size, egg quality, and breeding date to the mother's survival and future reproduction are of interest in that, together with their effects on fecundity, they set the relative investment in each.

Considerable effort has gone into measuring the various coefficients illustrated in figure 12.3 (as well as paths to the mother's future reproductive success and survival), and sections 12.3–12.4 provide a summary of the results. Almost all the work has been in the temperate region. I first concentrate on clutch size and timing of breeding. These two traits have been extensively studied and have large effects on offspring fitness. This is followed by a consideration of maternal and paternal influences on offspring fitness that arise after the clutch has been laid. Wherever possible I refer to reviews or references that provide entry points into a vast literature.

12.3   Maternal Condition and Maternal Fecundity

Mothers in high condition should invest more in reproduction. If maternal condition is assumed independent of those metric traits that cause trade-offs, then the simple correlation between the reproductive parameters (clutch size, lay date, egg quality) and condition measures the direct effect of condition on those parameters (figure 12.3). The chief difficulty has been in finding an adequate measure of condition of individuals (Drent and Daan 1980). Condition is generally used in reference to the health and body reserves of an individual, but influences of the external environment, such as the projected availability of food on a territory at the time of rearing offspring (Drent and Daan 1980; Daan et al. 1990), may have a similar effect on reproductive strategies (Daan et al. 1990). The distinction between internal and external energy reserves is rarely made, and the consequences of these alternatives need to be studied.

Some early attempts to measure body condition are summarized by Drent and Daan (1980). A few more recent studies are described here. Murphy (1986) found a correlation between flight muscle mass and egg quality in eastern kingbirds, *Tyrannus tyrannus*. Winkler and Allen (1996) found a correlation between muscle and fat stores and both laying date and clutch size in barn swallows, but the correlation between clutch size and lay date was still present when these measures of condition were statistically held constant; this could perhaps reflect variation in food availability or harvesting rate. Andersson and Gustafsson (1995) have introduced a new method for measuring condition, based on levels of glycosylated hemoglobin in the blood, which holds much promise. In collared flycatchers, *Ficedula albicollis*, this measure correlates with both laying date and clutch size.

Table 12.1 Summaries of experimental studies on reproductive parameters in wild bird populations.

Experimental manipulation	Clutch size	Lay date	Egg size/quality
Supplemental feeding[a]	18/25[b]	21/26[b]	5/8[c]
Feather removal[d]	1/1[e]	1/1[e]	—
Effect on mother[f]	7/8[g]	1/2[h,i]	—
Effect on offspring[j]	28/53[g]	6/6[h,k]	3/9[l]

Number of studies demonstrating an association in the expected direction is reported as a fraction of the total number of studies. In none of the studies was there a significant effect in the direction opposite to that predicted.

[a]The fraction of studies in which supplemental feeding increased the reproductive measure.
[b]Meijer et al. (1990).
[c]Magrath (1992).
[d]The fraction of studies in which feather removal decreased the reproductive measure.
[e]Slagsvold and Lifjeld (1990).
[f]The fraction of studies in which experimental increase of the reproductive trait defined in the column heading decreased mother future reproductive success (and/or experimental decrease of the reproductive trait increased mother future reproductive success).
[g]Dijkstra et al. (1990a).
[h]Verhulst et al. (1995).
[i]Nilsson (1994), Nilsson and Svensson (1996).
[j]As for note f regarding offspring rather than mother fitness, except for egg size, which refers to examples where swaps of eggs between nests have shown to have an influence of egg size on offspring survival.
[k]Norris (1993), Wiggins et al. (1994).
[l]Williams (1994).

Measures of female condition are also positively correlated with egg size and quality (e.g., in six studies reported by Murphy 1986, discussed in section 12.5). In general, where egg quality, laying date, and clutch size are all measured in the same study, it is found that they are positively correlated, and correlated with age and presumably condition: older, higher condition birds breed earlier, with larger clutches and more nutritious (larger) eggs (e.g., Arnold et al. 1991; Perdeck and Cave 1992; Brinkhof et al. 1993). Although most studies imply positive effects of condition on each trait, such effects are not universal, and higher condition great tits (those that breed earlier and have large clutches) tend to lay small eggs (Perrins 1996). This suggests that in this species the effect of condition on egg quality is probably negative, and high-condition birds divert relatively more of their resources to clutch size and early breeding.

Experimental studies where adult condition is increased by supplemental feeding have often found positive effects on clutch size, laying date (supplemental feeding leads to earlier breeding), and egg size and quality (table 12.1). Clutch size has often been increased by supplemental feeding, but rarely significantly. In the two cases where clutch size was significantly increased, laying date was not significantly affected, suggesting a trade-off in apportioning condition between traits (see Winkler and Allen 1996). Slagsvold and Lifjeld (1990) reduced condition by plucking a few feathers while at the same time removing clutches to force females to renest. They found lowered clutch size and later nesting compared to controls.

12.4 Influence of Lay Date and Clutch Size on Fitness

Female condition influences clutch size, lay date, and egg quality. The effects of lay date and clutch size on both the mother's and the offspring's fitness have been widely studied, and often found to be large. In this section I summarize the results for these two traits and our understanding of how they are jointly optimized.

Lay Date

Chicks fledging early in the breeding season typically have high survival and recruitment into the breeding population (Perrins 1970; Daan et al. 1990; Rowe et al. 1994), but experimental manipulations are needed to separate possible direct effects of parental condition on egg quality and parental care, since these may also increase survival. A number of recent studies have attempted to manipulate fledging date independently of condition, using several methods that include inducing females to relay by removing the first clutch (Hatchwell 1991; Verhulst et al. 1995), holding eggs in a refrigerator (Wiggins et al. 1994), exchanging clutches among early and late breeders (Brinkhof et al. 1993; Norris 1993), and manipulating of photoperiod of adults held in captivity (Daan et al. 1990). All have shown that a delay in fledging date has negative effects on offspring survival. The causes of the decline in offspring survival with date are not well understood. They likely include competition from earlier fledged young and increased predation later in the season (e.g., Hatchwell 1991; Norris 1993; Verhulst et al. 1995). However, in many species there is a decline in fledgling weight and structural size through the season (e.g., Perrins 1970; Price 1991; Saino et al. 1997), which may reflect both the lower condition of late-breeding parents (Saino and Moller in press) and the need for parents to divert relatively more of their resources toward overwinter survival as the season draws to a close (Nilsson and Svensson 1996). Effects of parental care on offspring survival are discussed more fully in section 12.5.

There is some evidence that very early laying is detrimental to chick survival. This has occasionally been manipulated independently of condition (Daan et al. 1990; Brinkhof et al. 1993; Norris 1993), and in addition, supplemental feeding experiments have advanced breeding into periods of inclement weather, with detrimental effects on offspring survival (Davies and Lundberg 1985; Clamens and Isenmann 1989; Nilsson 1994).

Three studies have searched for an effect of experimentally manipulated fledging date on parental survival or future reproduction. Verhulst et al. (1995) were unable to detect an effect of delayed fledging date on the parents, but Nilsson and Svensson (1996) showed that in blue tits, *Parus caeruleus*, parents with experimentally delayed lay dates had both reduced overwinter survival and subsequent reduced fecundity. They showed that delayed birds had greater energy expenditure overnight, which they attributed to lower plumage quality arising out of a delayed molt. Nilsson (1994) showed that female blue tits whose lay date was advanced had lower survival than controls, implying that early breeding is costly to parents. This appears to be the only information on the effects of early breeding on parental survival, which is unfortunate because such effects are crucial to models of the joint optimization of lay date and clutch size (Daan et al. 1990).

Effects on parental fitness of early breeding have so far mostly had to be inferred from studies of the energetic costs of searching for food (Daan et al. 1990; Deerenberg et al. 1995; Perrins 1996). Food typically increases during the nesting season and is in short supply early on. Perrins (1970) suggested that tits (*Parus* species) were breeding as early as physiologically possible, which implies that earlier breeding would be physiologically detrimental and hence have mortality or reproductive consequences. Perrins's suggestion seems to have been interpreted as placing an absolute constraint on breeding time (e.g., Rowe et al. 1994; Winkler and Allen 1996), but it is perhaps better treated as a trade-off against survival and future reproduction (Daan et al. 1990). Considerable energy may be expended in egg formation: for a tit Perrins (1996) estimates that 40% of basal metabolic energy is spent every day in forming an egg. In addition, foraging costs for the special requirements of the egg can be high. In particular, calcium may be difficult to acquire, and a tit needs much more calcium for its clutch than it has in its own body. Recently, a decline in snails, and hence calcium in the environment, has been linked to severe reproductive difficulties for some tit populations (Graveland et al. 1994).

Clutch Size

Many experimental manipulations of clutch size have been carried out, and they often demonstrate negative effects of increased clutch size on offspring survival (table 12.1) and, in one of the rare occasions when it could examined, offspring fecundity (Gustafsson and Sutherland 1988; Schluter and Gustafsson 1993). In addition, there are often negative effects of increased clutch size on parental survival and future reproduction (Dijkstra et al. 1990a; see table 12.1).

Joint Optimization of Lay Date and Clutch Size

Early breeding in the season and a large clutch both increase a female's fecundity. Early breeding increases an individual offspring's fitness, but a large clutch reduces it. Controls on the joint determination of clutch size and lay date go a long way toward explaining maternal influences on offspring quality. Both clutch size and lay date have large effects on offspring fitness, and before discussing other sources of maternal variance in offspring fitness I discuss how these two traits are jointly optimized. I consider clutch size and lay date in isolation partly because they may explain much variation in offspring fitness, partly because they have been studied in depth, and partly because the principles have yet to be applied to other maternal influences on offspring fitness.

There is often a striking negative correlation of clutch size with lay date (Klomp 1970; Meijer et al. 1990; Winkler and Allen 1996; see figure 12.4), presumed to reflect the lower condition of later breeding birds. To show how this might arise, I consider the joint optimization of clutch size and lay date in the European kestrel (Daan et al. 1990). Clutch size decreases by an average of 0.037 eggs/day throughout the breeding season (figure 12.4). The analysis is based on >38,000 birds.

Daan et al. (1990) used brood manipulation experiments to examine the effect of brood size on both parental and offspring survival. More young fledged from enlarged broods (with lower weights, but no detectable influence on subsequent survival). Parental survival was reduced. The net result (product of parental survival and fecundity) was that control broods had highest fitness, but there was a considerable range over which the detrimental effects of increased broods on parental survival were nearly balanced by the positive effects on par-

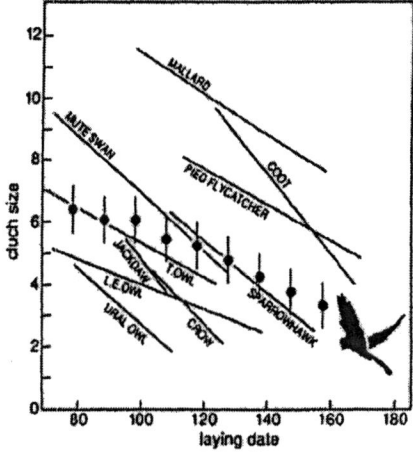

Figure 12.4 Associations of clutch size with lay date for several species of birds (from Meijer et al. 1990, p. 118). Circles ± standard error represent the values for the European kestrel.

ental fecundity (figure 12.5). Daan et al. point out that these are not complete measures of fitness for either the parents or the offspring, but they may be a reasonable approximation.

In the absence of experiments, the costs and benefits of laying on a particular date were assessed by the correlation of date with the reproductive value of both the offspring produced and with their parents. There is a steep decline in offspring fitness with laying date but almost no effect of lay date on the future reproductive success of the parents (figure 12.5). The relationship between offspring survival and lay date is assumed causal and not directly affected by parental quality or strategy. Food availability increases through the season, so there should be an optimal date where the benefits of waiting (in terms of less work needed to harvest the same quantity of food) are countered by the costs of waiting (reduced juvenile survival), just as there is an optimal clutch size. The costs and benefits of waiting given a measured changing food resource were worked out by Daan et al. (1990) using the results from the clutch manipulation experiment.

Optimization of laying date and clutch size for an individual bird thus depends on the balance between adult survival (decreased by large clutch and early breeding), juvenile survival (decreased by large clutch and late breeding), and adult fecundity (increased by large clutch size). In fact, the balance appears to result in a rather large number of clutches and lay dates with more or less equivalent fitnesses for a given mother.

On poor territories, where food is less available, adults will be selected to delay their breeding and breed with smaller clutches, as compared with good territories where there is more food. The combined influence of the benefits to the adult of delaying breeding (increased clutch size) and the costs (decreased juvenile survival) will mean that adults apportion their effort among both clutch size and lay date, and hence there will be a decline in clutch size with lay date (figure 12.6; Daan et al. 1990; see also Drent and Daan 1980; Rowe et al. 1994). Measurements of clutch size and lay date on territories of different quality agree well with the predictions.

There are two important conclusions from Daan et al.'s (1990) study. First, the path from parental condition to lay date is large. Variation among parents in the quality of their territory strongly affects breeding date and hence offspring survival. Second, even among individuals of the same condition, the trade-off between offspring quality and number results in a

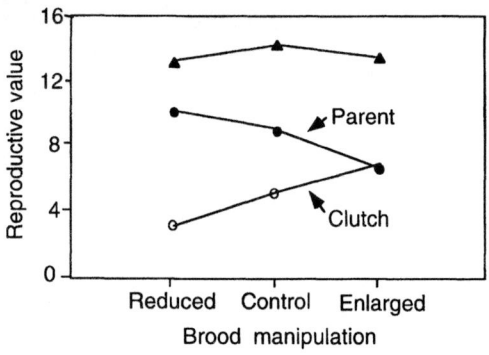

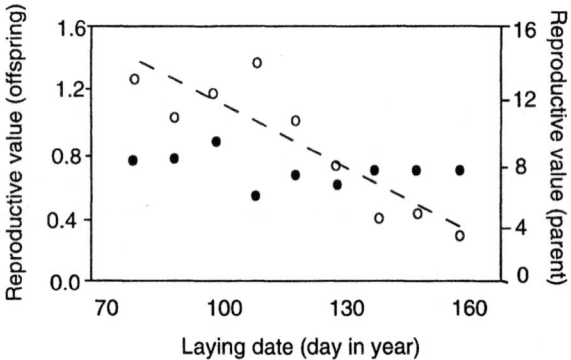

Figure 12.5 Top, reproductive value of parents and the brood they raise (both estimated as a product of survival and reproductive success) from experimentally reduced, control, and enlarged broods. Net fitness across all three classes is very similar (top curve), with a slight peak in the control class. Bottom, reproductive value of a single parent (●) and a single offspring (O) as a function of lay date. Reproductive value declines for offspring, but hardly varies for parents (from Daan et al. 1990).

wide range of clutches with approximately equivalent parental fitnesses. From the point of view of parental effects, this means there can be large variation among parents in their influences on offspring fitness, with few detrimental effects on the parent's own fitness. The summed effect of variation in condition and of the trade-off within condition classes implies that parental effects are large. These conclusions are similar to those found in a quantitative genetic study on a passerine bird, the collared flycatcher (Schluter and Gustafsson 1993; summarized in section 12.6), implying that large parental effects are widespread.

12.5 Additional Parental Influences on Offspring

In this section I review other effects of parents on offspring besides those acting through clutch size and lay date. Some of these affect the whole brood (as do clutch size and lay date), whereas others are directed toward individual offspring.

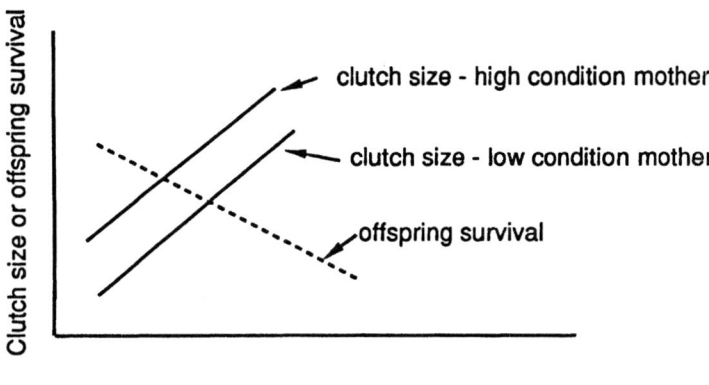

Laying date: early -> late

Figure 12.6 Explanations for the decline in clutch size with lay date. The figure shows the clutch size a given female can produce as a function of lay date (continuous lines), for the same cost. The clutch size increases because condition and/or resources increase through the season. Individuals in low condition are assumed to accumulate the resources needed for eggs less quickly than individuals in high condition. Juvenile survival (dashed line) is written in terms of units of fitness to the mother, scaled to be equivalent to the clutch size measure of fitness. If juvenile survival declines through the season, individuals in low condition are expected to breed later, and with smaller clutch sizes. The optima for individuals in different condition are at the points where the lines intersect (modified from Drent and Daan 1980; Rowe et al. 1994; see also figure 12.2).

Egg Size and Quality

It is thought that influences of egg size on offspring fitness are less strong than those of lay date (Perrins 1991). Nevertheless, egg size can vary greatly in a population (e.g., some eggs are 70% larger than others in song sparrows, *Melospiza melodia*; Arcese and Smith 1988). Larger eggs may contain more water, protein, or lipid, all of which should benefit the hatchling. Lipids and water seem to account for most of the differences (Williams 1994).

Some studies have demonstrated an association between egg quality and offspring survival over the first weeks after hatching, after controlling for influences of parental feeding and chick brooding by swapping eggs or recently hatched young between nests (table 12.1). Williams (1994, p. 42) suggests that the effect of egg size on offspring survival may be underestimated because mortality that occurs soon after hatch is not examined. In grouse and ptarmigan, chick survival from eggs taken into captivity was correlated with feeding conditions of the mother and hence inferred condition of the mother (Moss and Watson 1984; Moss et al. 1984). In contrast to studies on clutch size and lay date, no one appears to have experimentally investigated the effects of egg size on offspring survival in later life (e.g., after fledging).

In addition to nutrients, females differ in the quantity of antibodies and hormones they transfer into the egg. In the gold and blue macaw, *Ara ararauna*, maternal antibodies persist in the chick until at least 6 weeks of age; the decline after that is correlated with the ability of chicks to develop their own immune responses (Lung et al. 1996). Experimental studies in chickens have shown that maternal transfer of antibodies provides the chick with resistance related to infections to which the mother has been exposed (Smith et al. 1994).

Females transfer estradiol and testosterone to eggs (Schwabl 1993; Adkins-Regan et al. 1995). In a study on Japanese quail, *Coturnix japonica*, injections of estradiol into the mother permanently altered the morphology of the oviducts in their female offspring (Adkins-Regan et al. 1995). In canaries, *Serinus canaria*, testosterone levels in the egg correlated with dominance behavior in later life (Schwabl 1993). Schwabl (1996) injected testosterone into eggs and showed that it increased chick begging behavior. Since parents feed their young by responding to the one that most actively begs (although this is not true of all species; see Ricklefs 1993), high testosterone levels result in high chick growth rates. Within clutches, Schwabl (1993) demonstrated that testosterone levels are up to three times higher in the last laid egg than the first. He interpreted this as an adaptation to compensate for the disadvantage that later-hatching chicks experience in competition with their earlier-hatching nest mates (Schwabl 1993, 1996).

It is often found that the last egg is consistently either larger or smaller than the average size of the clutch (Ojanen et al. 1981), and this has been given an adaptive interpretation by Slagsvold et al. (1984). They suggest that a small last egg allows rapid elimination of small chicks in times of food scarcity, whereas a large last egg prevents accidental elimination of a chick through sibling competition, using a similar argument to that adopted by Schwabl (see also Magrath 1992). Variation in egg size is also affected by feeding conditions (e.g., Nilsson and Svennson 1993, who showed that relatively small first eggs were not present among females given supplemental food) or temperature (Jarvinen and Ylimaunu 1986; Nager and Zandt 1994), and this suggests that the future reproductive success of the parent affects egg size. Alternatively, direct selection on current fecundity may affect egg size. A fascinating example comes from those species that are subject to parasitism by cuckoos. Some species are able to reject a cuckoo's egg apparently by comparing its size to that of the size of the mother's own eggs, and these species seem to have low within-clutch egg variation (Marchetti 1992).

Incubation

An important means of varying quality among individual chicks within a clutch is to stagger hatching time, by beginning incubation before the clutch is complete. There is a vast literature on the adaptive significance of the ensuing hatching asynchrony. Lack's original hypothesis (see Magrath 1990) was that it enables easy elimination of latehatched chicks during times of food shortage, and this has found strong support in some experimental studies (Magrath 1988) but weaker support in others (e.g., Amundsen and Slagsvold 1991).

Incubation periods vary significantly among females within a species (Ricklefs and Smeraski 1983; Price and Jamdar 1991). Ricklefs and Smeraski (1983) showed by using a cross-fostering experiment on starlings that this is almost entirely a property of the foster mother, although interestingly, there is also a slight effect of the egg. Incubation periods have obvious effects on a mother's reproductive fitness. For example, Price and Jamdar (1991) demonstrated a correlation of incubation period with ambient temperature among yellow-browed leaf warblers, *Phylloscopus humei*, with mothers not incubating the clutch at all on some cold days. Short incubation periods have many advantages for the mother's survival and fecundity, including reduced predation on the whole clutch, but some species have exceptionally long incubation periods, implying that some advantages accrue (Ricklefs 1993).

Ricklefs suggests that these are advantages to the offspring, in that it may allow them to develop their own immune system before being exposed to the environment. In support of this suggestion, he shows that incubation period and parasite load are negatively correlated across species (Ricklefs 1992).

Parental Care

In many species the young are entirely dependent on the parents for food when they first hatch, giving ample opportunity for parental effects. Clutch swap experiments often demonstrate an effect of parental care on offspring growth rate (Bolton 1991; Smith and Wettermark 1995). However, heritability estimates of skeletal traits following cross-fostering of chicks are consistently as high as those from nonexperimental nests, implying little lasting effect of parental care on these traits (Smith and Dhondt 1980; Alatalo and Lundberg 1986; Wiggins 1989; Smith 1993). Under good conditions it appears that any differences among chicks (due to late hatching or to hatching from small eggs) are largely eliminated in the nest, with smaller chicks catching up to their larger siblings (Schifferli 1973; Price and Grant 1985). Under poor conditions parents bring less food to the nest, and then there is often a chick × environment interaction, with the larger chicks suffering relatively more than the smaller chicks (van Noordwijk et al. 1988; Price 1991). Undernourished young may be selectively eliminated either before or soon after fledging (Alatalo and Lundberg 1986; van Noordwijk et al. 1988).

For many temperate birds there is a decline in tarsus length and mass of chicks (measured when the chicks are about to leave the nest) through the season (Perrins 1970; Price 1991; Saino and Moller in press) attributable to lower parental provisioning by later breeding birds. It has been shown in the barn swallow, *Hirundo rustica*, that, in addition, there is a decline in immunocompetence (Saino and Moller in press). Both a clutch swap experiment and direct observations of feeding rates have established that much of the variation in immunocompetence is attributable to variation in food provisioning by the parents.

There is direct evidence that parental care affects the fitness of offspring from brood enlargement studies. Increased broods often reduce survival of offspring (e.g., table 12.1). Recently a few studies have begun to investigate longer term effects of parental care on daughter fecundity. In the collared flycatcher there is a large effect of parental care on the clutch size of the female offspring (Gustafsson and Sutherland 1988, discussed below), and in the great tit growth rate of a female chick in the nest is correlated with her clutch size the next year (Haywood and Perrins 1992).

There are also lasting effects of parental care on reproductive success of male offspring. Gustafsson et al. (1995) report that in the collared flycatcher, a male secondary sexual trait, the size of the white patch on the forehead, is affected by nestling conditions. Males coming from experimentally enlarged broods tend to have smaller patches, whereas males from experimentally reduced broods have larger patches. Patch size correlates with mating success, in particular, the chances of being polygamous. T. Price, K. Marchetti, A. Suarez and S. Bensch (unpublished data) found that the size of the wing patch in offspring leaf warblers, *Phylloscopus humei*, is strongly affected by environmental conditions, with smaller patches being produced late in the season and in years when conditions are generally poor. The size of the wing patch has been experimentally demonstrated to affect mating success (Marchetti 1993). In this case, however, it is possible that patch size is determined by nutritional content of the egg rather than parental care, because the correlation of patch size with the mother is higher than that with the father (T. Price, K. Marchetti, A. Suarez, and S. Bensch, unpublished observations), and feather patterns are laid down early in development.

Cultural Inheritance

Lasting maternal effects on gosling size have been demonstrated in the barnacle goose, *Branta leucopsis* (Larsson and Forslund 1992). In this species parental body size correlates with egg size (and this probably correlates with hatchling size), but egg size does not correlate with final size, when parental body size is controlled for. Instead, Larsson and Forslund argue that the maternal effect is the result of cultural transmission of grazing sites, with some sites providing more nutritious food than others.

In many communally breeding species, young birds often remain on their natal territory and eventually attain breeding status. In the acorn woodpecker, *Melanerpes formicivorus*, territory quality varies substantially because of variation in the number of stored acorns (Koenig and Stacey 1990). The acorns are used as a resource throughout the year, and stocks are maintained by the whole group; an important factor limiting the size of the store is the availability of suitable storage sites. Offspring that hatch in good territories have a high chance of inheriting those territories, and it appears that it can be to their advantage to remain as a nonbreeder on the territory for more than 2 years, rather than dispersing to a poorer quality territory and breeding earlier. In the Florida scrub jay, *Aphelocoma coerulescens*, most males that establish new territories do so by expanding the area of their natal territory with the assistance of parents and other family members (Woolfenden and Fitzpatrick 1990, p. 247). Once a territory is acquired, a female from outside the group joins him, and his former group members retreat to some or all of the original territory.

Apart from song, the evidence for cultural transmission of behavior in birds is weak, and evidence for parent-offspring transmission even weaker. Most cultural transmission of song appears to be lateral, especially between neighbors, and there have been a number of adaptive explanations for why song should be learned in this manner (Kroodsma and Miller 1996). Darwin's ground finches, *Geospiza* spp., provide the best example of a strong father-son transmission of song. In three species there is a correlation between characteristics of the father's song and that of his son (Grant 1984; Gibbs 1990; Grant and Grant 1996), although occasionally males seem to learn the song of another male, and more rarely that of another species. The fitness consequences of this have been examined. First, there is a weak tendency for females to avoid mating with males that sing the same song as their fathers, and hence this might provide a mechanism to avoid inbreeding (Grant 1984; Grant and Grant 1996). Second, misimprinting on other species' songs sometimes results in hybridization; learning from the father will prevent this (Grant and Grant 1996). The ease with which other species' songs are learned in nature seems to be the main difference between Darwin's finches and other species and may account for paternal transmission of song in this group.

12.6 Estimates of the Strength of Maternal Effects in Birds

The review in sections 12.3–12.5 implies that maternal and paternal effects are likely to be large. In some studies the magnitude of these effects can be quantified, and I discuss the methods and estimates here. Kirkpatrick and Lande (1989) introduced a matrix of maternal effect coefficients, which are needed to model the change in the mean value of a trait from one generation to the next. The relationship between mother and offspring is inherently multivariate, so, for example, selection on a mother's condition may lead to a change in the daughters' clutch size. However, the relationship is often studied univariately, for example, as the relationship between a mother's clutch size and her daughters' clutch size, ignoring the contributions of other maternal traits (such as condition) to clutch size. The univariate description can be a misleading guide to the process of evolution under selection, and a complete matrix of partial regression coefficients is preferable, describing the effect of each trait on each other, when all other sources are held constant (Lande and Price 1989). This is usually not possible to obtain, even in theory (because there are more coefficients than there are independent ways of estimating them). Instead, a few associations are estimated, and it is hoped they summarize the important patterns in the data. Experimental approaches of the kind listed in table 12.1 are probably the best way to estimate maternal effects, but many traits are difficult to manipulate experimentally.

While Kirkpatrick and Lande's (1989) formulation is designed specifically to model evolution given selection, these same coefficients can be used to gauge the strength of maternal influences when compared with other factors affecting the phenotype (genes, random environmental effects). In the univariate case, a partial regression coefficient, m, measures the effect of the mother's phenotype on the offspring when genetic transmission is held constant (m is termed the maternal effect coefficient; Kirkpatrick and Lande 1989). If the trait being studied is the same in mother and offspring (e.g., adult body size), m is equivalent to a partial correlation coefficient, and it is bounded by ± 1.0; m^2 then measures the fraction of the variance in the offspring attributable to variance in the mother: large absolute values of m^2 imply large maternal effects. When the offspring and mother trait differ, partial regressions are not bounded in this way and need to be standardized before they can be used as an indicator of the relative importance of maternal effects in affecting the offspring phenotype.

If we assume that a single maternal trait affects a single offspring trait and that paternal effects are absent, m can be estimated by subtracting the regression of offspring on father from that of offspring on mother (Lande and Price 1989). Maternal effects have most often been estimated in this way for body size. It is often found that there is a strong maternal effect on young chicks just after they hatch. For example, $m \cong 0.6$ for the influence of various measures of adult structural size on the homologous measures of chick size at hatching in Darwin's medium ground finch, and $m \cong 0.3$ for a similar measure in the great tit (Lande and Price 1989). This is apparently a consequence of egg size determining hatchling size (Ricklefs et al. 1978; Williams 1994). Egg size varies considerably among broods, with up to 98% of the variance being between different mothers (Vaisanen et al. 1972; Ojanen et al. 1981; Grant 1982; Ricklefs 1984), and egg size is correlated with the mother's body size in many species (Williams 1994).

The maternal effect on chick size rapidly disappears, and lasting maternal effects on adult morphology may be rare (chicken literature is reviewed in Schifferli 1973; Price and Grant 1985). Direct demonstration of this in nature suffers from the difficulty that fathers are not always known. A lower regression of offspring on putative father than on mother is a very common observation, but this has usually been attributed to the effects of extrapair copulations (Moller and Birkhead 1992). In some studies extrapair copulations have been shown to be absent (using DNA fingerprinting; Hasselquist et al. 1995) or thought to be unimportant

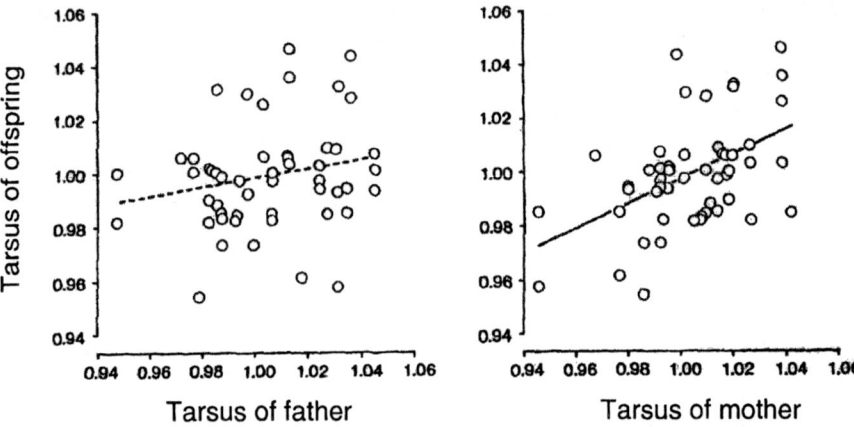

Figure 12.7 Regression of tarsus length of offspring on father and on mother in a natural population of great reed warblers, *Acrocephalus arundinaceus* (Hasselquist et al. 1995). The regression on fathers is $b = 0.17 \pm 0.12$ ($P > 0.1$), and on mothers is $b = 0.47 \pm 0.12$ ($P < 0.001$). The difference implies a maternal effect of $m = 0.3$.

(Larsson and Forslund 1992), and in these cases differences between regressions on mother and on father are more convincingly attributed to a maternal effect. For the great reed warbler, *Acrocephalus arundinaceus*, $m = 0.3$ for the effect of mother tarsus length on full-grown offspring tarsus length (Hasselquist et al. 1995; figure 12.7), and for a similar measure in the barnacle goose, *Branta leucopsis*, $m = 0.3$ (Larsson and Forslund 1992). In both these studies, regressions of offspring on mother are significantly higher than the corresponding regression of offspring on father (using one-tailed tests). Because the same trait is being measured in the parents and offspring in these studies, the estimates imply that at least 9% of the variance in full-grown tarsus length is attributable to direct maternal influences (the estimate is a lower bound, because it assumes an absence of paternal effects, which are likely to be present).

Many maternally influenced traits are sex limited in their expression, making comparisons between regression of offspring on mother and father impossible. Lande and Price (1989) suggested one way around this. If trait X (e.g., body size) in the mother is thought to affect the expression of trait Y (e.g., tarsus length) in the offspring, but trait Y in the parent is not thought to affect expression of trait X in the offspring, comparisons of the two correlations (X in parent, Y in offspring; Y in parent, X in offspring) can potentially be used to estimate the maternal effect. This has never been done, because of a reluctance to make the necessary assumptions, although Schluter and Gustafsson (1993) used a similar approach (see below).

Another method to estimate maternal effects for sex-limited traits is to compare regression of granddaughter on paternal grandmother with the regression of granddaughter on maternal grandmother. An association of granddaughter with paternal grandmother is attributed to genetic transmission, whereas an association with maternal grandmother is attributed to both genetic and maternal transmission. Van Noordwijk et al. (1981) used this method in a study of clutch size. The regression of granddaughter clutch size on maternal (0.16 ± 0.07 SE) and paternal grandmother's clutch size (0.12 ± 0.07 SE) differed slightly, although standard errors are large. Comparing the two regressions does depend on there being no cultural

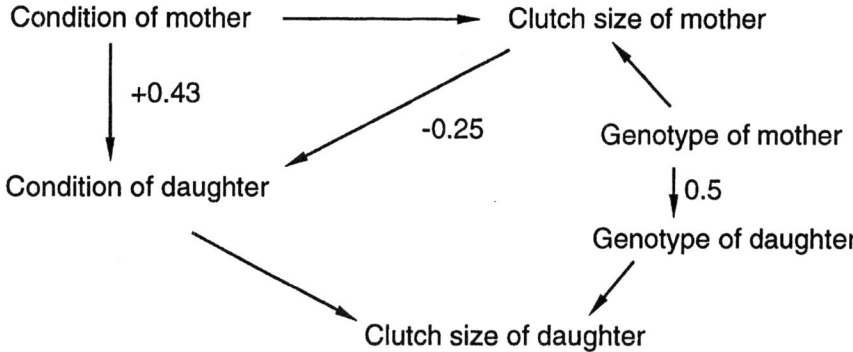

Figure 12.8 Path model for the main effects assumed by Schluter and Gustafsson (1993) to affect resemblance between mothers and daughters for clutch size. Unlabeled path coefficients are set to 1.0, with the variance in condition and genotype adjusted accordingly. The direct effect of mother's clutch size on daughter's clutch size (and hence condition) was estimated by experiment. The path from genotype to mother is not simply related to heritability, because when maternal effects are present, genetic and maternal influences become correlated (Kirkpatrick and Lande 1989). Nevertheless, the two remaining parameters (heritability and maternal influences of condition) can be estimated through comparisons of mother with daughters of different ages, given some assumptions.

transmission through the male line, and van Noordwijk et al. (1981) searched for influences of territory quality by comparing close neighbors and found none. The difference between the two regressions implies a small maternal effect coefficient of about $m = 0.06$ (using equations modified from Lande and Price 1989). This is somewhat low and inconsistent with the results of experimental studies (table 12.1). If the low value is real, and not a result of sampling, a possible explanation is that maternal influences on clutch size include influences both of maternal condition and of maternal clutch size per se, and these effects tend to cancel (Schluter and Gustafsson 1993). If this is the case, it demonstrates that the univariate approach can be misleading.

Grant and Grant (1996) used regressions of grandson on grandfather to examine paternal transmission of song characteristics in Darwin's medium ground finch, *Geospiza fortis* (using a summary statistic of five measurements taken from sonograms). They found a regression of $b = 0.54 \pm 0.08$ ($P < 0.001$) for the regression of grandson on paternal grandfather, but $b = -0.04 \pm 0.10$ for the regression on maternal grandfather. The latter regression implies an absence of genetic variance, and the paternal effect is thus estimated to be $p = 0.725$ (the square root of the regression on paternal grandfather) or 0.723 (from the regression of son on father).

The most comprehensive attempt to estimate the strength of maternal transmission in birds is that of Schluter and Gustafsson (1993) working with the collared flycatcher. They were solely concerned with the effects of maternal condition and clutch size on offspring clutch size. Clutch size of the daughter was defined as the sum of the effects of the daughter's condition and her genotype. It was assumed that there is no viability selection on condition of daughters (Gustafsson and Sutherland 1988) and no genetic variance in condition. To be able to estimate parameters, Schluter and Gustafsson further assumed that the condition of a daughter when laying a clutch depends on (1) the condition of the mother, (2) the size of the mother's clutch from which the daughter came, and (3) a random environmental effect. With these assumptions there are three parameters to estimate: the effect of mother's condition on daughter's condition (expected to be positive), the effect of mother's clutch size on daughter's condition (expected to be negative), and the effect of daughter's clutch size genotype (figure 12.8). Schluter and Gustafsson directly estimated the effect of mother's clutch size on daughter clutch size by adding and subtracting very small chicks from nests. They found that for every young added, clutch size of the daughter was reduced by 0.25 of an egg.

Two other variables remain to be estimated (heritability and direct effect of mother's condition), and the appropriate experiment to tease these apart has not been done. Instead Schluter and Gustafsson (1993) compared the resemblance between mothers and daughters breeding for the first time (i.e., when they were 1 year old), with the resemblance between mothers and daughters breeding for the second time (i.e., when they were 2 years old). Their reasoning was that maternal effects might have a strong effect on first-time breeders, but the strongest effect on second breeders was likely to be their previous reproductive effort. Thus, maternal effects are assumed to be absent in the second year, and a direct comparison of the two associations allows an estimate of both maternal transmission and the heritability of clutch size. By this means they obtained estimates of $m = 0.43$ for the effects of mother condition on daughter condition and a heritability of 0.33 for clutch size (figure 12.8). Transmission of condition from mother to daughter was also estimated more directly, where condition is defined as body mass corrected for skeletal size (tarsus length). The regression of daughter condition index on mother condition index is 0.26, which, in the assumed absence of genetic variance for condition, implies a maternal effect coefficient of 0.26.

A number of assumptions are always needed if one is to estimate maternal effects from resemblances among relatives, because of the potentially large number of influences of parents on their offspring (Lande and Price 1989), and it is necessary to find the equivalent number of relationships between relatives, where each can be used to estimate a different effect. Schluter and Gustafsson (1993) make several assumptions, but their results indicate a large effect of mother's condition on daughter condition: an estimated 34% of the variation in female condition (as it affects their clutch size) is attributable to maternal influences of condition or clutch size, and >25% of the variance in daughter clutch size is attributable to maternal effects. These estimates, although high, are consistent with other tests they employ and with the suggestion of large maternal influences on condition that have come from the many studies reviewed here. Their results also suggest that variance in condition among females accounts for about 80% of the maternal influences on variance in condition among daughters, whereas the trade-off between large clutches and high-quality offspring accounts for about 20% of the maternal influences on daughter condition (calculated from information in Schluter and Gustafsson 1993).

Conclusions

In this chapter I used the term "maternal effects" to describe the influence of the mother, or of both parents, without carefully distinguishing the two. Although the effects of the mother on offspring are expected to be larger than that of the father, there are many ways in which males directly influence the phenotype of their offspring. These need not occur only after the eggs have been laid. First, in many species males provide much food to the female during the time of egg formation (Nisbet 1973; Royama 1976), and in some species, such as the European kestrel, they provide all of the food (Daan et al. 1990). Second, it appears that females adjust their reproductive effort depending on the male they are mated with, and invest more when mated with preferred males (Burley 1988; De Lope and Moller 1993; Petrie 1993). A good example of male influences operating through the female comes from a study of sex ratio variation. It has been established in several species that females can vary the sex ratio of their brood. In some cases this correlates with laying date (e.g., Dijkstra et al. 1990b). However, Svensson and Nilsson (1996) show that in blue tits, females produce more male biased broods when they are mated to high-quality males. Among broods where the father survived, the proportion of males was 63%, whereas among broods where the father died the proportion of males was 50%.

The chief conclusion of this chapter is that parental effects on offspring are large and that they can last at least through the first year of life, and sometimes longer (Hasselquist et al. 1995; Grant and Grant 1996). Influences of parents are particularly strong on a young bird's condition and hence survival and/or reproduction. Lasting effects of poor rearing conditions on adult structural size have been demonstrated only rarely (Boag 1987), perhaps because offspring with stunted growth survive poorly in nature (van Noordwijk et al. 1988), but there are a few good examples where structural size of adults is affected by a maternal effect (Hasselquist et al. 1995).

Most parentally influenced traits are condition dependent. The best example where this is not the case is song, although other examples of imprinting, particularly as they may affect optimal levels of outbreeding, remain to be analyzed in detail (Immelman 1975; ten Cate and Bateson 1988). In the case of song, there is much evidence for cultural learning, but this is often from neighbors rather than parents (Kroodsma and Miller 1996). The main exception is in Darwin's finches, where substantial variation in songs within species (often as large as that between species) seems to generate a selection pressure to learn from parents and hence avoid misimprinting on heterospecific song (Grant and Grant 1996).

Identifying those offspring traits that are condition dependent and understanding how they affect offspring fitness are important tasks for future research. The word "condition" has been used loosely in this review and elsewhere, but might be defined in terms of those traits that are phenotypically plastic and that directly affect an individual's fitness. For example, Saino and Moller (1996) ask what this condition effect is and suggest that it has much to do with disease resistance in the form of the immune response (Saino et al. 1997). Gustafsson et al. (1995) demonstrate that a male secondary sexual trait is parentally influenced and affects mating success. The next step is to understand costs of the trait expression, how these costs are increased in low-condition birds, and the chain of causation from the materials supplied by the parents to the fitness of the offspring.

It is apparent that there is much variation in condition among individuals in nature. It is also apparent that investment in individual offspring is an evolved condition-dependent strategy on the part of the parents. Such parental effects in birds are adaptive, in that they are a consequence of individual optimization of reproductive fitness. However, much of the variance in maternal effects depends on there being variance in condition among parents, and the remainder on the trade-off between investment in individual offspring and other components of fitness. Maternal effects are adaptations, but it should be remembered that they are adaptations for parents and not the offspring.

Acknowledgments I thank B. Galef, R. Grant, P. Grant, M. Katti, T. Langen, L. Molles, J.-A. Nilsson, and A. Suarez for discussion or comments.

References

Adkins-Regan, E., M. A. Ottinger, and J. Park. 1995. Maternal transfer of estradiol to egg yolks alters sexual differentiation of avian offspring. J. Exp. Zool. 271:466–470.

Alatalo, R. V., and A. Lundberg. 1986. Heritability and selection on tarsus length in the pied flycatcher (*Ficedula hypoleuca*). Evolution 40:574–583.

Amundsen, T., and T. Slagsvold. 1991. Asynchronous hatching in the pied flycatcher: An experiment. Ecology 72:797–804.

Andersson, M., and L. Gustafsson. 1995. Glycosylated haemoglobin—a new measure of condition in birds. Proc. R. Soc. Lond. (ser. B) 260:299–303.

Arcese, P., and J. N. M. Smith. 1988. Effects of population density and supplemental food on reproduction in song sparrows. J. Anim. Ecol. 57:119–136.

Arnold, T. W., R. T. Alisauskus, and C. D. Ankney. 1991. Egg composition of American coots in relation to habitat, year, laying date, clutch size and supplemental feeding. Auk 108:532–547.

Boag, P. T. 1987. Effects of nestling diet on growth and adult size of zebra finches (*Poephila guttata*). Auk 104:155–166.

Bolton, M. 1991. Determinants of chick survival in the lesser black-backed gull: Relative contributions of egg size and parental quality. J. Anim. Ecol. 60:949–960.

Brinkhof, M., A. Cave, F. Hage, and S. Verhulst. 1993. Timing of reproduction and fledging success in the coot *Fulica atra*: Evidence for a causal relationship. J. Anim. Ecol. 62:577–587.

Burley, N. 1988. The differential allocation hypothesis: An experimental test. Am. Nat. 132:611–628.

Charnov, E., and J. R. Krebs. 1974. On clutch size and fitness. Ibis 116:217–219.

Clamens, A., and P. Isenmann. 1989. Effect of supplemental food on the breeding of blue and great tits in Mediterranean habitats. Ornis Scand. 20:36–42.

Daan, S., C. Dijkstra, and J. Tinbergen. 1990. Family planning in the kestrel (*Falco tinnunculus*)—the ultimate control of covariation of laying date and clutch size. Behaviour 114:83–116.

Davies, N. B., and A. Lundberg. 1985. The influence of food on the time budgets and timing of breeding of the dunnock *Prunella modularis*. Ibis 127:100–110.

Deerenberg, C., I. Pen, C. Dijkstra, B.-J. Arkies, G. H. Visser, and S. Daan. 1995. Parental energy expenditure in relation to manipulated brood size in the European kestrel *Falco tinnunculus*. Zool. Anal. Compl. Syst. 99:39–48.

De Lope, F., and A. P. Moller. 1993. Female reproductive effort depends on the degree of ornamentation of their mates. Evolution 47:1152–1160.

Dijkstra, C., A. Bult, S. Bilsma, S. Daan, T. Meijer, and M. Zilstra. 1990a. Brood size manipulations in the kestrel (*Falco tinnunculus*): Effects on offspring and parent survival. J. Anim. Ecol. 59:269–286.

Dijkstra, C., S. Daan, and J. Buker. 1990b. Adaptive seasonal variation in the sex ratio of kestrel broods. Funct. Ecol. 4:143–147.

Drent, R., and H. Daan. 1980. The prudent parent: Energetic adjustments in avian breeding. Ardea 68:225–252.

Gibbs, H. L. 1990. Cultural evolution of male song types in Darwin's medium ground finches, *Geospiza fortis*. Anim. Behav. 39:253–263.

Grant, P. R. 1982. Variation in the size and shape of Darwin's finch eggs. Auk 99:15–23.

Grant, B. R. 1984. The significance of song variation in a population of Darwin's finches. Behaviour 89:90–116.

Grant, B. R., and P. R. Grant. 1996. Cultural inheritance of song and its role in the evolution of Darwin's finches. Evolution 50:2471–2487.

Graveland, J., R. van der Waal, J. van Balen, and A. van Noordwijk. 1994. Poor reproduction in forest passerines from decline of snail abundance on acidified soils. Nature 368:446–448.

Gustafsson, L., and W. Sutherland. 1988. The costs of reproduction in the collared flycatcher *Ficedula albicollis*. Nature 355:813–815.

Gustafsson, L., A. Qvarnstrŝm, and B. Sheldon. 1995. Trade-offs between life-history traits and a secondary sexual character in male collared flycatchers. Nature 375:311–313.

Hasselquist, D., S. Bensch, and T. von Schantz. 1995. Estimating cuckoldry in birds: The heritability method and DNA fingerprinting give different results. Oikos 72:173–178.

Hatchwell, B. J. 1991. An experimental study of the effects of timing of breeding on the reproductive success of common guillemots (*Uria aalge*). J. Anim. Ecol. 60:721–736.

Haywood, S., and C. M. Perrins. 1992. Is clutch size in birds affected by environmental conditions during growth? Proc. R. Soc. Lond. (ser.)B 249:195–197.

Immelman, K. 1975. Ecological significance of imprinting and early learning. Annu. Rev. Ecol. Syst. 6:15–37.

Jarvinen, A., and J. Ylimaunu. 1986. Intraclutch egg-size variation in birds: Physiological responses of individuals to fluctuations in environmental conditions. Auk 103:235–237.

Kirkpatrick, M., and R. Lande. 1989. The evolution of maternal characters. Evolution 43:485–503 (errata 46:284, 1992).

Klomp, H. 1970. The determination of clutch size in birds. Ardea 58:1–124.

Koenig, W., and P. Stacey. 1990. Acorn woodpeckers: group-living and food storage under contrasting ecological conditions. Pp. 413–454 in P. Stacey and W. Koenig (eds.), Cooperative Breeding in Birds. Cambridge University Press, Cambridge.

Kroodsma, D., and E. Miller, (eds.). 1996. Ecology and Evolution of Acoustic Communication in Birds. Cornell Press, Ithaca, N.Y.

Lack, D. 1947. The significance of clutch size in birds. Part 1. Ibis 89:302–352.

Lack, D. 1948. The significance of clutch size in birds. Part 2. Ibis 90:25–45.

Lack, D. 1954. The natural regulation of animal numbers. Clarendon Press, Oxford.

Lande, R., and T. Price. 1989. Genetic correlations and maternal effect coefficients obtained from offspring-parent regression. Genetics 122:915–922.

Larsson, K., and P. Forslund. 1992. Genetic and social inheritance of body and egg size in the Barnacle goose (*Branta leucopsis*). Evolution 46:235–244.

Lung, N. P., J. P. Thompson, G. V. Kollias, J. H. Olsen, J. A. Zdziarski, and P. A. Klein. 1996. Maternal transfer of IgG antibodies and development of IgG antibody responses by blue and gold macaw chicks (*Ara ararauna*). Am. J. Vet. Res. 57:1157–1161.

Magrath, R. D. 1988. Hatching asynchrony and reproductive success in the blackbird. Nature 339:536–538.

Magrath, R. D. 1990. Hatching asynchrony in altricial birds. Biol. Rev. 65:587–622.

Magrath, R. D. 1992. Seasonal changes in egg mass within and among clutches of birds: General explanations and a field study of the blackbird (*Turdus merula*). Ibis 134:171–179.

Marchetti, K. 1992. Costs to host defence and the persistence of parasitic cuckoos. Proc. R. Soc. Lond. (ser. B) 248:41–45.

Marchetti, K. 1993. Dark habitats and bright birds illustrate the role of the environment in species divergence. Nature 362:149–152.

Meijer, T., Daan, S., and Hall, M. 1990. Family planning in the kestrel (*Falco tinnunculus*) —

the proximate control of covariation of laying date and clutch size. Behaviour 114:117–136.

Moller, A. P., and T. R. Birkhead. 1992. Validation of the heritability method to estimate extra-pair paternity in birds. Oikos 64:484–488.

Moss, R., and A. Watson. 1984. Maternal condition, egg quality, and breeding success of Scottish ptarmigan *Lagopus mutus*. Ibis 126:212–220.

Moss, R., A. Watson, P. Rothery, and W. Glennie. 1981. Clutch-size, egg-size, hatch-weight and laying date in relation to early mortality in red grouse *Lagopus lagopus scoticus* chicks. Ibis 123:450–462.

Murphy, M. T. 1986. Body size and condition, timing of breeding, and aspects of egg production in eastern kingbirds. Auk 103:465–476.

Nager, R. G., and H. S. Zandt. 1994. Variation in egg size in great tits. Ardea 82:315–328.

Nilsson, J.-A. 1994. Energetic costs during breeding and the reproductive cost of being too early. J. Anim. Ecol. 63:200–208.

Nilsson, J.-A., and E. Svensson. 1993. Causes and consequences of egg mass variation between and within blue tit clutches. J. Zool. (Lond.) 230:469–481.

Nilsson, J.-A., and E. Svensson. 1996. The cost of reproduction: A new link between current reproductive effort and future reproductive success. Proc. R. Soc. Lond. (ser. B) 263:711–714.

Nisbet, I. C. T. 1973. Courtship feeding, egg size, and breeding success in common terns. Nature 241:141–142.

Norris, K. 1993. Seasonal variation in the reproductive success of blue tits: An experimental study. J. Anim. Ecol. 62:287–294.

Ojanen, M., M. Orell, and R. A. Vaisanen 1981. Egg size variation within passerine clutches: Effects of ambient temperature and laying sequence. Ornis Fenn. 58:93–108.

Perdeck, A., and A. Cave. 1992. Laying date in the coot: effects of age and mate choice. J. Anim. Ecol. 61:13–20.

Perrins, C. M. 1970. The timing of birds breeding seasons. Ibis 112:242–255.

Perrins, C. M. 1991. Tits and their caterpillar food supply. Ibis 133(suppl.):49–54.

Perrins, C. M. 1996. Eggs, egg formation and the timing of breeding. Ibis 138:2–15.

Petrie, M. 1993. Peacocks lay more eggs for peacocks with larger trains. Proc. R. Soc. Lond. (ser. B) 251:127–131.

Price, T. D. 1991. Environmental and genotype-by-environment influences on chick size in the yellow-browed leaf warbler *Phylloscopus inornatus*. Oecologia 86:535–541.

Price, T. D. 1984. The evolution of sexual size dimorphism in Darwin's finches. Am. Nat. 123:500–518.

Price, T. D., and P. R. Grant. 1985. The evolution of ontogeny in Darwin's finches: A quantitative genetic approach. Am. Natur. 125:169–188.

Price, T. D., and N. Jamdar. 1991. The breeding biology of the yellow browed leaf warbler in Kashmir. J. Bombay Nat. Hist. Soc. 88:1–19.

Price, T. D., and D. Schluter. 1991. On the low heritability of life history traits. Evolution 45:853-861.

Ricklefs, R. E. 1984. Egg dimensions and neonatal mass of shorebirds. Condor 86:7–11.

Ricklefs, R. E. 1992. Embryonic development period and the prevalence of avian blood parasites. Proc. Natl. Acad. Sci. USA 89:4722–4725.

Ricklefs, R. E. 1993. Sibling competition, hatching asynchrony, incubation period and lifespan in birds. Curr. Ornithol. 11:199–276

Ricklefs, R. E., and C. A. Smeraski. 1983. Variation in incubation period within a population of the European starling. Auk 100:926–931.

Ricklefs, R. E., D. C. Hahn, and W. A. Montevecchi. 1978. The relationship between egg size and chick size in the laughing gull and Japanese quail. Auk 95:228–237.

Rowe, L., D. Ludwig, and D. Schluter. 1994. Time, condition, and the seasonal decline of avian clutch size. Am. Nat. 143:698–722.

Royama, T. 1966. A re-interpretation of courtship feeding. Bird Study 13:116–129.

Saino, M., and A. P. Moller. 1996. Sexual ornamentation and immunocompetence in the barn swallow. Behav. Ecol. 7:227–232.

Saino, N., S. Calza, and A. P. Moller. 1997. Immunocompetence of nestling barn swallows in relation to brood size and parental effort. J. Anim. Ecol. 66:827–836.

Schifferli, L. 1973. The effect of egg weight on the subsequent growth of nestling great tits (*Parus major*). Ibis 115:549–558.

Schluter, D., T. Price, and L. Rowe. 1991. Conflicting selection pressures and life-history tradeoffs. Proc. R. Soc. Lond. (ser. B) 246:11–17.

Schluter, D., T. Price, and L. Rowe. and L. Gustafsson. 1993. Maternal inheritance of condition and clutch size in the collared flycatcher. Evolution 47:658–667.

Schwabl, H. 1993. Yolk is a source of maternal testosterone for developing birds. Proc. Natl. Acad. Sci. USA 90:11446–11450.

Schwabl, H. 1996. Maternal testosterone in the avian egg enhances postnatal growth. Comp. Biochem. Physiol. 114A:271–276.

Slagsvold, T., and J. Lifjeld. 1990. Influence of male and female quality on clutch size in tits (*Parus* species). Ecology 71:1258–1266.

Slagsvold, T., J. Sandvik, G. Rofstad, O. Lorentsen, and M. Husby. 1984. On the adaptive value of intraclutch egg-size variation in passerines. Auk 101:685–697

Smith, H. G. 1993. Heritability of tarsus length in cross-fostered broods of the European starling (*Sturnus vulgaris*). Heredity 71:318–322.

Smith, H. G., and K.-J. Wettermark. 1995. Heritability of nestling growth in cross-fostered European starlings *Sturnus vulgaris*. Genetics 141:657–665.

Smith, J. N. M., and A. A. Dhondt. 1980. Experimental confirmation of heritable morphological variation in a natural population of song sparrows. Evolution 34:1155–1158.

Smith, N., M. Wallach, M. Petracca, R. Braun, and J. Eckert. 1994. Maternal transfer of antibodies induced by infection with *Eimeria maxima* partially protects chickens against challenge with *Eimeria tenella*. Parasitology 109:551–557.

Svensson, E., and J.-A. Nilsson. 1996. Mate quality affects offspring sex ratio in blue tits. Proc. R. Soc. Lond. (ser. B) 263:357–361.

ten Cate, C. and P. Bateson 1988. Sexual selection: The evolution of conspicuous characteristics in birds by means of imprinting. Evolution 42:1355–1358.

Vaisanen, R. A., O. Hilden, M. Soikkeli, and S. Vuolanto. 1972. Egg dimension variation in five wader species: The role of heredity. Ornis Fennica 49:25–44.

VanderWerf, E. 1992. Lack's clutch size hypothesis: An examination of the evidence using meta-analysis. Ecology 73:1699–1705.

van Noordwijk, A. J., and G. de Jong. 1986. Acquisition and allocation of resources: Their influence on variation in life-history tactics. Am. Nat. 128:137–142.

van Noordwijk, A. J., A. J. van Balen, and W. Scharloo. 1981. Genetic and environmental variation in the clutch size of the great tit *Parus major*. Neth. J. Zool. 31:342–372.

van Noordwijk, A. J., A. J. van Balen, and W. Scharloo. 1988. Heritability of body size in a natural population of the great tit (*Parus major*) and its relation to age and environmental conditions during growth. Genet. Res. (Cambridge) 51:149–162.

Verhulst, S., J. H. van Balen, and J. M. Tinbergen. 1995. Seasonal decline in reproductive success of the great tit: Variation in time or quality? Ecology 76:2392–2403.

Wiggins, D. A. 1989. Heritability of body size in cross-fostered tree swallow broods. Evolution 43:1808–1811.

Wiggins, D. A., T. Part, and L. Gustafsson. 1994. Seasonal decline in collared flycatcher *Ficedula albicollis* reproductive success: an experimental approach. Oikos 70:359–364.

Williams, T. D. 1994. Intra-specific variation in egg size and egg composition in birds: Effects on offspring fitness. Biol. Rev. 68:35–59.

Winkler, D. W., and P. E. Allen. 1996. The seasonal decline in clutch size: Physiological constraint or strategic adjustment? Ecology 77:922–932.

Woolfenden, G., and J. Fitzpatrick. 1990. Florida scrub jays: A synopsis after 18 years of study. Pp. 239–266 in P. Stacey and W. Koenig (eds.), Cooperative Breeding in Birds. Cambridge University Press, Cambridge.

13

Maternal Influences on Larval Competition in Insects

FRANK J. MESSINA

Maternal effects can occur whenever variation among maternal phenotypes provides an additional (usually nongenetic) source of variation among offspring phenotypes (Mousseau and Dingle 1991). For many insects, an important maternal character is the choice of an oviposition site. If juvenile stages are fairly sedentary, a female's egg-laying preferences can largely or completely determine the quality of her offspring's environment, sometimes for the duration of its development. Oviposition choices influence a larva's food quality and its susceptibility to natural enemies or abiotic stresses, as well as its probability of experiencing intra- or interspecific competition (Mitchell 1983; Thompson 1983). Here I consider maternal influences on the degree of competition among offspring in insects. I describe genetic and nongenetic sources of variation in egg-laying behavior and discuss how genetic variation in oviposition behavior can persist in populations despite its apparent link to fitness.

Although I focus on the relationship between egg-laying behavior and larval competition, it should be remembered that selection will produce oviposition behavior that improves the overall quality of the larval environment, rather than any particular aspect. A female's egg-laying preferences may in fact reflect a compromise between maximizing the nutritional quality of her offspring's food and minimizing its susceptibility to competition, natural enemies, or abiotic stress. It is also worth noting that females influence competition among offspring in ways other than by their choice of oviposition sites (e.g., by controlling the quantity of nutrients per egg; Parker and Begon 1986; Kawecki 1995; Fox and Mousseau, this volume; Rossiter, this volume). In contrast to avian and mammalian systems, in insects maternal effects on offspring competition are usually prenatal, since females abandon eggs after depositing them.

I use the example of the cowpea seed beetle, *Callosobruchus maculatus*, to illustrate how larval performance can jointly depend on a larva's own competitive ability and the pattern of oviposition displayed by its mother (Smith and Lessells 1985). This bruchid may thus be subject to *maternal selection*, in which an interaction between maternal and offspring phenotypes determines offspring fitness (Kirkpatrick and Lande 1989). Because *C. maculatus* has attacked human stores of legume seeds for thousands of years, it is well suited to laboratory conditions, and its short generation time makes it a good subject for disentangling genetic and environmental effects on oviposition behavior. Traits mediating intraspecific competition may be unusually important in *C. maculatus* because the storage environment provides a relatively stable, benign habitat in which competitive ability contributes more to fitness than resistance to abiotic stress (Anderson and Löfqvist 1996; Parsons 1996).

13.1 Avoidance of Occupied Hosts by Female Insects

Nonrandom egg-laying preferences of insects are generally expected to reflect variation in the quality of potential oviposition sites for offspring (Thompson 1988). Females in many species avoid ovipositing on hosts that already bear conspecific eggs or larvae (Prokopy et al. 1984). The gain in fitness from this behavior will be substantial if larvae feed on small, discrete hosts (such as seeds, fruits, or other insects) and must complete their development on the natal host (Roitberg and Mangel 1988). At the population level, avoidance of occupied hosts can produce a uniform (regular) dispersion of juvenile stages among hosts (Averill and Prokopy 1989; Wajnberg et al. 1989). Female insects usually detect conspecific eggs or larvae on occupied hosts by means of so-called marking pheromones. Identification of these pheromones has proven difficult (Hurter et al. 1987), in part because compounds that mediate avoidance of occupied hosts in nature are not easily distinguished from a wider array of compounds that deter egg laying in laboratory choice tests (Messina et al. 1987; Credland and Wright 1990; Blaakmeer et al. 1994).

Avoidance of occupied hosts is sometimes weak, intermittent, or absent in insects that would otherwise be expected to respond to conspecific density, that is, species with sedentary larvae that feed on small, discrete hosts. In some cases, this can be explained by a nonlinear relationship between conspecific density and fitness, often referred to as an Allée effect. An Allée effect may be observed if the presence of multiple larvae improves the quality or accessibility of the resource (e.g., Breden and Wade 1987; Ruiz-Dubreuil et al. 1994) or if the per capita risk to natural enemies declines with increasing numbers of eggs or larvae (Stamp 1980; Lawrence 1990). In parasitoids, for example, provisioning a host with multiple eggs can increase the probability that at least one larva avoids encapsulation by the host immune system (van Alphen and Visser 1990).

Even if larval fitness declines monotonically with increasing larval density, it may nevertheless be adaptive for a female to add an egg to an occupied host (Charnov and Skinner 1984; Iwasa et al. 1984). A rich theoretical literature has emerged to account for such "superparasitism" in solitary parasitoids, which attack hosts that can support only one parasitoid larva (e.g., Parker and Courtney 1984; Mangel 1989, 1992). The decision to accept an occupied host (and how many eggs to lay per host) should depend on several aspects of the maternal phenotype, including her age, size, egg load, and handling time (Parker and Begon 1986; Weisser and Houston 1993; Wilson and Lessells 1994). For parasitoids and fruit flies, it has been suggested that egg-laying females may prefer occupied hosts because of the savings in time or energy associated with using an entry hole created by an earlier-arriving female (Takasu and Hirose 1991; Papaj et al. 1992; Lalonde and Mangel 1994). Assessment of the "best" oviposition strategy must thus recognize a potential parent-offspring conflict regarding optimal clutch sizes and rates of host acceptance (Godfray et al. 1991; Rosenheim 1993).

Finally, egg-laying behavior may simply be suboptimal with respect to offspring competition because of information-processing constraints or a lack of underlying genetic variation (Ward 1992; Heard 1994; Adamo et al. 1995; but see Bouskila et al. 1995).

Females of *C. maculatus* provide a relatively straightforward example of the advantages gained by discrimination between occupied and unoccupied hosts. Females attach eggs singly to the surfaces of legume seeds, and the hatching, legless larva chews its way into the seed at the oviposition site. Seeds of many legume hosts can support the development of only one or a few larvae. By avoiding egg-laden seeds, females consistently produce uniform dispersions of eggs and larvae among seeds (Utida 1943; Mitchell 1975). Once all available seeds bear eggs, females can discriminate between seeds bearing few eggs versus many eggs (Messina and Renwick 1985; Wilson 1988) and thus maintain uniform egg dispersions even as egg densities increase to several eggs per seed. Detection of eggs is accomplished by receptors on the beetle's mouthparts, which are in frequent contact with the seed surface (Messina et al. 1987). Females do not appear to distinguish seeds bearing her own eggs from those bearing eggs of other females (as observed in some parasitoids; Messina and Tinney 1991). The advantages of recognizing "self" and "non-self" eggs may be small, however, because females lay eggs singly and probably rarely revisit seeds in storage or field environments.

The survival of *C. maculatus* larvae and the sizes of emerging adults decline monotonically with increasing numbers of larvae per seed, and adult size is positively related to fecundity and life span (Credland et al. 1986; Colegrave 1993). By spreading eggs uniformly among seeds, females in one population achieved an average fitness (estimated as the net replacement rate) that was up to 70% higher than the expected fitness of random female, that is, one that produces a Poisson distribution of eggs among seeds (Mitchell 1975, 1991; Mitchell and Thanthianga 1990). In the closely related *C. chinensis*, larval competition caused a linear, inverse relationship between the number of eggs per seed (from one to six) and the net replacement rate of progeny (Ryoo and Chun 1993).

13.2 Genetic Variation in Oviposition Behavior

Although there is ample evidence that egg-laying insects avoid occupied hosts to reduce offspring competition, little attention has been given to variation in this behavior as a function of maternal genotype. Some indirect evidence is provided by studies of true fruit flies (Tephritidae) that "mark" fruits after inserting eggs in them (Prokopy 1972). Prokopy and colleagues (Prokopy et al. 1976, 1978; Papaj et al. 1990) have noted that the potency of marking pheromones or the response to them can decline in laboratory populations. This observation implies variation within the wild populations from which laboratory flies were derived (and relaxed selection for avoidance of occupied hosts in the laboratory), but differences between culture flies and wild flies were not clearly attributed to genetic causes. Comparisons among wild populations revealed minor (Boller et al. 1994) to moderate (Boller and Aluja 1992) variation in the tendency of flies to avoid occupied hosts. The latter study compared flies on different host species, but it was not apparent why flies attacking one host should be less discriminating than those attacking another. Differences among populations are also suggested in a few other insect species (Credland and Dendy 1992). For example, genetic differences may explain conflicting results as to whether *Battus philenor* butterflies avoid ovipositing on egg-laden plants (compare Rauscher 1979; Tatar 1991).

Evidence for genetic variation in behavior within populations is similarly limited. In *Drosophila melanogaster*, conditioning of a resource (such as a rotting fruit) by early-colonizing larvae may improve the performance of subsequent colonists, thus producing an Allée effect (e.g., Chess and Ringo 1985; Chess et al. 1990). Distributions of eggs tend to be aggregated, but high levels of aggregation reduce larval survival on small hosts (Ruiz-Dubreuil et al. 1994). Oviposition behavior of *D. melanogaster* females responded significantly to bidirectional selection for low versus high aggregation, and analyses of selected lines indicated additive, polygenic inheritance of the trait (Ruiz-Dubreuil and del Solar 1986). Diallel crosses of divergent, inbred strains revealed an additional component of dominance toward higher aggregation, as well as a possible maternal (cytoplasmic) effect (Ruiz-Dubreuil and del Solar 1993). Differences in egg dispersion between high- and low-selected lines were apparently mediated by differing levels of female movement (Ruiz-Dubreuil et al. 1994).

In species that are not as amenable to genetic analyses as *Drosophila* (such as arrhenotokous and thelytokous parasitoids), comparisons of isofemale lines have been useful for detecting genetic variation in oviposition behavior (Boulétreau 1986). *Trichogramma* wasps insert their eggs into the eggs of other insects and have been used frequently for the biological control of crop pests. Comparisons among isofemale lines indicated significant genetic variation in both the propensity to superparasitize (i.e., lay eggs in already parasitized hosts) and in the ultimate dispersion of eggs among available hosts (Wajnberg et al. 1989; Chassain and Boulétreau 1991). Cronin and Strong (1996) recently reported genetic variation in traits that affect how a fairyfly parasitoid distributes her eggs among host planthoppers. Both the time a female spends in a patch of hosts and the number of hosts parasitized per patch were as genetically variable as two morphological traits in the same population.

For *Callosobruchus maculatus*, genetic variation in oviposition behavior can be assessed under seminatural conditions. Messina and Mitchell (1989) developed an index that quantifies how uniformly females distribute eggs among seeds. Because this U score is independent of the number of eggs laid (unlike the conventional variance-to-mean ratio), it can be used to compare individuals or populations that differ in realized fecundity. Egg distributions varied substantially among seven geographic strains (Messina and Mitchell 1989). Females from an Indian strain frequently spread their eggs as evenly as possible among seeds, whereas females from a Brazilian strain produced egg distributions that were often close to (but not quite) random. Electrophoretic analyses indicated that these two strains were highly divergent in allozyme frequencies as well as in oviposition behavior (Berg and Mitchell 1993).

Reciprocal crosses between Indian and Brazilian strains of *C. maculatus* suggested polygenic inheritance of egg laying behavior, with dominance toward more uniform egg-laying and possible epistasis (Messina 1989). Even females derived from backcrosses to the "sloppier" Brazilian strain showed a dominance deviation toward the Indian strain. A factorial experiment was used to examine whether the stronger avoidance of occupied hosts by Indian-strain females was caused by a greater deterrence associated with Indian-strain eggs (perhaps because of a more potent marking pheromone) or by a higher sensitivity to egg-laden seeds on the part of Indian-strain females. Differences in both female behavior and in the eggs themselves appeared to contribute to the divergent egg dispersions in the two populations (figure 13.1; Messina et al. 1991). Females from the two strains exhibited similar foraging behavior in standardized arrays of egg-free seeds (Messina and Dickinson 1993) and thus do not appear to differ in their tendency to "sample" the local environment for unoccupied hosts.

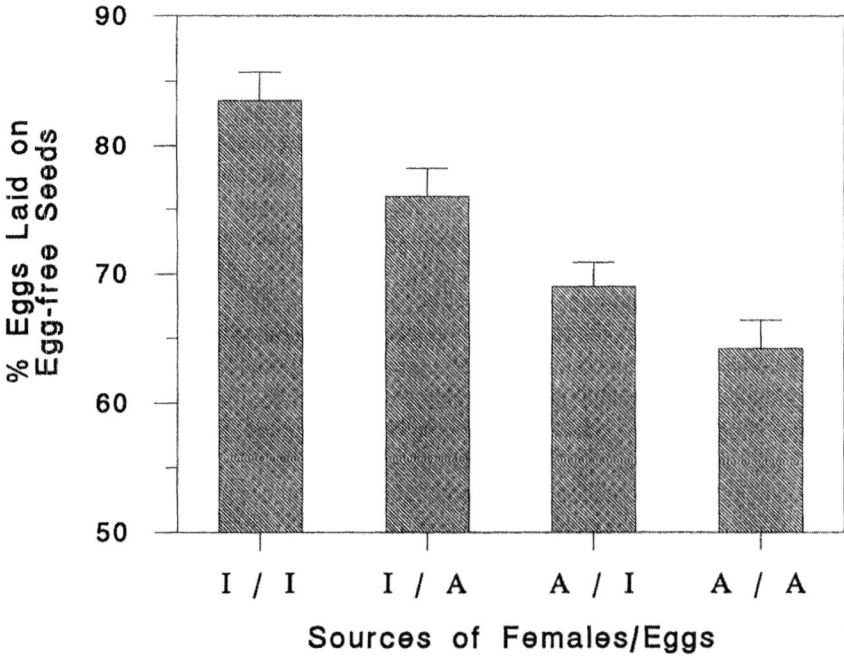

Figure 13.1 The tendency of *Callosobruchus maculatus* females to avoid occupied hosts as a function of the female's strain of origin (Indian or African) and the source of eggs (Indian or African) on egg-laden seeds. The source of females is listed first under each column. $N =$ 45–70 females/treatment.

Quantitative-genetic analysis was used to assess the amount of additive-genetic variation within *C. maculatus* populations in the tendency to distribute eggs evenly among hosts. Resemblances among half-siblings indicated nonzero heritabilities for U scores (i.e., egg dispersions) in each of two outbred populations from West Africa (Messina 1993). Moreover, the evolvability of egg-laying behavior, as estimated by the coefficient of additive-genetic variation (Houle 1992), was slightly higher than that for a morphological trait (wing length) in both populations. Heritabilities based on between-dam variance components were comparable to those based on between-sire components, which suggests an absence of maternal, cytoplasmic effects on egg-laying behavior. A different approach was used to examine variation in oviposition behavior within a population of the closely related *C. chinensis*. By measuring the repeatability of the tendency to avoid the putative marking pheromone (which sets an upper limit for the trait's heritability; Falconer 1989), Tanaka (1991) concluded that about half of the total phenotypic variation in this trait was due to among-individual variation.

13.3 Nongenetic Effects and Genotype × Environment Interactions

In contrast to the small number of genetic studies, there are abundant examples of how nongenetic factors, such as a female's age, physiological condition, and experience, can modify her tendency to avoid laying eggs on occupied hosts. An important physiological variable is the number of mature oocytes that a female carries (Minkenberg et al. 1992). Female insects are well known to increase their acceptance of less-preferred host species with an increase in the time elapsed since their last oviposition bout (Singer 1982; Courtney and Kibota 1990). A similar process is probably responsible for increased acceptance of occupied hosts and increased clutch sizes following a period of host deprivation (Pilson and Rausher 1988; Mangel 1989; Tatar 1991). The probability of accepting a host can also depend on the types of hosts that a female has recently encountered (Papaj and Prokopy 1989), but distinguishing the effects of experience and physiological state may be difficult in many insects (Rosenheim and Rosen 1991).

The tephritid fly *Rhagoletis pomonella* illustrates how a suite of nongenetic factors can affect oviposition choices. Experience with marking pheromone enhances a fly's response to it, but the deterrent effect of the pheromone is greatly diminished if she has not recently laid an egg (Roitberg and Prokopy 1981, 1983). On the stimulus side, the amount of pheromone deposited can depend on a female's age, size, or level of hunger, as well as the size of the host fruit (Averill and Prokopy 1988). Because host-deprived females tend to ignore marking pheromone, we might expect random or aggregated distributions of larvae when few fruits are available. This was indeed observed in a natural population of *R. pomonella*; the dispersion of the larvae among fruits shifted progressively from aggregated to random to regular as the season progressed (Averill and Prokopy 1989) and presumably more fruits became penetrable by ovipositing flies (Messina and Jones 1990),

Host deprivation modifies the oviposition behavior of *Callosobruchus* females, but the magnitude of the response can vary among populations. Females emerge from seeds with about eight mature oocytes and, in the absence of hosts, accumulate mature oocytes in their abdomen for about 2 days (Credland and Wright 1989; Wilson and Hill 1989). A 10-hour period of host deprivation increased the frequency at which Indian-strain females committed at least one oviposition "mistake" from 19% to 50%, where a mistake is defined as placing a second egg on a seed when an egg-free seed is available (Messina et al. 1992). Host deprivation had no effect on Brazilian-strain females, but even nondeprived females from this strain invariably committed at least one mistake. Thus, the populations differed not only in their absolute probabilities of accepting an egg-laden host, but also in their population-level, reaction norms in response to host deprivation.

Plastic oviposition behavior is also evident if *C. maculatus* females experience a shortage of egg-free seeds. When females from the Indian strain were provided only 10 seeds on which to lay their lifetime supply of eggs, their realized fecundities were approximately half of those of females given 40 seeds (figure 13.2; see also Messina 1991b). This is because females simply ceased laying eggs once each of the 10 seeds bore two or more eggs. In contrast, realized fecundities of females from an African strain were unaffected by the same levels of seed availability (figure 13.2). The two populations again exhibited different reaction norms, in this case for the relationship between seed availability and lifetime fecundity. By affecting the intensity of larval competition, population differences in the response to a shortage of seeds will translate into variation in the dynamics and economic impact of beetle populations (Dick and Credland 1984; Mitchell 1990).

Although the African strain exhibited a flat response to seed availability in the experiment depicted in figure 13.2, a subsequent experiment showed that lifetime fecundity does decrease when the shortage of seeds becomes severe, such as when females are provided no seeds (figure 13.3). This experiment also revealed substantial variation among individuals. Whereas

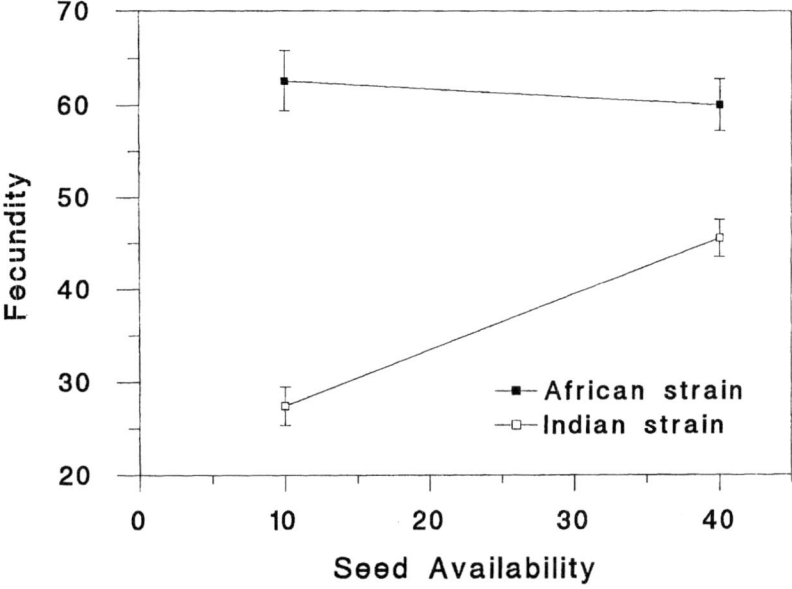

Figure 13.2 Mean lifetime fecundities of Indian- and African-strain females of *Callosobruchus maculatus* presented 10 or 40 seeds. $N = 25$ females/treatment.

the fecundities of females provided ample seeds followed the expected Gaussian distribution, the fecundities of females provided no seeds were strikingly bimodal (figure 13.4). Some females laid few or no eggs; others "dumped" more than half of their complement of eggs on the container in which they were confined (see also Wilson and Hill 1989). It should be possible to determine if this bimodality reflects genotypic variation (e.g., by splitting families into seed-poor and seed-rich environments), and thus determine whether fecundity within populations is controlled by the same genotype × environment interaction observed at the population level (figure 13.2).

The role of experience in the egg-laying decisions of *C. maculatus* is not as easy to demonstrate as the role of physiological condition (Horng 1994). Mitchell (1975) suggested that females achieve uniform egg dispersions by comparing the current seed with the last seed encountered, and a simulation model based on this decision rule fit his observed dispersions reasonably well. Alternatively, females may respond only to the number of eggs on the current seed but are less likely to accept a seed with each increase in egg density (Messina and Renwick 1985). Wilson (1988) constructed models incorporating both "absolute" or "relative" rules and concluded that the choices of females presented seeds with different egg densities were best explained by an absolute rule (i.e., one assuming no effect of experience). However, subsequent simulations by Mitchell (1990) suggested that none of Wilson's (1988) models could generate the extreme hyperdispersion of eggs observed in the Indian strain.

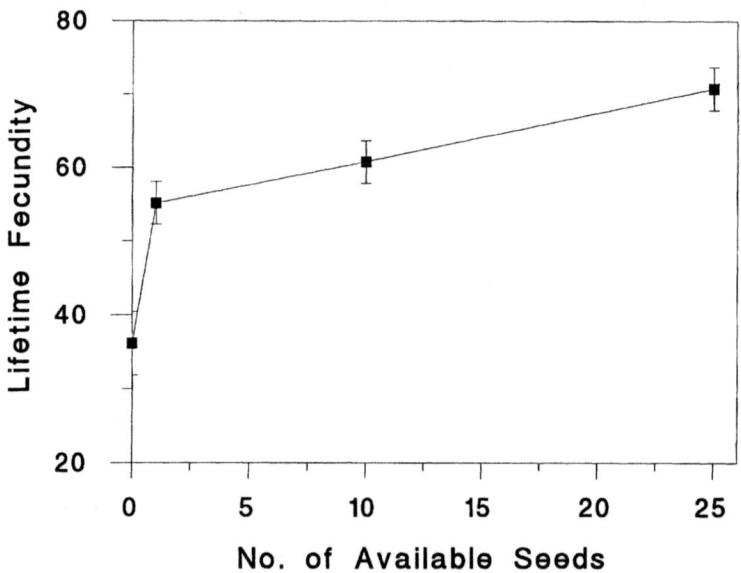

Figure 13.3 Mean lifetime fecundities of African-strain females of *Callosobruchus maculatus* presented 0, 1, 10, or 25 seeds. $N = 35$ females/treatment.

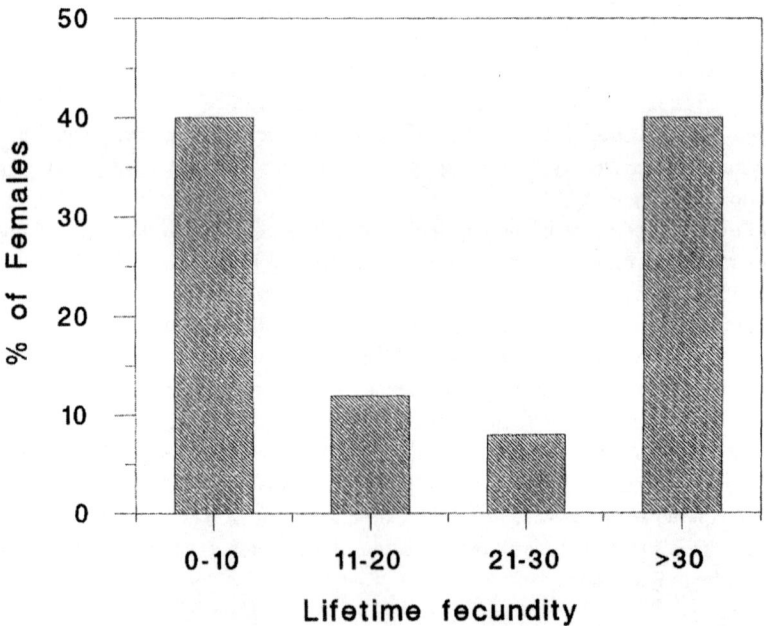

Figure 13.4 Frequencies of *Callosobruchus maculatus* females in different fecundity classes when females were confined without seeds ($N = 50$).

13.4 Maintenance of Genetically Variable Behavior

The above examples illustrate how an egg-laying female's local environment or physiological condition will influence her response to densities of conspecific individuals on potential hosts (Mangel 1992). From an evolutionary perspective, it is equally relevant to ask why populations might harbor genetic variation in the tendency to avoid occupied hosts, given the obvious consequences for larval fitness. One explanation is that reducing larval competition entails a cost through some other fitness component. Deposition of marking pheromone by tephritid flies enhances detection of the egg by parasitoids (Prokopy and Webster 1978). The apparent trade-off between minimizing the effects of competition and natural enemies may lead to the stable coexistence of marking and nonmarking flies in a population (Roitberg and Lalonde 1991). A similar trade-off could affect clutch size if predation risk were inversely density dependent; larvae from small clutches would experience less competition but higher predation. Although egg-laying preferences of some insects appear to have evolved in response to natural enemies rather than in response to intraspecific competition (Atsatt 1981; Damman 1991), too few studies have considered both factors simultaneously (Siemens and Johnson 1992; Ryoo and Chun 1993).

Fitness trade-offs might also arise because of the pleiotropic effects of alleles controlling oviposition behavior. Lines of *Drosophila melanogaster* selected for high versus low aggregation of eggs also differed in adult size and in the feeding rates and foraging behavior of larvae (Ruiz-Dubreuil et al. 1996). Before they reached the critical mass needed for pupation, larvae from the low-aggregation line fed more and developed faster than those from the high-aggregation line. This difference would make larvae from the low-aggregation line superior competitors when food is limiting. On the other hand, flies from the high-aggregation line attained a larger adult size and were therefore more fecund. Many genes undoubtedly affect larval and adult behaviors simultaneously in insects (e.g., the "rover-sitter" polymorphism of *Drosophila;* Periera and Sokolowski 1993). Antagonistic pleiotropy might also occur within the adult stage. For example, genotypes that invest more time or energy locating unoccupied hosts may deposit fewer eggs before they die (Iwasa et al. 1984). We did not obtain negative genetic correlations between lifetime fecundity and U scores (i.e., egg dispersion) in two wild populations of *C. maculatus*, but confinement of females within a small arena may have artificially reduced search costs associated with producing the most uniform egg dispersions (Messina 1993).

A second explanation for variation in oviposition behavior is that it reflects underlying variation in the degree to which fitness is reduced when larvae are forced to compete (Zimmerman 1982). In some insects, the tendency to superparasitize or the number of eggs laid per clutch depends on host size (and thus the number of larvae that can develop per host; Mitchell 1990; Fox and Mousseau 1995). Depending on the temporal and spatial scales over which hosts are encountered, variation in host size might lead to plastic oviposition behavior or the maintenance of genotypic variation. Apple and hawthorn "races" of *Rhagoletis pomonella* have diverged genetically in their preferences for the two hosts (Prokopy et al. 1988); we might expect them to differ in the tendency to avoid occupied fruits as well, because the number of larvae that can develop per fruit is quite different for hawthorns and apples (Averill and Prokopy 1988).

Variation in the cost of larval competition can also be caused by variation in the way co-occurring larvae interact. Populations of *C. maculatus* show striking differences in the num-

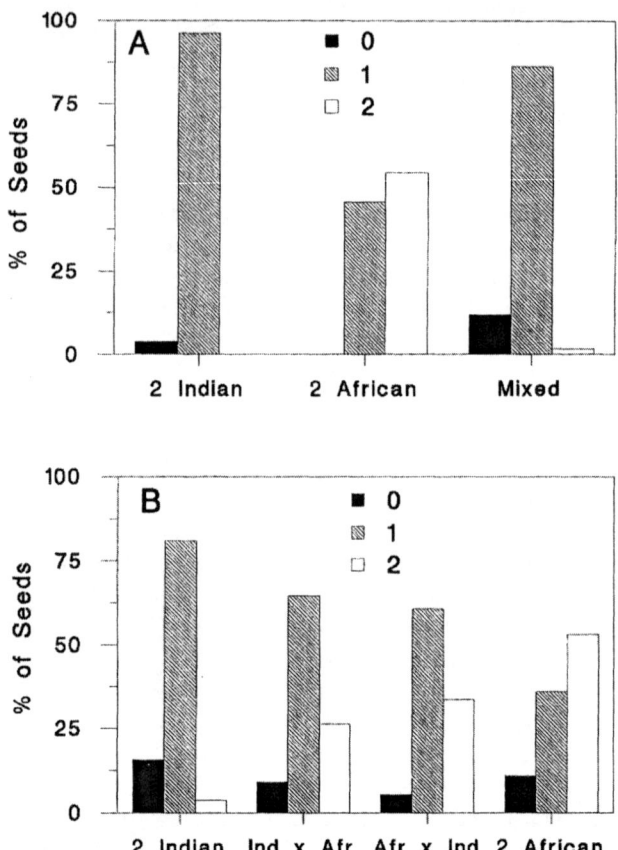

Figure 13.5 A, frequency of seeds yielding 0, 1, or 2 emerging adults of *Callosobruchus maculatus* depending on whether seeds contained two Indian-strain larvae, two African-strain larvae, or one larva from each strain. $N = 100-110$ seeds/treatment. B, frequency of seeds yielding 0, 1, or 2 emerging adults depending on whether seeds contained two Indian-strain larvae, two African-strain larvae, or two hybrid larvae from reciprocal crosses (the maternal source is listed first for each cross). $N = 75-100$ seeds/treatment.

ber of adults that can emerge from seeds infested by a given number of larvae (Dick and Credland 1984; Credland et al. 1986). Particularly strong competition occurs between larvae of the Indian strain, whose females produce highly uniform egg dispersions and cease ovipositing if few seeds are available (Thanthianga and Mitchell 1987). No seed yielded two adults when two Indian-strain larvae co-occurred in a small host (mung bean), yet over half of the seeds bearing two African-strain larvae did so (figure 13.5A; Messina 1991b). When seeds were manipulated to receive one larva from each strain, the reduction in survival was as large as it was when two Indian-strain larvae shared a seed, and 70% of the beetles from single-survivor seeds were from the Indian strain (figure 13.5A). A hybridization experiment (using an intermediate-sized host) suggested that these differences in competitive ability were inherited additively, as the frequencies of seeds yielding two adults were intermediate when two hybrid larvae shared a seed (figure 13.5B; Messina 1991a). Additive inheritance was also observed when the effects of larval competition were investigated over a wide range of larval densities (Toquenaga and Fujii 1991; Toquenaga 1993; Toquenaga et al. 1994). In behavior trials, Indian-strain larvae exhibited more persistent biting behavior and were more likely to inflict fatal wounds.

Differences in larval competitiveness may help explain differences in oviposition behavior among populations, but why should larvae from interfertile populations compete in such different ways? For *C. maculatus*, both empirical and theoretical investigations have implicated the evolutionary importance of host size. Because the optimal behavior of a larva should depend on the frequencies of alternative phenotypes, Smith and Lessells (1985; also Smith 1990) used game-theory to predict levels of larval aggression among granivorous beetles. They concluded that an aggressive phenotype (*attack*) would be a superior competitor in a small host but a passive strategy (*avoid*) would be the evolutionary stable strategy in a large host, mostly because the cost of exploitation competition would be low in a seed with ample resources. Experiments using different *Callosobruchus* species or divergent populations of *C. maculatus* tend to confirm the importance of seed size in determining the most successful larval strategy (Toquenaga 1993; Toquenaga et al. 1994; Colegrave 1995). When data from figure 13.5 were used as parameters for the Smith-Lessells model, the evolutionarily stable strategy depended on initial frequencies; neither *attack* nor *avoid* could invade a population primarily composed of the other strategy (Messina 1991b). It is noteworthy that the Indian strain of *C. maculatus* was obtained from a region where most hosts are small seeded, and other populations in this region exhibit strong competitive ability (Mitchell 1991).

By influencing the costs of multiple larvae per seed, the level of aggression between *C. maculatus* larvae should affect the evolution of oviposition behavior in this species. Conversely, the distribution of eggs among seeds affects the probability of larval interactions, and hence the optimal level of larval aggression (Smith and Lessells 1985; Godfray 1987). Because the fitness of a larva depends jointly on its aggressiveness and on the oviposition behavior of its mother, larval competitive ability should be under maternal selection (Kirkpatrick and Lande 1989). It may be possible to examine the coevolution of larval and maternal characters using selection experiments similar to those of Toquenaga et al. (1994), who looked strictly at microevolutionary changes in larval competitiveness.

Conclusions

The oviposition choices of female insects have lasting effects on the performance of their offspring. Some aspects of egg-laying behavior appear to have evolved to minimize the likelihood that offspring will have to compete for resources. Female responses to conspecific densities on potential hosts should be most evident in species with endophagous larvae that cannot move among hosts. A large fraction of all insect species possess these characteristics, particularly when one considers the large but poorly known diversity of parasitoid wasps.

The examples discussed in this review suggest that the tendency to avoid occupied hosts can vary substantially within and among insect populations, in some cases as a result of genotypic differences among host-seeking females. This variation may have important consequences at both the individual and population level. In tephritid flies, for example, seasonal changes in female behavior may make offspring of early-emerging females more susceptible to intraspecific competition than those of late-emerging females (Averill and Prokopy 1989; Lalonde and Mangel 1994), and this pattern may in turn produce microevolutionary change in fly phenology. The degree of avoidance of occupied hosts can affect population dynamics and stability (Smith and Lessells 1985; Toquenaga et al. 1994) as well as diet breadth (if females switch to a less preferred host when a favored host already bears eggs or larvae; Wasserman 1981). Among parasitoids, the tendency to superparasitize will influence the efficacy of biological control (Chassain and Boulétreau 1991), as genotypes with a strong aversion to superparasitism would be expected to inflict more host mortality than those that readily superparasitize. Similarly, the spatial distribution of eggs or larvae in a population of herbivorous insects can affect the extent of injury to an associated crop (Credland and Dendy 1992).

Understanding the evolution of nonrandom oviposition behavior would be enhanced by more information on the degree of genetic variation for the trait in natural populations and on the typical modes of inheritance. Particularly interesting are possible constraints imposed by genetic correlations between egg-laying behavior and other fitness-related characters, including those expressed by juvenile stages (Ruiz-Dubreuil et al. 1996). In general, little is known regarding the genetic architecture surrounding most behavioral traits in insects. Because of the potential link between the optimal oviposition strategies of adults and the behavioral interactions of larvae (Smith and Lessells 1985; Godfray 1987; see also Wade, this volume), intra- and interspecific variation in the tendency to avoid occupied hosts may prove useful for understanding the coevolution of suites of characters that mediate intraspecific competition.

References

Adamo, S. A., D. Robert, J. Perez, and R. R. Hoy. 1995. The response of an insect parasitoid, *Ormia ochracea* (Tachinidae), to the uncertainty of larval success during infestation. Behav. Ecol. Sociobiol. 36:111–118.

Anderson, P., and J. Löfqvist. 1996. Asymmetric oviposition behaviour and the influence of larval competition in the two pyralid moths *Ephestia kuehniella* and *Plodia interpunctella*. Oikos 76:47–56.

Atsatt, P. R. 1981. Ant-dependent food plant selection by the mistletoe butterfly *Ogyris amaryllis* (Lycanenidae). Oecologia 48:60–63.

Averill, A. L., and R. J. Prokopy. 1988. Factors influencing release of host-marking pheromone by *Rhagoletis pomonella* flies. J. Chem. Ecol. 14:95–111.

Averill, A. L., and R. J. Prokopy. 1989. Distribution patterns of *Rhagoletis pomonella* (Diptera: Tephritidae) eggs in hawthorn. Ann. Entomol. Soc. Am. 82:38–44.

Berg, D. J., and R. Mitchell. 1993. Associations of allozyme variation and behavior in the cowpea weevil *(Callosobruchus maculatus)*. Entomol. Exp. Appl. 69:215–220.

Blaakmeer, A., D. Hagenbeek, T. A. van Beek, A. E. de Groot, L. M. Schoonhoven, and J. J. A. van Loon. 1994. Plant response to eggs vs. host marking pheromone as factors inhibiting oviposition by *Pieris brassicae*. J. Chem. Ecol. 20:1657–1665.

Boller, E. F., and M. Aluja. 1992. Oviposition deterring pheromone in *Rhagoletis cerasi* L. J. Appl. Entomol. 113:113–119.

Boller, E. F., C. Hippe, R. J. Prokopy, W. Enkerlin, B. I. Katsoyannos, J. S. Morgante, S. Quilici, D. Crespo de Stilinovic, and M. Zapater. 1994. Response of wild and laboratory-reared *Ceratitis capitata* Wied. (Dipt., Tephritidae) flies from different geographic origins to a standard host marking pheromone solution. J. Appl. Entomol. 118:84–91.

Boulétreau, M. 1986. The genetic and coevolutionary interactions between parasitoids and their hosts. Pp. 169–200 in J. Waage and D. Greathead (eds.), Insect Parasitoids. Academic Press, London.

Bouskila, A., I. C. Robertson, M. E. Robinson, B. D. Roitberg, B. Tenhumberg, A. J. Tyre, and E. vanRanden. 1995. Submaximal oviposition rates in a mymarid parasitoid: Choosiness should not be ignored. Ecology 76:1990–1993.

Breden, F., and M. J. Wade. 1987. An experimental study of the effect of group size on larval growth and survivorship in the imported willow leaf beetle, *Plagiodera versicolora* (Coleoptera: Chrysomelidae). Envir. Entomol. 16:1082–1086.

Charnov, E. L., and S. W. Skinner. 1984. Evolution of host selection and clutch size in parasitoid wasps. Fla. Entomol. 67:5–21.

Chassain, C., and M. Boulétreau. 1991. Genetic variability in quantitative traits of host exploitation in *Trichogramma* (Hymenoptera: Trichogrammatidae). Genetica 83:195–202.

Chess, K. F., and J. M. Ringo. 1985. Oviposition site selection by *Drosophila melanogaster* and *Drosophila simulans*. Evolution 39:869–877.

Chess, K. F., J. M. Ringo, and H. B. Dowse. 1990. Oviposition by two species of *Drosophila* (Diptera: Drosophilidae): Behavioral responses to resource distribution and competition. Ann. Entomol. Soc. Am. 83:717–724.

Colegrave, N. 1993. Does larval competition affect fecundity independently of its effect on adult weight? Ecol. Entomol. 18:275–277.

Colegrave, N. 1995. The cost of exploitation competition in *Callosobruchus* beetles. Funct. Ecol. 9:191–196.

Courtney, S. P., and T. T. Kibota. 1990. Mother doesn't know best: selection of hosts by ovipositing insects. Pp. 161–188 in E.A. Bernays (ed.), Insect-Plant Interactions. Vol. 2 CRC Press, Boca Raton, Fla.

Credland, P. F., and J. Dendy. 1992. Intraspecific variation in bionomic characters of the Mexican bean weevil, *Zabrotes subfasciatus*. Entomol. Exp. Appl. 65:39–47.

Credland, P. F., and A. W. Wright. 1989. Factors affecting female fecundity in the cowpea seed beetle, *Callosobruchus maculatus* (Coleoptera: Bruchidae). J. Stored Prod. Res. 25:125–136.

Credland, P. F., and A. W. Wright. 1990. Oviposition deterrents of *Callosobruchus maculatus* (Coleoptera: Bruchidae). Physiol. Entomol. 15:285–298.

Credland, P. F., K. M. Dick, and A. W. Wright. 1986. Relationships between larval density, adult size and egg production in the cowpea seed beetle, *Callosobruchus maculatus*. Ecol. Entomol. 11:41–50.

Cronin, J. T., and D. R. Strong. 1996. Genetics of oviposition success of a thelytokous fairyfly parasitoid, *Anagrus delicatus*. Heredity 76:43–54.

Damman, H. 1991. Oviposition behaviour and clutch size in a group-feeding pyralid moth, *Omphalocera munroei*. J. Anim. Ecol. 60:193–204.

Dick, K. M., and P. F. Credland. 1984. Egg production and development of three strains of *Callosobruchus maculatus* (F.) (Coleoptera: Bruchidae). J. Stored Prod. Res. 20:221–227.

Falconer, D.S. 1989. Introduction to Quantitative Genetics. 3rd ed. Longman Scientific-Wiley, New York.

Fox, C. W., and T. A. Mousseau. 1995. Determinants of clutch size and seed preference in a seed beetle, *Stator beali* (Coleoptera: Bruchidae). Envir. Entomol.1557–1561.

Godfray, H. C. J. 1987. The evolution of clutch size in parasitic wasps. Am. Nat. 129:221–233.

Godfray, H. C. J., L. Partridge, and P. H. Harvey. 1991. Clutch size. Annu. Rev. Ecol. Syst. 22:409–429.

Heard, S. B. 1994. Imperfect oviposition decisions by the pitcher plant mosquito (*Wyeomyia smithii*). Evol. Ecol. 8:493–502.

Horng, S. 1994. What is the oviposition decision rule by bean weevil, *Callosobruchus maculatus*? Zool. Studies 33:278–286.

Houle, D. 1992. Comparing evolvability and variability of quantitative traits. Genetics 130:195–204.

Hunter, J. E. F. Boller, E. Stadler, B. Blattman, J. Buser, N. U. Bosshard, L. Damm, M. W. Kozlowski, R. Schoni, F. Raschdorf, R. Dahinden, E. Schlumpf, H. Fritz, W. J. Richter, and J. Schreiber. 1987. Ovipositron-deterring pheremone in Rhagoletic cerasi L.: purification and determination of the chemical constitution. Experientia 43:157–164.

Iwasa, Y., Y. Suzuki, and H. Matsuda. 1984. Theory of oviposition strategy of parasitoids. I. Effect of mortality and limited egg number. Theor. Pop. Biol. 26:205–227.

Kawecki, T. J. 1995. Adaptive plasticity of egg size in response to competition in the cowpea weevil, *Callosobruchus maculatus* (Coleoptera: Bruchidae). Oecologia 102:81–85.

Kirkpatrick, M., and R. Lande. 1989. The evolution of maternal characters. Evolution 43:485–503.

Lalonde, R. G., and M. Mangel. 1994. Seasonal effects on superparasitism by *Rhagoletis completa*. J. Anim. Ecol. 63:583–588.

Lawrence, W. S. 1990. The effects of group size and host species on development and survivorship of a gregarious caterpillar *Halisidota caryae* (Lepidoptera: Arctiidae). Ecol. Entomol. 15:53–62.

Mangel, M. 1989. Evolution of host selection in parasitoids: does the state of the parasitoid matter? Am. Nat. 133:688–705.

Mangel, M. 1992. Descriptions of superparasitism by optimal foraging theory, evolutionarily stable strategies and quantitative genetics. Evol. Ecol. 6:152–169.

Messina, F. J. 1989. Genetic basis of variable oviposition behavior in *Callosobruchus maculatus* (Coleoptera: Bruchidae). Ann. Entomol. Soc. Am. 82:792–796.

Messina, F. J. 1991a. Competitive interactions between larvae from divergent strains of the cowpea weevil (Coleoptera: Bruchidae). Envir. Entomol. 20:1438–1443.

Messina, F. J. 1991b. Life-history variation in a seed beetle: Adult egg-laying vs. larval competitive ability. Oecologia 85:447–455.

Messina, F. J. 1993. Heritability and 'evolvability' of fitness components in *Callosobruchus maculatus*. Heredity 71:623–629.

Messina, F. J., and J. A. Dickinson. 1993. Egg-laying behavior in divergent strains of the cowpea weevil (Coleoptera: Bruchidae): Time budgets and transition matrices. Ann. Entomol. Soc. Am. 86:207–214.

Messina, F. J., and V. P. Jones. 1990. Relationship between fruit phenology and infestation by the apple maggot (Diptera: Tephritidae) in Utah. Ann. Entomol. Soc. Am. 83:742–752.

Messina, F. J., and R. Mitchell. 1989. Intraspecific variation in the egg-spacing behavior of the seed beetle *Callosobruchus maculatus*. J. Insect Behav. 2:727–741.

Messina, F. J., and J. A. A. Renwick. 1985. Ability of ovipositing seed beetles to discriminate between seeds with differing egg loads. Ecol. Entomol. 10:225–230.

Messina, F. J., and T. R. Tinney. 1991. Discrimination between 'self' and 'non-self' eggs by egg-laying seed beetles: A reassessment. Ecol. Entomol. 16:509–512.

Messina, F. J., J. L. Barmore, and J. A. A. Renwick. 1987. Oviposition deterrent from eggs of *Callosobruchus maculatus*: Spacing mechanism or artifact? J. Chem. Ecol. 13:219–226.

Messina, F. J., S. L. Gardner, and G. E. Morse. 1991. Host discrimination by egg-laying seed beetles: Causes of population differences. Anim. Behav. 41:773–780.

Messina, F. J., J. L. Kemp, and J. A. Dickinson. 1992. Plasticity in the egg-spacing behavior of a seed beetle: Effects of host deprivation and seed patchiness (Coleoptera: Bruchidae). J. Insect Behav. 5:609–621.

Minkenberg, O. P. J. M., M. Tatar, and J. A. Rosenheim. 1992. Egg load as a major source of variability in insect foraging and oviposition behavior. Oikos 65:134–142.

Mitchell, R. 1975. The evolution of oviposition tactics in the bean weevil, *Callosobruchus maculatus* (F.). Ecology 56:696–702.

Mitchell, R. 1983. Effects of host-plant variability on the fitness of sedentary herbivorous insects. Pp. 343–370 in R. F. Denno and M. S. McClure (eds.), Variable Plants and Herbivores in Natural and Managed Systems. Academic Press, New York.

Mitchell, R. 1990. Behavioral ecology of *Callosobruchus maculatus*. Pp. 317–330 in: K. Fujii, A. M. R. Gatehouse, C. D. Johnson, R. Mitchell, and Y. Yoshida (eds.), Bruchids and Legumes: Economics, Ecology, and Coevolution. Kluwer Academic, Dordrecht.

Mitchell, R. 1991. The traits of a biotype of *Callosobruchus maculatus* (F.) (Coleoptera: Bruchidae) from South India. J. Stored Prod. Res. 27:221–224.

Mitchell, R., and C. Thanthianga. 1990. Are the oviposition traits of the South India strain of *Callosobruchus maculatus* maintained by natural selection? Entomol. Exp. Appl. 57:143–150.

Mousseau, T. A., and H. Dingle. 1991. Maternal effects in insect life histories. Annu. Rev. Entomol. 36:511–534.

Papaj, D. R., and R. J. Prokopy. 1989. Ecological and evolutionary aspects of learning in phytophagous insects. Annu. Rev. Entomol. 34:315–350.

Papaj, D. R., B. D. Roitberg, S. B. Opp, M. Aluja, R. J. Prokopy, and T. T. Y. Wong. 1990. Effect of marking pheromone on clutch size in the Mediterranean fruit fly. Physiol. Entomol. 15:463–468.

Papaj, D. R., A. L. Averill, R. J. Prokopy, and T. T. Y. Wong. 1992. Host-marking pheromone and use of previously established oviposition sites by the Mediterranean fruit fly (Diptera: Tephritidae). J. Insect Behav. 5:583–598.

Parker, G. A., and M. Begon. 1986. Optimal egg size and clutch size: Effects of environment and maternal phenotype. Am. Nat. 128:573–592.

Parker, G. A., and S. P. Courtney. 1984. Models of clutch size in insect oviposition. Theor. Pop. Biol. 26:27–48.

Parsons, P. A. 1996. Competition versus abiotic factors in variably stressful environments: Evolutionary implications. Oikos 75:129–132.

Periera, H. S., and M. B. Sokolowski. 1993. Mutations in the larval foraging gene affect adult locomotory behavior after feeding in *Drosophila melanogaster*. Proc. Natl. Acad. Sci. USA 90:5044–5046.

Pilson, D., and M. D. Rausher. 1988. Clutch size adjustment by a swallowtail butterfly. Nature 333:361–363.

Prokopy, R. J. 1972. Evidence for a marking pheromone deterring repeated oviposition in apple maggot flies. Envir. Entomol. 1:326–332.

Prokopy, R. J., and R. P. Webster. 1978. Oviposition deterring pheromone of *Rhagoletis pomonella*: A kairomone for its parasitoid *Opius lectus*. J. Chem. Ecol. 4:481–494.

Prokopy, R. J., W. H. Reissig, and V. Moericke. 1976. Marking pheromones deterring repeated oviposition in *Rhagoletis* flies. Entomol. Exp. Appl. 20:170–178.

Prokopy, R. J., J. R. Ziegler, and T. T. Y. Wong. 1978. Deterrence of repeated oviposition by fruit-marking pheromone in *Ceratitis capitata*. J. Chem. Ecol. 4:55–63.

Prokopy, R. J., B. D. Roitberg, and A. L. Averill. 1984. Resource partitioning. Pp. 301–330 in W. J. Bell and R. T. Carde (eds.), Chemical Ecology of Insects. Sinauer, Sunderland, Mass .

Prokopy, R. J., S. R. Diehl, and S. S. Cooley. 1988. Behavioral evidence for host races in *Rhagoletis pomonella* flies. Oecologia 76:138–147.

Rausher, M. D. 1979. Egg recognition: Its advantage to a butterfly. Anim. Behav. 27:134–1040.

Roitberg, B. D., and R. G. Lalonde. 1991. Host marking enhances parasitism risk for a fruit-infesting fly *Rhagoletis basiola*. Oikos 61:389–393.

Roitberg, B. D., and M. Mangel. 1988. On the evolutionary ecology of marking pheromones. Evol. Ecol. 2:289–315.

Roitberg, B. D., and R. J. Prokopy. 1981. Experience required for pheromone recognition by the apple maggot fly. Nature 292:540–541.

Roitberg, B. D., and R. J. Prokopy. 1983. Host deprivation influence on response of *Rhagoletis pomonella* to its oviposition deterring pheromone. Physiol. Entomol. 8:69–72.

Rosenheim, J. A. 1993. Single-sex broods and the evolution of nonsiblicidal parasitoid wasps. Am. Nat. 141:90–104.

Rosenheim, J. A., and D. Rosen. 1991. Foraging and oviposition decisions in the parasitoid *Aphytis lingnanensis*: Distinguishing the influences of egg load and experience. J. Anim. Ecol. 60:873–893.

Ruiz-Dubreuil, D. G., and E. Del Solar. 1986. Effect of selection on oviposition site preference in *Drosophila melanogaster*. Austral. J. Biol. Sci. 39:155–160.

Ruiz-Dubreuil, D. G., and E. del Solar. 1993. A diallel analysis of gregarious oviposition in *Drosophila melanogaster*. Heredity 70:281–284.

Ruiz-Dubreuil, G., B. Burnet, and K. Connolly. 1994. Behavioural correlates of selection for oviposition by *Drosophila melanogaster* females in a patchy environment. Heredity 73:103–110.

Ruiz-Dubreuil, G., B. Burnet, K. Connolly, and P. Furness. 1996. Larval foraging behaviour and competition in *Drosophila melanogaster*. Heredity 76:55–64.

Ryoo, M. I., and Chun, Y. S. 1993. Oviposition behavior of *Callosobruchus chinensis* (Coleoptera: Bruchidae) and weevil population growth: Effects of larval parasitism and competition. Envir. Entomol. 22:1009–1015.

Siemens, D. H., and C. D. Johnson. 1992. Density-dependent egg parasitism as a determinant of clutch size in bruchid beetles (Coleoptera: Bruchidae). Envir. Entomol. 21:610–619.

Singer, M. C. 1982. Quantification of host preference by manipulation of oviposition behavior in the butterfly *Euphydryas editha*. Oecologia 52:224–229.

Smith, R. H. 1990. Adaptations of *Callosobruchus* species to competition. Pp. 351–360 in K. Fujii, A. M. R. Gatehouse, C. D. Johnson, R. Mitchell, and Y. Yoshida (eds.), Bruchids and Legumes: Economics, Ecology, and Coevolution. Kluwer Academic, Dordrecht.

Smith, R. H., and C. M. Lessells. 1985. Oviposition, ovicide and larval competition in granivorous insects. Pp. 423–448 in R. M. Sibly and R. H. Smith (eds.), Behavioral Ecology. Blackwell, Oxford.

Stamp, N. E. 1980. Egg deposition patterns in butterflies: Why do some species cluster their eggs rather than deposit them singly? Am. Nat. 115:367–380.

Takasu, K., and Y. Hirose. 1991. The parasitoid *Ooencyrtus nezarae* (Hymenoptera: Encyrtidae) prefers hosts parasitized by conspecifics over unparasitized hosts. Oecologia 87:319–323.

Tanaka, Y. 1991. Individual variation of scent avoiding oviposition behavior in azuki bean weevil *Callosobruchus chinensis*. J. Ethol. 9:31–33.

Tatar, M. 1991. Clutch size in the swallowtail butterfly, *Battus philenor*: The role of host quality and egg load within and among seasonal flights in California. Behav. Ecol. Sociobiol. 28:337–344.

Thanthianga, C., and R. Mitchell. 1987. Vibrations mediate prudent resource exploitation by competing larvae of the bruchid bean weevil *Callosobruchus maculatus*. Entomol. Exp. Appl. 44:15–21.

Thompson, J. N. 1983. Selection pressures of phytophagous insects feeding on small host plants. Oikos 40:438–444.

Thompson, J. N. 1988. Evolutionary ecology of the relationship between oviposition preference and performance of offspring in phytophagous insects. Entomol. Exp. Appl. 47:3–14.

Toquenaga, Y. 1993. Contest and scramble competitions in *Callosobruchus maculatus* (Coleoptera: Bruchidae) II. Larval competition and interference mechanisms. Res. Pop. Ecol. 35:57–68.

Toquenaga, Y., and K. Fujii. 1991. Contest and scramble competitions in *Callosobruchus maculatus* (Coleoptera: Bruchidae) I. Larval competition curves and resource sharing patterns. Res. Pop. Ecol. 32:199–211.

Toquenaga, Y., M. Ichinose, T. Hoshino, and K. Fujii. 1994. Contest and scramble competitions in an artificial world: genetic analysis with genetic algorithms. Pp. 177–199 in C. G. Langdon (ed.), Artificial Life III. Addison-Wesley, Reading, Mass.

Utida, S. 1943. Studies on the experimental population of the azuki bean weevil, *Callosobruchus chinensis* (L.). VIII. Statistical analysis of the frequency distribution of the emerging weevils on beans. Mem. Coll. Agric. Kyoto Imp. Univ. 54:1–22.

van Alphen, J. J. M., and M. E. Visser. 1990. Superparasitism as an adaptive strategy for insect parasitoids. Annu. Rev. Entomol. 35:59–79.

Wajnberg, E., J. Pizzol, and M. Babault. 1989. Genetic variation in progeny allocation in *Trichogramma maidis*. Entomol. Exp. Appl. 53:177–187.

Ward, D. 1992. The role of satisficing in foraging theory. Oikos 63:312–317.

Wasserman, S. S. 1981. Host-induced oviposition preferences and oviposition markers in the cowpea weevil, *Callosobruchus maculatus*. Ann. Entomol. Soc. Am. 74:242–245.

Weisser, W. W., and A. I. Houston. 1993. Host discrimination in parasitic wasps: When is it advantageous? Funct. Ecol. 7:27–39.

Wilson, K. 1988. Egg laying decisions by the bean weevil *Callosobruchus maculatus*. Ecol. Entomol. 13:107–118.

Wilson, K., and C. M. Lessells 1994. Evolution of clutch size in insects. I. A review of static optimality models. J. Evol. Biol. 7:339–363.

Wilson, K., and L. Hill. 1989. Factors affecting egg maturation in the bean weevil *Callosobruchus maculatus*. Physiol. Entomol. 14:115–126.

Zimmerman, M. 1982. Faculative deposition of an oviposition-deterring pheromone by *Hylemya*. Envir. Entomol. 11:519–522.

14

Maternal Effects, Developmental Plasticity, and Life History Evolution

An Amphibian Model

ROBERT H. KAPLAN

In many frogs and salamanders the largest investment made in progeny is in the provisioning of eggs with material and energy. At the time a female makes this substantial contribution, the environment that her offspring will experience may be unpredictable. How much influence does uncertainty in the larval environment exert over the way in which this maternal investment is allocated? Much progress was made in answering this question by using optimality techniques and their accompanying assumptions (e.g., Roff 1992; Stearns 1992). These techniques were firmly based on neo-Darwinian principles emphasizing natural selection among phenotypic variants that result directly from the action of discrete genes. This is apparent in the amphibian literature where models were developed to explain different patterns of reproductive allocation with the major assumptions being that for a given amount of reproductive expenditure, a female may produce many small offspring or few large offspring, and that offspring fitness is positively related to investment in offspring (Smith and Fretwell 1974; Brockelman 1975; Wilbur 1977).

The work on amphibians in the 1970s and 1980s resulted in a solid empirical foundation that can now be reevaluated in light of an increasing departure from the more narrowly defined aspects of neo-Darwinian thinking. This departure takes a variety of forms, but a very natural extension of life history strategy research incorporates an emphasis on the evolution of phenotypic response patterns of genotypes, or the norms of reaction (Stearns 1989; Sultan 1995). When this emphasis is applied to reproductive traits, it naturally leads to a consideration of maternal phenotype variation. When such variation has ramifications for the development of offspring, it is a maternal effect.

Figure 14.1 shows a simplified conceptual framework for the amphibian model considered here. Even before it became productive to think about maternal effects and their impact on the amphibian life cycle, there was a great deal of work on the interaction of various stages of the life cycle with the variation in both abiotic (e.g., temperature) and biotic (e.g., competition and density) features of the environment. This environmental variation is ubiquitous in studies of the evolution and ecology of amphibian reproduction. Each of the effects shown by

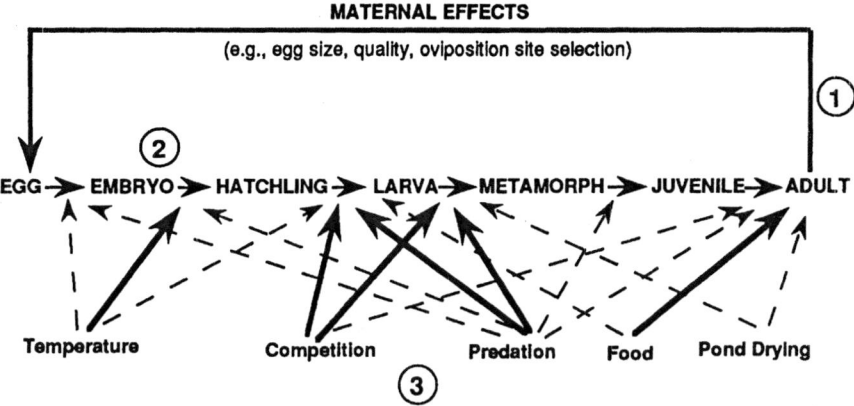

Figure 14.1 A model of environmental influences (including maternal effects) on the life cycle of a typical amphibian. Solid lines indicate effects addressed in the text, and numbers indicate the specific text sections.

the arrows in figure 14.1 has numerous studies addressing it (too many to be reviewed), and not all effects are shown. The phenotypic variation that accumulates within the life cycle as a result of proximate influences from the external environment was found to play a major role in subsequent development. For example, the pioneering work of Wilbur and Collins (1973; and many studies that followed) showed that characteristics of larvae such as size and growth rate were readily influenced by crowding. These characteristics became incorporated into the cycle and were shown to influence subsequent characteristics, such as the time to and size at metamorphosis. This orientation (i.e., the external environment resulting in developmental flexibility) can be applied to other stages of the life cycle. If one focuses on the egg and considers sources of variation in egg size, and the ramifications of the variation once incorporated into the life cycle, one is grappling directly with the issue of the developmental plasticity and evolution of maternal effects. The environment plays the dual role of affecting the developmental process and setting the fitness function (Scheiner 1993).

Within this framework, this chapter is divided into four sections. The first deals with why there is so much interest in ovum size variation in amphibians. Section 14.2 deals with the main effects of this maternal effect variation for offspring development and fitness. Section 14.3 follows with how maternal effects variation goes on to interact with variation in the larval environment, and section 14.4 discusses the implications of these interactive effects for the evolution of adaptive maternal effects and developmental plasticity in the reproductive traits of females.

14.1 Ovum Size Variation in Amphibians

In many organisms that lack parental care, egg size can be considered a direct measure of maternal investment in individual offspring. Studies in a wide variety of taxa have found that larger propagule sizes and greater levels of investment per offspring can have positive effects on offspring growth, development, and survivorship (e.g., Janzen 1977; Marsh 1986; McGinley et al 1987; Meffe 1987; Petranka et al. 1987; Tessier and Consolatti 1989; Amundsen and Stokland 1990; Rossiter 1991; Parichy and Kaplan 1992a; Bernardo 1996a). Recently, attention has focused on how these maternal effects influence offspring performance. For example, the consequences of maternal investment for lizard morphology and locomotor ability have been examined by experimentally manipulating egg sizes (Sinervo 1990; Sinervo and Huey 1990). These studies have shown that larger hatchlings (those that developed from larger eggs) have faster sprint speeds.

Amphibians are ideal for the study of ovum size variation as maternal effects (Bernardo 1991; Kaplan 1991). In addition to the historical reasons described in the introductory remarks, there are some significant practical reasons. For example, sources of variation in birth weight in viviparous organisms can be partitioned into factors associated with the genotype and the maternal environment of offspring (Falconer 1989). But, in amphibians with external fertilization, the offspring's genotype has not expressed itself at the time of oviposition, and therefore all variation in the offspring's phenotype at this time (e.g., egg size and quality) is the result of maternal effects. These maternal effects can be further partitioned into those that result from the genotype of the female and those that result from two types of maternal plasticity. The first effects result from environmental variation associated with the female during her development, and the second from environmental variation that immediately affects the female during the egg-making process. Arrows leading to the adult female in figure 14.1 place these effects within the context of the amphibian model. In addition, other factors such as parity and age can play a role.

Until fairly recently, the optimality approach to the study of life histories diverted attention from considering variation in maternal investment in offspring because it was the average of the trait that was felt to be optimized and the variance was uninteresting noise. But, as data accumulated showing large amounts of variation, the notion of an optimal egg size became untenable. This is also due to examples of offspring fitness having complex relationships with investment per offspring (see below). Currently, with more work focusing on maternal effects and developmental plasticity, it is the optimization of the variance in egg size that is moving into a more central position. Numerous studies have grappled with this problem, and some of this work is reviewed below. While it might seem unnecessary to spend a lot of time showing that ovum size variation is ubiquitous in amphibians, there are surprisingly few reports that provide significant statistical detail of variances to make meaningful comparisons within or among species (Bernardo 1996b).

Variation in ovum size has been commonly observed among species of amphibians (e.g., Crump and Kaplan 1979; Kaplan and Salthe 1979; Nussbaum 1985; Beachy 1993) and among populations within species (e.g., Dushane and Hutchinson 1944; Kozlowska 1971; Diaz et al. 1987). Such variation is to be expected even under optimality theories, because these groups may be subject to different environments and different selective pressures. However, there is also copious evidence of egg size variation among individuals within populations (e.g., Kuramoto 1978; Kaplan 1980; Travis 1983; Crump 1984; Williamson and Bull 1989, 1995), among successive clutches of individual females (Howard 1978; Kaplan 1987; Berven 1988), and even within single clutches (e.g., Crump 1981, 1984; Tejedo and Reques 1992; Beachy 1993). These findings are less readily explained by current optimality theory.

Figure 14.2 shows variation in egg size found in a population of oriental fire-bellied toads, *Bombina orientalis*, that is typical of numerous amphibian populations. Less well appreciated are the magnitude and susceptibility to temporal changes in the two types of variation cap-

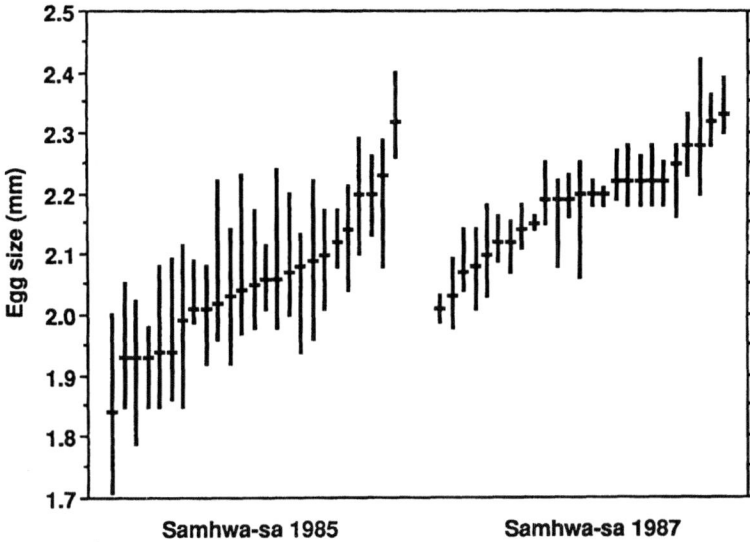

Figure 14.2 Each vertical line shows the range and mean in observed egg size for individual female *Bombina orientalis* within a single breeding season. Data are for the Samhwa-sa population from the Republic of Korea during two breeding seasons (1985 and 1987). It should be noted that these diameter measurements translate into a twofold volume (and energy) difference between the smallest and largest egg. From Kaplan and King (1997).

tured in the figure: mean egg size per clutch and intraclutch egg size variation. Figure 14.3 summarizes some of the results of an 8-year field study that compared patterns of temporal variation in egg size among three populations of the fire-bellied toad from the Republic of Korea (Kaplan and King 1997). This study provides evidence that egg size is a developmentally plastic trait in this species by describing substantial variation among populations, among individuals within populations, and among eggs within the clutches of single individuals as well as temporal changes in these characteristics.

The amount of variation observed in the field was substantial. The bold lines in figure 14.3 show the mean and range of the average egg size produced by individual females in two different years in the same population (Samhwa-sa). The coefficients of variation (CV) of mean egg size among females in a single season show a range of 3.8–5.7% in various years in the three populations. This compares to ranges of 5.7–8.2% in four populations of *Rana temporaria* (Kozlowksa 1971), 4.4–12.4% in nine species of frogs and three species of salamanders (Kuramoto 1978), 2.1–3.9% among ages and years in one population of *Rana sylvatica* (Berven 1988), 3.5–7.0% among months and years in an extended breeding frog, *Ranidella signifera* (Williamson and Bull 1995), and 3.3–12.7% in five species of salamanders (Beachy 1993). In addition, figure 14.3 shows (as an example) how the average shifted from 2.05 mm to 2.18 mm between 1985 and 1987. This observation is in accordance with prior studies that show yearly changes in egg size in recaptured females (Berven 1988), mean egg size changes over time in unmarked individuals (Cummins 1986), and monthly changes in mean egg size (Williamson and Bull 1995).

Figure 14.3 also shows the range of egg sizes produced within the clutch of the female

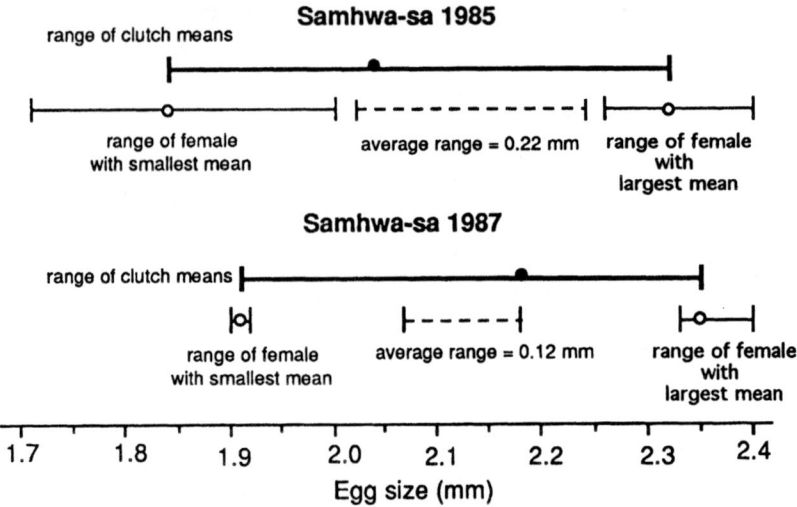

Figure 14.3 Representative summary data from an 8-year field study on *Bombina orientalis*. Data are from the same population during two different years. The bold lines show the limits and mean for the average egg size produced by females during a single breeding season. The plain solid lines show the limits of egg size produced within a single clutch and the average of that clutch for the females that produced both the smallest average egg and the largest average egg during each of the seasons. The dashed lines show the average intraclutch egg size variation for each of the years. From Kaplan and King (1997).

that produced the smallest eggs and the largest eggs in each of the representative years, as well as the average intraclutch egg size variation in each of the years. While no significant differences in intraclutch egg size variation were found among the three populations of *Bombina* studied (Kaplan and King 1997), we did find substantial variation among years in intraclutch CV, with the lowest mean being 1.9% in Samhwa-sa in 1987 and the highest mean being 3.4% in a population from Kangchon in 1985, and an overall range for individuals extending from 0.5% to 7.5%. As with mean egg size, the levels of intraclutch variation observed in these field studies are similar to those reported for other amphibians. Intraclutch CVs among five species of tropical hylids showed no significant differences, but values ranged from 2.2% to 4.8 % (Crump 1981). In a single population of *Hyla crucifer*, the intraclutch CV ranged from 2.1% to 5.4% (Crump 1984). Williamson and Bull (1995) found substantial intraclutch variability in *Ranidella signifera*, with intraclutch CV values changing over time and ranging from 0.88% to 5.37%, while Tejedo and Reques (1992) found values ranging from 2.9% to 5.9% in a population of *Bufo calamita*, and Beachy (1993) found significant differences among five salamander species, with mean values ranging from approximately 4.1% to 7.0%. Thus, as figure 14.3 shows, the intraclutch egg size variation of an individual can itself be a small to a substantial part of the overall egg size variation observed in a population.

Concomitant laboratory studies bolster the strength of the field results that imply a great deal of plasticity in ovum size (Kaplan and King 1997). For example, yearly shifts in egg size were positively associated with environmentally induced shifts in body mass in the field in a manner analogous to correlated responses to temperature- and food-induced changes in body size and egg size in the laboratory (Jørgensen 1982; Kaplan 1987). There were also significant differences in the magnitude of plasticity exhibited by individual females in each of three *Bombina* populations as measured by CVs of mean egg size and intraclutch egg size variation among females based on repeated breedings of individuals in the laboratory. Repeatability values for individual populations ranged from 0.14 to 0.27 for mean egg size, but were not significantly different from zero for intraclutch egg size variation. Other specific environmental influences on egg size are discussed in section 14.4.

Taken together, these results demonstrate that egg size variation in amphibians (which is close to the currency of natural selection) is highly subject to nongenetic sources of variation and that maternal investment in offspring shows a great deal of phenotypic plasticity (which can be defined here as equivalent to the norm of reaction of a particular genotype; Sultan 1995). The question of selection for levels of plasticity is dealt with in section 14.4.

14.2 The Main Effects of Maternal Effect Variation on Offspring Development and Implications for Offspring Fitness

The basic amphibian model presented in figure 14.1 can be dissected to ask how this maternal effect variation by itself can impact subsequent development of individual offspring. This question is first considered in the context of a constant offspring environment (i.e., excluding effects related to section 3 in figure 14.1) and then in relation to a more realistic (and complex) heterogeneous larval environment.

While interspecific variation and correlation with other life history traits have long been recognized (Salthe 1969; Salthe and Duellman 1973; Kaplan and Salthe 1979; Nussbaum 1985), published laboratory and field research provides evidence that intrapopulational ovum size variation has important ramifications for offspring development within many species of amphibians (e.g., see Kaplan 1985 for review). In *Bombina orientalis*, for example, studies have shown that offspring that develop from larger eggs are larger at the hatching stage, stage of first feeding, and metamorphosis (Kaplan 1985, 1989). In addition, studies have found that individuals developing from larger eggs can metamorphose in less time (Parichy and Kaplan 1992a; Semlitsch and Schmiedehausen 1994).

In terms of fitness ramifications, many studies of amphibian larvae have shown deleterious effects of small size. For instance, smaller larvae can be more vulnerable to predators if they have lower growth rates (Caldwell et al. 1980; Wilbur et al. 1983; Kusano et al. 1985) or reduced sprint speeds (Wassersug and Sperry 1977; Huey 1980; Richards and Bull 1990). Smaller larvae can also experience chemical or behavioral interference from larger larvae (Richards 1962; Wilbur and Collins 1973; Woodward 1987; C. K. Smith 1990), and they may be less able to exploit limited resources (Wilbur 1977; Steinwascher 1979; Travis 1984), resulting in slower developmental rates and delayed metamorphosis. This in turn can increase the risks associated with habitat drying or predation (Newman 1987; Petranka and Sih 1987). Moreover, individuals that are small at metamorphosis can experience greater risks of desiccation or starvation (Berven and Gill 1983) and can be more vulnerable to terrestrial predators because of reduced locomotor ability and stamina (Taigen and Pough 1981; John-Alder and Morin 1990). Also, small metamorphs may breed later (Smith 1987; Semlitsch et al.

1988; Berven 1990), have reduced fecundity (Kaplan and Salthe 1979), and have less mating success (Howard 1980).

14.3 The Interactive Effects of Maternal Effects Variation with Variation in the Larval Environment

The preceding examples suggest that offspring size is positively related to fitness. These observations in amphibians and other taxa (e.g., Bagenal 1969; Thomas 1983; McGinley et al. 1987; Tessier and Consolatti 1989; Amundsen and Stokland 1990; Rossiter 1991) have influenced the development of optimality models of parental investment as discussed in the introductory remarks and in section 14.1 (e.g., Smith and Fretwell 1974; Brockelman 1975; McGinley et al. 1987; Morris 1987; Winkler and Wallin 1987). Such models generally assume that (1) larger initial size results in greater offspring fitness, and (2) a necessary trade-off exists between energy expenditure per offspring and the number of offspring produced. Both of these assumptions are problematic (for some treatment of the second assumption, see Kaplan and Salthe 1979; Kaplan 1980; Lessels et al. 1989). There is a growing realization that a particular level of maternal investment in individual offspring can have a range of consequences in different environments (Capinera 1979; Berven and Chadra 1988; Venable and Brown 1988; Semlitsch and Gibbons 1990; Hutchings 1991; Kaplan 1992; Bernardo 1996b). Thus, what follows considers the shape of the fitness function that relates offspring fitness to maternal investment. The shape of this fitness function has implications for theoretical models that address the issue of the evolution of variance in propagule size. Some of these models are discussed more fully in section 14.4.

The maternal effect on offspring growth and development must interact with the environment in which the offspring finds itself (as shown in figure 14.1). This environment is notoriously heterogeneous, and the search for interactions between the maternal environment and subsequent larval environment on offspring growth, development, and ultimately offspring fitness is currently receiving much attention. A variety of studies have addressed this point, and interesting interactions have been uncovered (e.g., Kaplan 1985; Berven and Chadra 1988; Semlitsch and Gibbons 1990; Semlitsch and Schmiedehausen 1994). For example, in one study on *Bombina orientalis*, Parichy and Kaplan (1992a) found that the length of the larval period was unaffected by maternal investment in offspring unless larvae were injured by a simulated predator attack, in which case injured larvae that received less maternal investment required a greater length of time to metamorphose than those that received greater maternal investment.

Another study focused on the interaction between early larval size and rearing conditions in *B. orientalis* (Parichy and Kaplan 1992b). Conspecific density and resource availability have long been known to limit growth, development, and survivorship of anuran larvae (Wilbur and Collins 1973; Wilbur 1977; Semlitsch and Caldwell 1982; Newman 1987; Berven 1990; D. C. Smith 1990). To examine whether the effects of different levels of maternal investment vary among these kinds of environments, we reared larvae ($N = 360$) in one of two treatments representing different degrees of environmental quality. Snout-vent length at the feeding stage (stage 25; Gosner 1960) was used as a measure of maternal investment. In a "low-quality" environment, larvae were reared with two conspecific tadpoles and food was limited, whereas in a "high-quality" environment, larvae were reared individually and

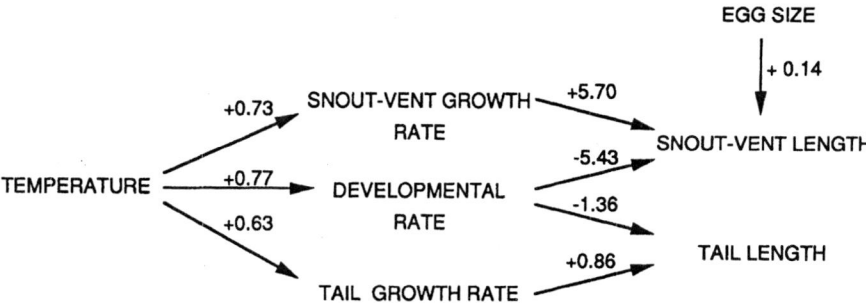

Figure 14.4 A path analysis of the effects of temperature during the prehatching period and maternal investment (egg size) on larval morphology. Standardized regression coefficients are shown and all are significant, $P < 0.01$. From Kaplan (1992).

were fed ad libitum. Among tadpoles reared in the low-quality environment, individuals that received less initial investment had smaller body sizes through metamorphosis and longer larval periods than individuals that received more. Among tadpoles reared in the high-quality environment, greater maternal investment had only a weak influence on later larval size and did not significantly affect metamorphic size or the duration of the larval period. This interaction between maternal investment and rearing conditions suggests that production of initially small offspring could be advantageous if these offspring develop in relatively benign environments (based on the assumption of an egg size-number trade-off) but disadvantageous in low-quality environments.

In the field, interactions between maternal investment in offspring and temperature in which larvae develop have also been shown to influence offspring fitness in ways not concordant with the assumptions of simple optimality models (Kaplan 1992). Offspring fitness was evaluated by exposing newly hatched *Bombina* larvae to interspecific tadpole predators and determining whether the large amount of thermal heterogeneity that is commonly found at a particular field site influences the relationship between maternal investment and survival. The path diagram in figure 14.4 shows the direct effects of temperature and egg size on larval characteristics. The outcome of these significant effects on offspring morphology and the probability of surviving predation by an interspecific tadpole are presented schematically in figure 14.5. The larger size that was induced by development in colder environments (as a result of increased tail size) seems to have afforded larvae significant protection from predation. However, the effects of increased maternal investment were not straightforward. Increased maternal investment and its concomitant larger size (mostly due to increased snout-vent length and associated mass) only provided protection from predation in colder environments. In warmer environments, increased maternal investment actually decreased offspring fitness.

Is there a functional basis for this morphology-fitness relationship that confounds the notion that increasing maternal investment increases offspring fitness? Studies in the laboratory (Parichy and Kaplan 1995) and field (Kaplan and Phillips in preparation) are elucidating the mechanisms for this interaction between maternal investment and temperature variability during development. It was found in the laboratory, as it had been in the field, that egg size had positive effects primarily on snout-vent length and to a lesser extent on tail

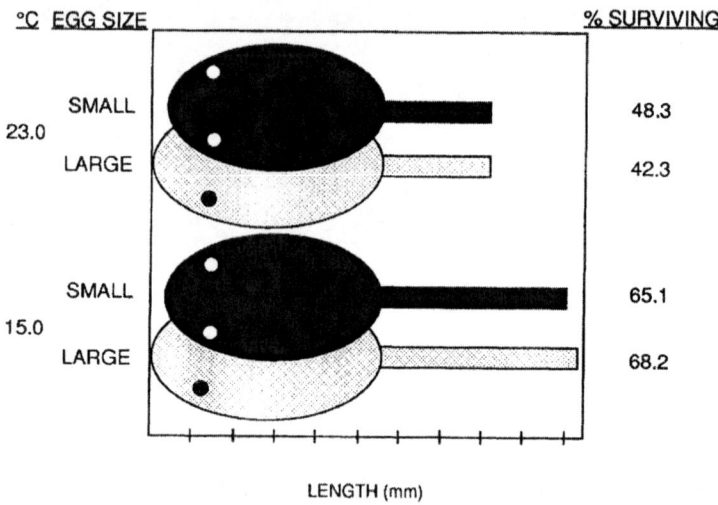

Figure 14.5 The relationship between maternal investment, developmental temperature, tadpole morphology (drawn to scale), and the probability of surviving predation based on the path analysis of figure 14.4 (from Kaplan 1992). See section 4.3 for explanation.

length, but temperature experienced during embryonic development had main effects at the time of hatching only on tail lengths, with individuals that developed at higher temperatures having shorter tails than individuals that developed at lower temperatures. These effects of egg size and developmental temperature combined to generate a range of larval morphologies that influenced sprint speed. The results of this study (Parichy and Kaplan 1995) are summarized in figure 14.6. The same interaction seen in the field, where increased maternal investment decreased offspring survival at warmer temperatures, was seen when locomotor performance (measured as sprint speed) was measured in the laboratory.

These results and those from the field (Kaplan 1992, described above) differ from earlier studies (cited in section 14.3) since larger eggs can increase or decrease both sprint speed and survival, depending on the stage at hatching (see also Sih and Moore 1993) and temperature experienced during development. These findings add to a growing body of empirical evidence indicating that the consequences of particular levels of maternal investment can vary significantly with offspring environment. In plants, for example, successful establishment can be independent of seed size in good conditions but positively related to seed size when conditions are less favorable (Venable and Brown 1988). In cladocerans, Tessier and Goulden (1987) found that smaller offspring survived better than larger offspring in environments of poorer quality. In frogs, individuals that initially receive lower levels of maternal investment can sometimes equal (Crump 1984; Parichy and Kaplan 1992a,b) or exceed (Kaplan 1985; Berven and Chadra 1988) the growth and developmental rates of individuals that receive greater levels of investment, depending on the quality of the larval environment. The effects of egg size on offspring viability can also differ across environments in salamanders (Semlitsch and Gibbons 1990), lizards (Sinervo et al. 1992), and fish (Hutchings 1991). Finally, offspring receiving greater levels of investment can be more susceptible to predation (Kaplan 1992) if they suffer reduced locomotor performance (as described above) or are more con-

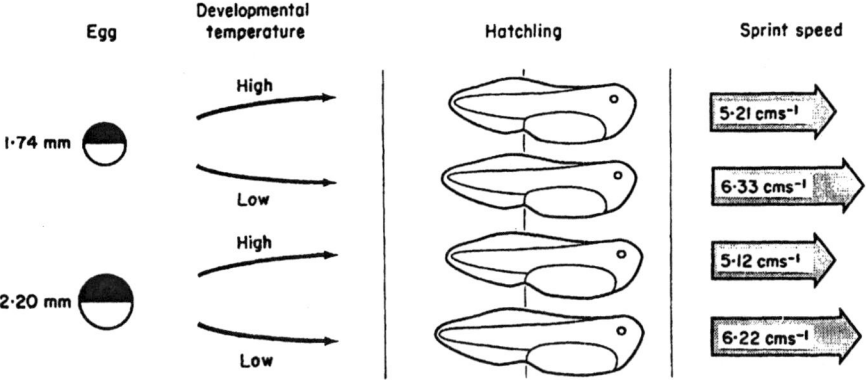

Figure 14.6 Schematic representation of the relationships among egg size, developmental temperature, hatchling morphology, and sprint speed. Figures are drawn to scale and are based on Parichy and Kaplan (1995).

spicuous (Capinera 1979), or if predators prefer large over small individuals (Woodward 1983; Crump and Vaira 1991).

14.4 Interactive Effects and the Evolution of the Sensitivity of Females to Fluctuations in their External Environment

Given that maternal investment in offspring in temperate zone amphibians is greatly influenced by environmental factors, the question arises as to whether such plasticity can be adaptive. It clearly has important consequences for offspring fitness.

An understanding of the interactions between initial offspring size (the maternal effect) and factors in the larval environment in which offspring develop is necessary to answer this question. The existence of such interactions has at least two implications. First, a female could maximize her fitness with different combinations of investment per individual offspring and total number of offspring produced, depending on the environment in which her offspring will develop. Under conditions favorable to all offspring, a female could increase her reproductive success by producing greater numbers of smaller offspring with the same total maternal investment. But, under more restrictive conditions, a female would be expected to maximize fitness by cuing on the environment and restricting her response to the most adaptive level of investment per offspring. These types of predictions have been made for other organisms (e.g., Venable and Brown 1988; Fox and Mousseau, this volume), and as discussed in the introduction and in section 14.1, such trade-offs between offspring size and number are implicit in optimality models of maternal investment (Smith and Fretwell 1974; Brockelman 1975; McGinley et al. 1987; Morris 1987; Winkler and Wallin 1987).

Second, the situation becomes particularly complex when considering organisms that occupy two or more distinct, spatially heterogeneous and temporally unpredictable environments during different stages of the life cycle (Wilbur 1980; Ebenman 1992). For example, breeding pools used by *B. orientalis* differ temporally and spatially in temperature, persistence, resource availability, and conspecific density (Kaplan 1989, 1992). As suggested above, different combinations of egg size and clutch size could be optimal in different pools. Yet, because maternal provisioning occurs in a terrestrial environment over an extended period prior to oviposition, females may not be able to predict the future aquatic environment their offspring will experience. Moreover, in ectothermic vertebrates the nutritional state and temperature experienced by a female during vitellogenesis (Shrode and Gerking 1977; Fraser 1980; Marsh 1984; Kaplan 1987), as well as female age (Berven 1988; Semlitsch and Gibbons 1990), female body size (Kaplan and Salthe 1979; Berven 1988), and geographic and altitudinal effects (Berven 1982; Takahashi and Iwasawa 1988), can all profoundly influence levels of investment and consequent initial offspring size. Hence, offspring might or might not have the appropriate phenotypes for the conditions they will encounter. Moreover, even if females selected oviposition sites (Crump 1991; Resetarits 1996) favorable for the particular sizes and numbers of eggs to be deposited, breeding ponds are ephemeral and conditions can change rapidly prior to metamorphosis (Newman 1987, 1992).

To reiterate, most theoretical investigations of optimal levels of maternal investment have assumed a monotonic, increasing relationship between levels of investment per offspring and offspring fitness (e.g., Smith and Fretwell 1974), and some have considered differences in the magnitude or convexity of this relationship (Brockelman 1975; Lloyd 1987; McGinley et al. 1987; Winkler and Wallin 1987; Venable and Brown 1988; Niewiarowski and Dunham 1994). But, conditions necessary for the evolution of within- and among-clutch variance in maternal investment have also been explored, and the work discussed here highlights the importance of considering not only optimal mean levels of investment but optimal variances (Gillespie 1977; Janzen 1977; Capinera 1979; Kaplan and Cooper 1984; Bull 1987; Philippi and Seger 1989; Schultz 1989, 1991; and see discussion in Bernardo 1996b).

Roff (1992, p. 389) discusses the problem of the evolution of optimal propagule size in a temporally stochastic environment and emphasizes that variance in egg size is most likely to be adaptive in such a situation. He presents several models including those of Kaplan and Cooper (1984) and McGinley et al. (1987). These models tend to disagree with each other, the former advocating greater likelihood of the evolution of adaptive variation in offspring size and the latter a lower likelihood. The disagreement very much hinges on the assumed shape of the fitness function. In Kaplan and Cooper's fitness function there is an optimal egg size in each environment, with fitness diminishing for that particular egg size as the environment changes (i.e., egg size × environment interaction). McGinley et al. (1987) argue that fitness is likely to be a monotonic function of egg size and hence of environmental quality. However, the data presented and reviewed in section 14.3 clearly show that, in a particular environment, increasing maternal investment does not necessarily increase offspring fitness and can, in fact, decrease offspring fitness contrary to the standard offspring size-fitness function assumption used again by McGinley et al. (1987). The point is that the shape of the fitness function is critical, and both models would be correct depending on that detail (see Bernardo 1996b for a further elaboration and concurring assessment of this controversy).

Conclusion

The studies discussed here suggest that impact of ovum size as a maternal effect can depend on a variety of factors in the offspring environment, and that the magnitude and occasionally even the sign of these impacts can change ontogenetically. Clearly, the evolutionary significance of such interactions will depend on their strength and timing during development (Cowley and Atchley 1992), though the outcome of selection is especially difficult to predict when both genetic and maternal environmental effects are considered (Kirkpatrick and Lande 1989). Nevertheless, observations such as these indicate that greater emphasis should be placed on evaluating the morphological and performance consequences of different levels of maternal investment per offspring across environments and ontogenetic stages.

The optimization of variance in offspring size, or the targeting of the reaction norm of this maternal effect by natural selection, is not without controversy (also see discussion in Bernardo 1996b). It is clear that much more work is needed both empirically and theoretically on the interactions between these maternal effects and the embryonic and larval environment in order to understand how they individually define the shape of the fitness function. Furthermore, such studies are needed to understand how the various interactions (arrows in figure 14.1) collect within the life cycle to exert an overall influence on the evolution of adaptive maternal effects.

Acknowledgments   I am grateful to David Parichy and Elizabeth King for contributing substantially to the several studies that are highlighted in this chapter. I also thank Tim Mousseau and several anonymous referees for helping to improve the quality of the manuscript.

References

Amundsen, T., and J. N. Stokland. 1990. Egg size and parental quality influence nestling growth in the shag. Auk 107:410–413.

Bagenal, T. B. 1969. Relationship between egg size and fry survival in brown trout *Salmo trutta* L. J. Fish. Biol. 1:349–353.

Beachy, C. K. 1993. Differences in variation in egg size for several species of salamanders (Amphibia: Caudata) that use different larval environments. Brimleyana 18:71–82.

Bernardo, J. 1991. Manipulating egg size to study maternal effects on offspring traits. Trends Ecol. Evol. 6:1–2.

Bernardo, J. 1996a. Maternal effects in animal ecology. Am. Zool. 36:83–105.

Bernardo, J. 1996b. The particular maternal effect of propagule size, especially egg size: Patterns, models, quality of evidence and interpretations. Am. Zool. 36:216–236.

Berven, K. A. 1982. The genetic basis of altitudinal variation in the wood frog *Rana sylvatica*. I. An experimental analysis of life history traits. Evolution 36:962–983.

Berven, K. A. 1988. Factors affecting variation in reproductive traits within a population of wood frogs (*Rana sylvatica*). Copeia 1988:605–615.

Berven, K. A. 1990. Factors affecting population fluctuations in larval and adult stages of the wood frog (*Rana sylvatica*). Ecology 71:1599–1608.

Berven, K. A., and B. G. Chadra. 1988. The relationship among egg size, density, and food level on larval development in the wood frog (*Rana sylvatica*). Oecologia 75:67–72.

Berven, K. A., and D. E. Gill. 1983. Interpreting geographic variation in life-history traits. Am. Zool. 23:85–97.

Brockelman, W. Y. 1975. Competition, the fitness of offspring, and optimal clutch size. Am. Nat. 109:677–699.

Bull, J. J. 1987. Evolution of phenotypic variance. Evolution 41:303–315.

Caldwell, J. P., Thorp J. H., and T. O. Jervey. 1980. Predator-prey relationships among larval dragonflies, salamanders, and frogs. Oecologia 46:285–289.

Capinera, J. L. 1979. Qualitative variation in plants and insects: Effect of propagule size on ecological plasticity. Am. Nat. 114:350–361.

Cowley, D. E., and W. R. Atchley. 1992. Quantitative genetic models for development, epigenetic selection, and phenotypic evolution. Evolution 46:495–518.

Crump, M. L. 1981. Variation in propagule size as a function of environmental uncertainty for tree frogs. Am. Nat. 117:724–737.

Crump, M. L. 1984. Intraclutch egg size variability in *Hyla crucifer* (Anura: Hylidae). Copeia 1984:302–308.

Crump, M. L. 1991. Choice of oviposition site and egg load assessment by a treefrog. Herpetologica 47:308–315.

Crump, M. L., and R. H. Kaplan. 1979. Clutch energy partitioning of tropical tree frogs (Hylidae). Copeia 1979:626–635.

Crump, M. L., and M. Vaira. 1991. Vulnerability of *Pleurodema borelli* tadpoles to an avian predator: Effect of body size and density. Herpetologica 47:316–321.

Cummins, C. P. 1986. Temporal and spatial variation in egg size and fecundity in *Rana temporaria*. J. Anim. Ecol. 55:303–316.

Diaz, N. F., M. Sallaberry, and J. Valencia. 1987. Microhabitat and reproductive traits in populations of the frog, *Batrachyla taeniata*. J. Herpetol. 21:317–323.

Dushane, G. P., and C. Hutchinson. 1944. Differences in size and developmental rate between eastern and midwestern embryos of *Ambystoma maculatum*. Ecology 25:414–423.

Ebenman, B. 1992. Evolution in organisms that change their niches during the life cycle. Am. Nat. 139:990–1021.

Falconer, D. S. 1989. Introduction to Quantitative Genetics. Longman, New York.

Fraser, D. T. 1980. On the environmental control of oocyte maturation in a plethodontid salamander. Oecologia 46:302–307.

Gillespie, J. H. 1977. Natural selection and variances in offspring numbers: A new evolutionary principle. Am. Nat. 111:1010–1014.

Gosner, K. L. 1960. A simplified table for staging anuran embryos and larvae with notes on identification. Herpetologica 16:183–190.

Howard, R. D. 1978. The influence of male-defended oviposition sites on early embryo mortality in bullfrogs. Ecology 59:789–798.

Howard, R. D. 1980. Mating behavior and mating success in wood frogs, *Rana sylvatica*. Anim. Behav. 28:705–716.

Huey, R. B. 1980. Sprint velocity of tadpoles (*Bufo boreas*) through metamorphosis. Copeia 1980:537–540.

Hutchings, J. A. 1991. Fitness consequences of variation in egg size and food abundance in brook trout *Salvelinus fontinalis*. Evolution 45:1162–1168.

Janzen, D. H. 1977. Variation in seed size within a crop of a Costa Rican *Micuna andreana* (Leguminosae). Am. J. Bot. 64:347–349.

John-Alder, H. B., and P. J. Morin. 1990. Effects of larval density on jumping ability and stamina in newly metamorphosed *Bufo woodhousii fowleri*. Copeia 1990:856–859.

Jørgensen, C. B. 1982. Factors controlling the ovarian cycle in a temperate zone anuran, the toad *Bufo bufo*: Food uptake, nutritional state, and gonadotropin. J. Exp. Zool. 224:437–443.

Kaplan, R. H. 1980. The implications of ovum size variability for offspring fitness and clutch size within several populations of salamanders (*Ambystoma*). Evolution 34:51–64.

Kaplan, R. H. 1985. Maternal influences on offspring development in the California newt, *Taricha torosa*. Copeia 1985:1028–1035.

Kaplan, R. H. 1987. Developmental plasticity and maternal effects of reproductive characteristics in the frog, *Bombina orientalis*. Oecologia 71:273–279.

Kaplan, R. H. 1989. Ovum size plasticity and maternal effects on the early development of the frog, *Bombina orientalis*, in a field population in Korea. Funct. Ecol. 3:597–604.

Kaplan, R. H. 1991. Developmental plasticity and maternal effects in amphibian life histories. Pp. 794–799 in E. C. Dudley (ed.), The Unity of histories. Evolutionary Biology. Dioscorides Press, Portland, Ore.

Kaplan, R. H. 1992. Greater maternal investment can decrease offspring survival in the frog *Bombina orientalis*. Ecology 73:280–288.

Kaplan, R. H., and W. S. Cooper. 1984. The evolution of developmental plasticity in reproductive characteristics: An application of the "adaptive coin-flipping" principle. Am. Nat. 123:393–410.

Kaplan, R. H., and E. G. King. 1997. Egg size is a developmentally plastic trait: evidence from long term studies in the frog *Bombina orientalis*. Herpetologica. 53:149–165.

Kaplan, R. H. and P. C. Phillips. The environmental and developmental context of natural selection: morphology, performance, and fitness in larval fire-bellied toads (*Bombina orientalis*) In preparation.

Kaplan, R. H., and S. N. Salthe. 1979. The allometry of reproduction: An empirical view in salamanders. Am. Nat. 113:671–689.

Kirkpatrick, M., and R. Lande. 1989. The evolution of maternal characters. Evolution 43:485–503.

Kozlowska, M. 1971. Differences in the reproductive biology of mountain and lowland common frogs, *Rana temporaria* L. Acta Biol. Cracov. 14:2–32.

Kuramoto, M. 1978. Correlations of quantitative parameters of fecundity in amphibians. Evolution 32:287–296.

Kusano, T., Kusano, H., and K. Miyashita. 1985. Size related cannibalism among larval *Hynobius nebulosus*. Copeia 1985:472–476.

Lessels, C. M., F. Cooke, and R. F. Rockwell. 1989. Is there a tradeoff between egg weight and clutch size in wild lesser snow geese (*Anser c. caerulescens*)? J. Evol. Biol. 2:457–472.

Lloyd, D. G. 1987. Selection of offspring size at independence and other size-versus-number strategies. Am. Nat. 129:800–817.

Marsh, E. 1984. Egg size variation in central Texas populations of *Etheostoma spectabile* (Pisces: Percidae). Copeia 1984:291–301.

Marsh, E. 1986. Effects of egg size on offspring fitness and maternal fecundity in the orangethroat darter, *Etheostoma spectabile* (Pisces: Percidae). Copeia 1986:18–30.

McGinley, M. A., D. H. Temme, and M. A. Geber. 1987. Parental investment in offspring in variable environments: Theoretical and empirical considerations. Am. Nat. 130:370–398.

Meffe, G. K. 1987. Embryo size variation in mosquitofish: Optimality vs. plasticity in propagule size. Copeia 1987:762–768.

Morris, D. W. 1987. Optimal allocation of parental investment. Oikos 49:332–339.

Newman, R. A. 1987. Effects of density and predation on *Scaphiopus couchi* tadpoles in desert ponds. Oecologia 71:301–307.

Newman, R. A. 1992. Adaptive plasticity in amphibian metamorphosis. Bioscience 42:671–678.

Niewiarowski, P. H., and A. E. Dunham. 1994. The evolution of reproductive effort in squamate reptiles: Costs, tradeoffs, and assumptions reconsidered. Evolution 48:137–145.

Nussbaum, R. A. 1985. The evolution of parental care in salamanders. Misc. Publ. Mus. Zool. Univ. Mich. 169:1–50.

Parichy, D. M., and R. H. Kaplan. 1992a. Developmental consequences of tail injury on larvae of the oriental fire-bellied toad, *Bombina orientalis*. Copeia 1992:129–137.

Parichy, D. M., and R. H. Kaplan. 1992b. Maternal effects on offspring growth and development depend on environmental quality in the frog *Bombina orientalis*. Oecologia 91:579–586.

Parichy, D. M., and R. H. Kaplan. 1995. Maternal investment and developmental plasticity: Functional consequences for locomotor performance of hatchling frog larvae. Funct. Ecol. 9:606–617.

Petranka, J. W., and A. Sih. 1987. Habitat duration, length of the larval period, and the evolution of a complex life cycle of a salamander, *Ambystoma texanum*. Evolution 41:1347–1356

Petranka, J. W., A. Sih, L. B. Kats, and J. R. Holomuzki. 1987. Stream drift, size specific predation, and the evolution of ovum size in an amphibian. Oecologia 71:624–630.

Philippi, T., and J. Seger. 1989. Hedging one's evolutionary bets, revisited. Trends Ecol. Evol. 4:41–44.

Resetarits, W. J. 1996. Oviposition site choice and life history evolution. Am. Zool. 36:205–215.

Richards, C. M. 1962. The control of tadpole growth by algal-like cells. Physiol. Zool. 35:285–296.

Richards, S. J., and C. M. Bull. 1990. Size-limited predation on tadpoles of three Australian frogs. Copeia 1990:1041–1046.

Roff, D. A. 1992. The Evolution of Life Histories. Chapman and Hall, New York.

Rossiter, M. C. 1991. Maternal effects generate variation in life history: Consequences of egg weight plasticity in the gypsy moth. Funct. Ecol. 5:386–393.

Salthe, S. N. 1969. Reproductive modes and the number and sizes of ova in the urodeles. Am. Midl. Nat. 81:467–490.

Salthe, S. N., and W. E. Duellman. 1973. Quantitative constraints associated with reproductive mode in anurans. Pp. 229–249 in J. L. Vial (ed.), Evolutionary Biology of the Anurans. University of Missouri Press, Columbia.

Scheiner, S. M. 1993. Genetics and evolution of phenotypic plasticity. Annu. Rev. Ecol. Syst. 24:35–68.

Schultz, D. L. 1989. The evolution of phenotypic variance with iteroparity. Evolution 43:473–475.

Schultz, D. L. 1991. Parental investment in temporally varying environments. Evol. Ecol. 5:415–427.

Semlitsch, R. D., and J. P. Caldwell. 1982. Effects of density on growth, metamorphosis, and survivorship in tadpoles of *Scaphiopus holbrooki*. Ecology 63:905–911.

Semlitsch, R. D., and J. W. Gibbons. 1990. Effects of egg size on success of larval salamanders in complex aquatic environments. Ecology 71:1789–1795.

Semlitsch, R. D., and S. Schmiedehausen. 1994. Parental contributions to variation in hatchling size and its relationship to growth and metamorphosis in tadpoles of *Rana lessonae* and *Rana esculenta*. Copeia 1994:406–412.

Semlitsch, R. D., D. E. Scott, and J. H. K. Pechmann. 1988. Time and size at metamorphosis related to adult fitness in *Ambystoma talpoideum*. Ecology 69:184–192.

Shrode, J. B., and S. D. Gerking. 1977. Effect of constant and fluctuating temperatures on reproductive performance of a desert pupfish, *Cyprinodon n. nevadensis*. Physiol. Zool. 50:1–10.

Sih, A., and R. D. Moore. 1993. Delayed hatching of salamander eggs in response to enhanced larval predation risk. Am. Nat. 142:947–960.

Sinervo, B. 1990. The evolution of maternal investment in lizards: an experimental and comparative analysis of egg size and its effects on offspring performance. Evolution 44:279–294.

Sinervo, B., and R. B. Huey. 1990. Allometric engineering: an experimental test of the causes of interpopulational differences in performance. Science 248:1106–1109.

Sinervo, B., P. Doughty, R. Huey, and K. Zamudio. 1992. Allometric engineering: A causal analysis of natural selection on offspring size. Science 258:1927–1930.

Smith, C. K. 1990. Effects of variation in body size on intraspecific competition among larval salamanders. Ecology 71:1777–1788.

Smith, D. C. 1987. Adult recruitment in chorus frogs: Effects of size and date at metamorphosis. Ecology 68:344–350.

Smith, D. C. 1990. Population structure and competition among kin in the chorus frog (*Pseudacris triseriata*). Evolution 44:1529–1541.

Smith, C. C., and S. D. Fretwell. 1974. The optimal balance between size and number of offspring. Am. Nat. 108:499–506.

Stearns, S. C. 1989. Evolutionary significance of phenotypic plasticity. Bioscience 39:436–445.

Stearns, S. C. 1992. The Evolution of Life Histories. Oxford University Press, New York.

Steinwascher, K. 1979. Competitive interactions among tadpoles: responses to resource level. Ecology 60:1172–1183.

Sultan, S. E. 1995. Phenotypic plasticity and plant adaptation. Acta Bot. Neerl. 44:363–383.

Taigen, T. L., and F. H. Pough. 1981. Activity metabolism of the toad (*Bufo americanus*): Ecological consequences of ontogenetic change. J. Comp. Physiol. 144:247–252.

Takahashi, H., and H. Iwasawa. 1988. Interpopulation variations in clutch size and egg size in the Japanese salamander, *Hynobius nigrescens*. Zool. Sci. (Tokyo) 5:1073–1081.

Tejedo, M., and R. Reques. 1992. Effects of egg size and density on metamorphic traits in tadpoles of the natterjack toad (*Bufo calamita*). J. Herpetol. 26:146–152.

Tessier, A. J., and N. L. Consolatti. 1989. Variation in offspring size in *Daphnia* and consequences for individual fitness. Oikos 56:269–276.

Tessier, A. J., and C. E. Goulden. 1987. Cladoceran juvenile growth. Limnol. Oceanogr. 32:680–686.

Thomas, C. S. 1983. The relationships between breeding experience, egg volume and reproductive success of the kittiwake *Rissa tridactyla*. Ibis 125:567–574.

Travis, J. 1983. Variation in development patterns of larval anurans in temporary ponds. I. persistent variation within a *Hyla gratiosa* population. Evolution 37:496–512.

Travis, J. 1984. Anuran size at metamorphosis: Experimental test of a model based on intraspecific competition. Ecology 65:1155–1160.

Venable, D. L., and J. S. Brown. 1988. The selective interactions of dispersal, dormancy, and seed size as adaptations for reducing risk in variable environments. Am. Nat. 131:360–384.

Wassersug, R. J., and D. G. Sperry. 1977. The relationship of locomotion to differential predation on *Pseudacris triseriata* (Anura: Hylidae). Ecology 58:830–839.

Wilbur, H. M. 1977. Propagule size, number, and dispersion pattern in *Ambystoma* and *Asclepias*. Am. Nat. 111:43–68.

Wilbur, H. M. 1980. Complex life cycles. Annu. Rev. Ecol. Syst. 11:67–93.

Wilbur, H. M., and J. P. Collins. 1973. Ecological aspects of amphibian metamorphosis. Science 182:1305–1314.

Wilbur, H. M., Morin, P. J., and R. N. Harris. 1983. Salamander predation and the structure of experimental communities: Anuran responses. Ecology 64:1423–1429.

Williamson, I., and C. M. Bull. 1989. Life history variation in a population of the Australian frog *Ranidella signifera*: Egg size and early development. Copeia 1989:349–356.

Williamson, I., and C. M. Bull. 1995. Life-history variation in a population of the Australian frog *Ranidella signifera*: Seasonal changes in clutch parameters. Copeia 1995:105–113.

Winkler, D. W., and J. Wallin. 1987. Offspring size and number: A life history model linking effort per offspring and total effort. Am. Nat. 129:708–720.

Woodward, B. 1983. Tadpole size and predation in the Chihuahuan desert. Southw. Nat. 28:470–471.

Woodward, B. D. 1987. Interactions between Woodhouse's toad tadpoles (*Bufo woodhousii*) of mixed sizes. Copeia 1987:380–386.

15

Perinatal Influences on the Reproductive Behavior of Adult Rodents

MERTICE M. CLARK & BENNETT G. GALEF, JR.

In litter-bearing rodent species, the location that a male or female fetus occupies relative to siblings of same or opposite sex influences its level of prenatal exposure to gonadal hormones. In particular, varying amounts of prenatal exposure to testosterone produce cascades of neurendocrine events that, in turn, result in variation in such biologically important characteristics as the timing of puberty, an individual's lifetime fecundity, the sex ratio of its offspring, and the magnitude of its parental investment. Consequently, studies of intrauterine effects on adult patterns of reproduction provide a means of examining the relationship between normally occurring variation in perinatal exposure to hormones and the variation in reproductive tactics often seen in mammalian populations. Dams can only indirectly influence the intrauterine position of their offspring by varying the size and sex ratio of their litters and the distribution of litter members within the uterus. However, because the same hormonal mechanisms that support indirect maternal effects on behavior also support direct maternal influences on the phenotype of offspring, examination of effects of intrauterine position increases understanding of direct maternal effects on phenotypes of adult mammals.

In mammals, as in other vertebrates (Schwabl 1993; Hews et al. 1994), an individual's reproductive performance can be profoundly affected by its level of exposure to gonadal hormones in the days surrounding its birth. Consequently, differences in perinatal exposure to hormones can be used to explain variation in reproductive tactics exhibited by individuals of the same species and sex.

Results of the relatively few studies examining effects of variation in perinatal exposure to hormones on adult reproductive behavior in natural populations of mammals have been less conclusive than have results of studies conducted in the laboratory with domesticated animals. Consequently, when discussing natural populations, we shall have to settle for clues as to effects of perinatal hormonal experiences on reproduction. Definitive evidence of such effects has been provided only in laboratory situations.

15.1 Rodent Models

House Mouse and Norway Rat

In many litter-bearing rodent species, such as the house mouse (*Mus domesticus*) and Norway rat (*Rattus norvegicus*), the intrauterine position (IUP) that a male or female fetus occupies relative to siblings of same or other sex influences the hormonal milieu in which that fetus matures (Gandelman et al. 1977; Clemens et al. 1978; Meisel and Ward 1981). For example, a male mouse fetus occupying an IUP between two male fetuses (a 2M male) has greater blood concentrations of testosterone than does its brother maturing in an IUP between two female fetuses (2F males). Similarly, female fetuses located between males (2M females) have higher testosterone titers than do their sisters located between females (2F females). (At different times during the exploration of IUP effects, different classification schemes have been used to assign fetuses to IUPs. To simplify matters here, we discuss all the data as though fetuses had been classified in terms of the number of adjacent male and female fetuses, though that is not always the case in the original research.)

There has been some controversy as to how steroid hormones travel between fetuses and, consequently, as to whether the sex of immediate intrauterine neighbors determines the level of exposure to gonadal hormones that a fetus receives (Meisel and Ward 1981; Richmond and Sachs 1984; Houtsmuller and Slob 1990; Houtsmuller et al. 1994). However, recent studies of both dye transport within the uteri of pregnant rats and the movement of radioactively labeled testosterone between fetal rats and fetal mice indicate that, at least in these two species, androgens secreted by the gonads of males late in gestation diffuse through the amniotic fluid and cross fetal membranes to adjacent fetuses (Even et al. 1992; vom Saal and Dahr 1992). Such diffusion causes fetuses located between males to receive greater prenatal exposure to exogenous testosterone than do fetuses located between females.

In house mice, a range of androgen-sensitive, biologically meaningful characteristics are correlated with IUP: 2M females have first estrous at a later age, a longer estrous cycle, a shorter reproductive life, fewer litters during their lifetimes, and a greater percentage of males/litter and are both less attractive to males and more aggressive than are their 2F sisters (vom Saal 1984, 1989; Vandenbergh and Huggett 1995). Male mice from 2M IUPs are more aggressive and (perhaps unexpectedly) less sexually active, less infanticidal, and more paternal than are their 2F brothers (vom Saal 1984, 1989).

Some laboratories have failed to observe some of these effects of IUP on the phenotypes of female house mice (Simon and Cologer-Clifford 1991; Juliban and Nyby 1992). However, observations of robust effects of IUP on both the morphology and reproductive behavior of females of another litter-bearing rodent species (the Mongolian gerbil, *Meriones unguiculatus*), quite similar to those observed in mice, suggest that effects of IUP on mice reported in the literature are real.

Below, we first review our studies of effects of IUP on reproduction in Mongolian gerbils, then discuss briefly the effects of stressors applied to rodent dams on the phenotypes of their offspring. A more comprehensive general review of maternal effects on mammalian development can be found in Moore (1995).

Mongolian Gerbils

Females.   Our interest in effects of IUP on reproduction in Mongolian gerbils arose from a serendipitous discovery made while studying the development of sexual behavior in female Mongolian gerbils. We discovered that age at vaginal introitus (which, in Mongolian gerbils, is significantly correlated with age at first parturition; Clark and Galef 1988) was bimodally distributed: some gerbils exhibited vaginal perforation before their eyes opened (day 16 post-partum), and others only after they weaned (day 25 postpartum); there also was a period from day 22 to day 27 postpartum when essentially no vaginal opening occurred (Clark et al. 1986).

When we examined the reproductive life histories of samples of such early- and late-maturing female gerbils, we found marked differences in their patterns of reproduction. (1) Early-maturing females were less likely than were late-maturing females to behave aggressively toward, and more likely to be impregnated by, strange males with whom they were randomly paired. (2) Early-maturing females reproduced for the first time at an earlier age, had significantly more litters during their lifetimes, produced slightly more young in each litter, and, consequently, had a lifetime fecundity more than twice as great as that of late-maturing females. (3) Litters delivered by early-maturing females contained a greater proportion of daughters and a greater proportion of early-maturing daughters than did litters delivered by late-maturing females. (4) Finally, early-maturing females directed less maternal behavior toward their young than did late-maturing females; the former animals spent less time nursing their young and were less likely to retrieve offspring displaced from the nest than were the latter (Clark and Galef 1986; Clark et al. 1986).

Some of these differences between early- and late-maturing females reflected differences in their ages at first parturition or responses to litters of varying size and sex ratio. However, when we controlled such factors postnatally, differences in the sex ratios of litters produced by early- and late-maturing females persisted, as did differences in ages at sexual maturation of their daughters (Clark and Galef 1986). Presumably, differences in the prenatal experiences of daughters of early- and late-maturing mothers affected the daughters' patterns of reproduction when adult.

As already noted, early-maturing female gerbils tended to be born as members of relatively large, female-biased litters, while late-maturing females came from relatively small, male-biased litters. The size and sex composition of a litter affect the probability that any fetus in it will occupy an IUP between either two male or two female fetuses: as the proportion of males in a litter increases, the probability that a fetus in that litter will occupy an IUP between males increases; as the size of a litter decreases, the probability of a fetus being located at the end of a line of pups in a uterine horn increases, and the probability of its being between two fetuses of either sex decreases (Clark et al. 1993b). Consequently, it seemed reasonable to ask whether the correlations between age at sexual maturation of a mother, the sex ratios of her litters, and the reproductive behaviors of her daughters might be mediated by different levels of prenatal exposure to gonadal hormones experienced by females that matured either early or late.

We found that, regardless of their respective dam's age at sexual maturation, daughters from 2F IUPs were almost sure to be early maturing, while daughters from 2M IUPs produced slightly more late- than early-maturing daughters (Clark and Galef 1988). Further, within both sexes, fetuses from 2M IUPs had higher circulating levels of testosterone than did fetuses from 2F IUPs (Clark et al. 1991), a result consistent with the hypothesis that differences in prenatal exposure to testosterone, resulting from IUP, might mediate some of the observed differences in the reproductive profiles of female gerbils born to early- and late-maturing mothers.

We could not entirely explain the difference in ages at vaginal introitus of daughters born to early- and late-maturing mothers by reference to the IUPs that mothers had occupied as fetuses. Daughters of early-maturing mothers located in 2M IUPs were twice as likely to mature early as were 2M daughters of late-maturing mothers (Clark and Galef 1988). Fetuses located in utero between one male and one female fetus and gestated by late-maturing mothers had higher blood titers of testosterone than did fetuses in similar IUPs gestated by early-maturing mothers (Clark et al. 1991). Thus, while IUP accounted for much variation in both circulating levels of testosterone in gerbil fetuses and age at vaginal opening in female gerbils, there were additional direct maternal effects on the endocrinology of young and their consequent rates of maturation.

Yalcinkaya et al. (1993) have proposed that the malelike aggressive behavior and genitalia of female spotted hyenas (*Crocuta crocuta*) result from exposure of fetal female hyenas to high levels of androstenedione produced in the ovaries of hyena mothers and converted to testosterone in the placenta. Possibly, some effects on female gerbils gestated by 2M mothers also result from exposure of gerbil fetuses to the higher circulating levels of testosterone to be found in 2M than in 2F gerbil dams (Clark et al. 1991).

Whatever the hormonal mechanisms that mediate effects of IUP on the reproductive morphology and behavior of Mongolian gerbils, such effects are profound (Clark and Galef, 1995a, 1995b). As previously noted, because of differences in the sex ratios of litters gestated by 2M and 2F gerbils (and mice; Vandenbergh and Huggett 1995), daughters born to 2M females are more likely to themselves be 2M females than are daughters of 2F females. Conversely, daughters born to 2M mothers are more likely than are daughters born to 2F mothers to be gestated in 2F IUPs. Such influence of the IUP in which a mother matured on the IUP of her daughters results in hormonally mediated transmission between gerbil dams and their daughters of those characteristics affected by levels of prenatal exposure to testosterone. Gerbil daughters resemble their mothers not only because they have in common a large proportion of genes, but also because, as a result of their statistical tendency toward congruence in IUP, mothers and daughters tend to have similar histories of fetal exposure to steroids (Clark et al. 1993b; Clark and Galef 1994).

Males.   Given the variable and demanding steppe environment in which Mongolian gerbils evolved, it is not difficult to suggest reasons why natural selection might maintain the reproductive patterns exhibited by both 2M and 2F gerbil females (Clark et al. 1986). Observed effects of IUP on the reproductive profiles of male gerbils are considerably more difficult to understand because male Mongolian gerbils gestated in 2F IUPs appear to be at a distinct reproductive disadvantage.

Adult male gerbils from 2M and 2F IUPs paired for 3 weeks with each of a succession of virgin females differed in the number and size of litters that the females paired with them produced. Litters sired by 2M males were slightly (but not significantly) larger than those sired by 2F males, and 2F males were five times more likely than were 2M males to fail to impregnate a female during a 3-week period of cohabitation (Clark et al. 1992, 1996).

Observation of the copulatory patterns of males from different IUPs revealed consistent inadequacies in the copulatory performance of 2F males: they had longer latencies to intromit, longer latencies to ejaculate, and, of greatest importance, were significantly less likely than were 2M males to achieve ejaculation when paired with an unfamiliar virgin female (Clark et al. 1990, 1996).

As one might expect, given the lower potency of 2F than of 2M males, female gerbils can discriminate between them: females scent mark more frequently in response to scent marks of 2M males than to those of 2F males (Clark and Galef 1994) and, when in estrous (but not at other times during the estrous cycle), prefer to affiliate with the 2M rather than 2F males (Clark et al. 1992).

Differences in both the copulatory behavior of males from different IUPs and the response of females to them may be mediated by differences in circulating levels of testosterone and responsiveness to testosterone found in adult males that were gestated in 2M and 2F IUPs: adult males from 2M IUPs have higher circulating levels of testosterone than do those from 2F IUPs (Clark et al. 1992b), and castrate 2M males are more responsive to fixed levels of exogenous testosterone than are 2F males (Clark et al. 1993a).

The reduced attractiveness to females and lower sexual competence of 2F male gerbils are characteristics one might expect natural selection to have acted vigorously to suppress, and there is some evidence that such selection has been at work. By segregating male and female fetuses in different uterine horns, a gerbil mother could protect her sons from prenatal contact with females and the reduced copulatory success that such contact entails. In fact, female gerbils tend to gestate sons in the right uterine horn and daughters in the left (Clark and Galef 1990), thus producing greater numbers of 2M male and 2F female offspring than would otherwise be expected.

The right ovaries of female gerbils produce a greater proportion of male-destined eggs than do the left ovaries (Clark et al. 1994). When we surgically exchanged right and left ovaries within female gerbils, they produced more male fetuses in their left than in their right uterine horns. On the other hand, female gerbils whose left and right ovaries we removed and reimplanted in their original locations continued to produce more males in right than in left uterine horns (Clark et al. 1994).

There is second way that gerbil mothers might influence the future potency of their sons. It has been known for some time that the amount of anogenital licking that a male rat pup receives can influence its pattern of copulatory behavior when adult: male rats receiving greater amounts of anogenital licking as infants have shorter ejaculatory latencies and shorter interintromission intervals than do brothers receiving less anogenital grooming as infants, though it is not known whether the amount of anogenital stimulation that a pup has received has any affect on its reproductive success (Moore 1983; C. L. Moore, personal communication).

We have found that the greater the number of male intrauterine neighbors a gerbil pup has while a fetus, the more time its mother spends grooming its anogenital area (Clark et al. 1989). And, as already mentioned, there is a correlation between the IUP that a male gerbil fetus occupies and its reproductive success when adult: male gerbils from 2F IUPs that receive relatively little maternal anogenital grooming (like male rats that received relatively little maternal anogenital grooming) have longer ejaculation latencies and longer postejaculatory intervals than do male gerbils from 2M IUPs. Consequently, effects of IUP on copulatory performance may be mediated by differences in maternal anogenital grooming.

Whatever the underlying mechanism, the selective pressures maintaining the copulatory patterns of 2F males are difficult to understand. Fitness costs of reduced potency are obvious;

the benefits of impotence are not. However, in a recent review of effects of testosterone on the reproductive behavior of birds, Ketterson and Nolan (1994) described several cases in which elevated levels of testosterone have two effects, both increasing sexual behavior and decreasing parental behavior. Such negative correlations among testosterone-sensitive behavioral traits might be design constraints that limit adaptation or trade-offs that persist because they provide a simple mechanism by which animals adjust their reproductive tactics in response to variations in environmental conditions.

We have recently begun to explore the possibility that adult 2F male gerbils (with low circulating levels of testosterone), though clearly less potent than their 2M brothers (with high circulating levels of testosterone), may increase their reproductive success by increasing their investment in the relatively few young they sire. The data are very promising. At least in the laboratory, 2F male Mongolian gerbils are consistently more attentive to young than are 2M males. Perhaps in natural circumstances, the increased investment 2M male gerbils make in their offspring compensates for their decreased copulatory success.

15.2 Field Evidence

Because of the difficulty both of identifying rodents from different IUPs in the field and of manipulating early experience in natural circumstances, almost all work on perinatal effects on the behavior of mammals has been carried out in laboratory populations. However, in a few instances, animals have been released into natural environments in experiments studying effects of perinatal experience on fitness. Ims (1987) placed 74, individually marked, laboratory-bred voles (*Clethrionomys rufocanus*) onto a 1.8-hectare island (thus doubling the density of voles there) and live trapped the population at 3-week intervals. Laboratory-bred females that succeeded in establishing themselves on the island came predominantly from litters with a female-biased sex ratio (64% female), while females that failed to establish themselves came from male-biased litters (65% male). Ims, citing vom Saal and Bronson (1980), suggested that voles from female-biased litters were less likely to mature in IUPs adjacent to males and were therefore more likely to be docile and socially tolerant in high-density conditions than were presumably more aggressive female voles from male-biased litters. Of course, it is also possible that intrauterine exposure of female voles to males simply increased their probability of migrating. Indeed, Holekamp et al. (1984) found that experimental perinatal exposure of female Belding's ground squirrels (*Spermophilus beldingi*) to testosterone significantly increased their rate of dispersal.

In an experiment conceptually similar to that of Ims (1987), Zielinski et al. (1992) monitored the movement patterns and reproductive success of laboratory-reared *Mus musculus* from known IUPs after they were released onto highway islands. As predicted from the outcomes of staged encounters between 2F and 2M female mice in the laboratory, 2M females in the wild had significantly larger territories than did 2F females. However, no effect of IUP on either probability of survival or number of uterine scars was found in the released population.

Jacquot and Vessey (1995) examined the relationship between litter composition and dispersal of white-footed mice (*Peromyscus leucopus*) in an oak-hickory woodlot and found that females with many brothers dispersed farther than did females with few brothers.

While such studies provide useful first steps in identifying possible effects of perinatal experience on survival and reproduction in natural circumstances, they have, as yet, failed to provide compelling evidence of an impact of IUP on reproductive behavior in natural conditions. Technical limitations have made it impossible to monitor the reproductive profiles of animals from different IUPs in natural circumstances, so it is not too surprising that progress has been limited.

Recent evidence both from our laboratory (Forger et al. 1996) and that of Vandenbergh (Vandenbergh and Huggett 1994, 1995) suggest that it may be possible to determine the perinatal hormonal experience of adults by examining the morphology of their genitals. Vandenbergh and Huggett (1994) have described a weight-corrected index of the distance between the anal and genital opening in female house mice that permits discrimination, at weaning, of females exposed to high and low levels of androgens while in utero. Forger et al. (1996) have provided evidence of marked differences in the relative weight of the genital musculature of adult male Mongolian gerbils from 2M and 2F IUPs. In the future it may be possible to correlate differences in the reproductive tactics observed in free-living rodents with morphological characteristics indicative of their early exposure to gonadal hormones and, consequently, to understand some of the variability in individual reproductive tactics exhibited by members of natural populations.

15.3 Effects of Perinatal Stress on Adult Reproduction

The laboratory experiments described above demonstrate that experiences of rodents during the perinatal period, especially experience of prenatal exposure to steroid hormones originating in uterine neighbors, can have important effects on subsequent patterns of reproductive behavior. We focus here on the impact of IUP on adult reproduction because exploration of effects of IUP on adult phenotypes has been the focus of much recent research on effects of perinatal exposure to hormones in mammals. Of course, it has been known for some time that variables other than IUP influence both exposure of fetuses to gonadal hormones and expression of hormone-sensitive aspects of adult phenotypes. Physical stress applied to pregnant rodents affects many of the same anatomical, physiological, and behavioral traits that are affected by IUP: age at vaginal opening, length of estrous cycle, and litter sex ratios are all greater in daughters of stressed than of unstressed females. Daughters of physically stressed mothers are both less fertile and less fecund than are daughters of unstressed mothers (Herrenkohl and Politch 1978; Herrenkohl 1979; Ward and Weisz 1980; Politch and Herrenkohl 1984; Kinsley and Svare 1988). Similarly, daughters born to pregnant house mice housed at high densities (a potential social stressor) have greater anogenital distances and reduced copulatory receptivity relative to controls (Allen and Haggett 1977; Zielinski et al. 1991). Sons of stressed rat mothers are less willing to copulate with females in estrous than are sons of unstressed mothers (Ward 1972), and the impaired sexual behavior of sons of stressed mothers is believed to reflect a shift away from a sensitive period for central nervous system development in the age at which fetal sons of stressed mothers experience a species-typical surge in circulating levels of testosterone (Ward and Weisz 1980).

Given the hormonal basis of both effects of stress and IUP on adult reproductive phenotypes, it is not surprising to find that effects on reproductive behavior of stressors and IUP interact in important ways. For example, Vom Saal et al. (1990) have reported that physical stress applied to mice during the last week of their pregnancies results in (1) higher circulating levels of testosterone in both male and female mouse fetuses and (2) increases in anogenital distance and estrous cycle lengths in 2F, but not 2M, female mice. Direct effects of the internal state of a dam and effects of neighboring fetuses interact to produce the adult phenotype.

Conclusion

During prenatal life, the environment in which fetal mammals develop is open to manipulation by their dams. Variation in the hormonal state of dams, as well as in the sex of the uterine neighbors a dam provides for each of her fetuses, can have profound impact on the prenatal microenvironments to which her young are exposed from conception to delivery. As we have seen, the intrauterine environment in which young develop affects biologically important features of their reproductive behavior when adult.

Some maternal effects on adult phenotype are relatively direct. For example, stressors applied to a dam cause changes in the level of prenatal exposure of her young to gonadal hormones, which in turn cause changes in their reproductive tactics when adult: females from 2M IUPs produce daughters with different reproductive profiles than do females from 2F IUPs, and preferential anogenital licking of some sons increases their copulatory efficiency. Other maternal effects are less direct. Females produce litters of varying size and sex ratio and can influence some parameters of the distribution of fetuses in the uterus. In such cases, a mother influences the uterine environment in which her young mature by affecting the probability that they will occupy different IUPs.

Because direct and indirect maternal effects on mammalian development are mediated by similar hormonal mechanisms, study of IUP effects on adult behavior are not merely inherently interesting. Such studies also provide a means of exploring the hormonal mechanisms that underlie direct maternal effects on development of reproductive behaviors.

Perinatal effects on the reproductive behaviors of domesticated rodents are now well established. However, much difficult work remains to determine whether and how variation in conditions of perinatal life affect reproductive behavior of animals living in natural circumstances. Recent discoveries of markers of differential exposure to gonadal hormones in infancy in adult rodents of both sexes offer promise of progress in future studies of the fitness consequences of early exposure to hormones. Should such experiments prove successful, studies of effects on the adult reproductive behavior of naturally occurring variation in perinatal exposure to hormones will provide an opportunity for integrative investigations of mammalian behavior from molecular to population levels of analysis. Analysis of both direct and indirect effects of dams on the reproductive phenotypes of their offspring is prerequisite for development of such innovative research programs.

References

Allen, T., and B. Haggett. 1977. Group housing of pregnant female mice reduces copulatory receptivity of female progeny. Physiol. Behav. 19:61–68.

Clark, M. M., and B. G. Galef Jr. 1986. Postnatal effects on reproduction and maternal care in early- and late-maturing gerbils. Physiol. Behav. 36:997–1003.

Clark, M. M., and B. G. Galef Jr. 1988. Effects of uterine position on rate of sexual development in female Mongolian gerbils. Physiol. Behav. 42:15–18.

Clark, M. M., and B. G. Galef Jr. 1990. Sexual segregation in the left and right horns of gerbil uterus: "The male embryo is usually on the right and the female on the left" (Hippocrates). Dev. Psychobiol. 23:29–38.

Clark, M. M., and B. G. Galef Jr. 1994. Sex-ratio and inheritance. Nature 367:327–328.

Clark, M. M., and B. G. Galef Jr. 1995a. A gerbil mother's fetal intrauterine position affects the sex ratios of the litters she gestates. Physiol. Behav. 57:297–299.

Clark, M. M., and B. G. Galef Jr. 1995b. Prenatal influences on reproductive life-history strategies. Trends Ecol. Evol. 10:151–153.

Clark, M. M., C. A. Spencer, and B. G. Galef Jr. 1986. Reproductive life history correlates of early and late sexual maturation in Mongolian gerbils (*Meriones unguiculatus*). Anim. Behav. 34:551–560.

Clark, M. M., S. Bone, and B. G. Galef Jr. 1989. Intrauterine positions and schedules of urination: Correlates of differential maternal anogenital stimulation. Dev. Psychobiol. 22:389–400.

Clark, M. M., S. A. Malenfant, D. A. Winter, and B. G. Galef Jr. 1990. Fetal uterine position affects copulation and scent marking by adult male gerbils. Physiol. Behav. 47:301–305.

Clark, M. M., D. Crews, and B. G. Galef Jr. 1991. Concentrations of sex steroid hormones in pregnant and fetal Mongolian gerbils. Physiol. Behav. 49:239–243.

Clark, M. M., L. Tucker, and B. G. Galef Jr. 1992a. Stud males and dud males: Intrauterine position effects on the reproductive success of male gerbils. Anim. Behav. 43:215–221.

Clark, M. M., F. S. vom Saal, and B. G. Galef Jr. 1992b. Foetal intrauterine position correlates with endogenous testosterone levels of adult male Mongolian gerbils. Physiol. Behav. 51:957–960.

Clark, M. M., A. M. Bishop, F. S. vom Saal, and B. G. Galef Jr. 1993a. Responsiveness to testosterone of male gerbils from known intrauterine positions. Physiol. Behav. 53:1183–1187.

Clark, M. M., P. Karpiuk, and B. G. Galef Jr. 1993b. Hormonally mediated inheritance of acquired characteristics in Mongolian gerbils. Nature 364:712

Clark, M. M., M. Ham, and B. G. Galef Jr. 1994. Differences in the sex ratios of offspring originating in the left and right ovaries of Mongolian gerbils (*Meriones unguiculatus*). J. Reprod. Fertil. 101:393–396.

Clark, M. M., J. M. Vonk, and B. G. Galef Jr. 1996. Reproductive profiles of adult Mongolian gerbils gestated as the sole fetuses in a uterine horn. Physiol. Behav. 61:71–81.

Clemens, L. G., B. A. Gladue, and L. P. Coniglio. 1978. Prenatal endogenous androgenic influences on masculine sexual behavior and genital morphology in male and female rats. Horm. Behav. 10:40–53.

Even, M. D., M. G. Dhar, and F. S. vom Saal. 1992. Transport of steroids between fetuses via amniotic fluid in relation to the intrauterine position phenomenon in rats. J. Reprod. Fertil. 96:709–716.

Forger, N., B. G. Galef Jr. and M. M. Clark. 1996. Intrauterine position affects motoneuron number and muscle size in a sexually dimorphic neuromuscular system. Brain Research, 735, 119–124.

Gandelman, R., F. S. vom Saal, and J. M. Reinisch. 1977. Contiguity to male foetuses affects morphology and behavior of female mice. Nature 266:722–724.

Herrenkohl, L. 1979. Prenatal stress reduces fertility and fecundity in female offspring. Science 206:1097–1099.

Herrenkohl, L., and J. Politch. 1978. Effects of prenatal stress on the estrous cycle of female offspring as adults. Experientia 34:1240–1241.

Hews, D. K., R. Knapp, and M. C. Moore. (1994). Early exposure to androgens affects expression of alternative male types in tree lizards. Horm. Behav. 28:96–115.

Holekamp, K. E., L. Smale, H. B. Simpson, and N. E. Holekamp. 1984. Hormonal influences on natal dispersal in free-living Belding's ground squirrels (*Spermophilus beldingi*). Horm. Behav. 18:465–483.

Houtsmuller, E. J., and A. K. Slob. 1990. Masculanization and defeminization of female rats by males located caudally in the uterus. Physiol. Behav. 48:555–560.

Houtsmuller, E. J., J. Juranek, C. E. Gebauer, A. K. Slob, and D. L. Rowland. 1994. Males located caudally in the uterus affect sexual behavior of male rats in adulthood. Behav. Brain Res. 62:119–125.

Ims, R. A. 1987. Determinants of competitive success in *Clethrionomys rufocanus*. Ecology 68:1812–1818.

Jacquot, J. J., and S. H. Vessey. 1995. Influence of the natal environment on dispersal of white-footed mice. Behav. Ecol. Sociobiol. 37:407–412.

Jubilan, B. M., and J. G. Nyby. 1992. The intrauterine position phenomenon and precopulatory behaviors of house mice. Physiol. Behav. 51:857–872.

Ketterson, E. D. and V. Nolan. 1994. Hormones and life histories: An integrative approach. Pp. 327–353 in L. A. Real (eds.), Behavioral Mechanisms in Evolutionary Ecology. University of Chicago Press, Chicago.

Kinsley, C., and B. Svare. 1988. Prenatal stress alters maternal aggression in mice. Physiol. Behav. 42:7–13.

Meisel, R. L., and I. Ward. 1981. Fetal female rats are masculinized by male littermates located caudally in the uterus. Science 213:239–241.

Moore, C. L. 1983. Maternal contributions to the development of masculine sexual behavior in laboratory rats. Dev. Psychobiol. 17:347–356.

Moore, C. L. 1995. Maternal contributions to mammalian reproductive development and the divergence of males and females. Adv. Stud. Behav. 24:47–118.

Politch, J., and L. Herrenkohl. 1984. Effects of prenatal stress on reproduction in male and female mice. Physiol. Behav. 32:95–99.

Richmond, G., and B. D. Sachs. 1984. Further evidence for masculanization of female rats by males located caudally *in utero*. Horm. Behav. 18:484–490.

Schwabl, H. 1993. Yoke is a source of maternal testosterone for developing birds. Proc. Natl. Acad. Sci. USA 90:11446–11450.

Simon, N. G., and A. Cologer-Clifford. 1991. *In utero* contiguity to males does not influence morphology, behavioral sensitivity to testosterone of hypothalamic androgen binding in CF-1 female mice. Horm. Behav. 25:518–530.

Vandenbergh, J. G., and C. L. Huggett. 1994. Mother's prior intrauterine position affects the sex ratio of her offspring in house mice. Proc. Natl. Acad. Sci. USA 91:11055–11059.

Vandenbergh, J. G., and C. L. Huggett. 1995. The anogenital distance index, a predictor of the intrauterine position effects on reproduction in female house mice. Lab. Anim. Sci. 45:567–573.

vom Saal, F. S. 1984. The intrauterine position phenomenon: Effects on physiology, aggressive behavior and population dynamics in house mice. Pp. 135–179 in K. Flannelly, R. Blanchard, and D. Blanchard (eds.), Biological Perspectives on Aggression. Liss, New York.

vom Saal, F. S. 1989. Sexual differentiation in litter-bearing mammals: Influence of sex of adjacent fetuses in utero. J. Anim. Sci. 67:1824–1840.

vom Saal, F. S., and F. H. Bronson. 1980. Sexual characteristics of adult female mice are correlated with their blood testosterone levels during prenatal development. Science 208:597–599.

vom Saal, F. S. and M. G. Dahr. 1992. Blood flow in the uterine loop artery and loop vein is bidirectional in the mouse: Implications for transport of steroids between fetuses. Physiol. Behav. 52:163–171.

vom Saal, F. S., D. M. Quadagno, M. D. Even, L. W. Keisler, D. H. Keisler, and S. Kahn. 1990. Paradoxical effects of maternal stress on fetal steroids and postnatal reproductive traits in female mice from different intrauterine positions. Biol. Reprod. 43:751–761.

Ward, I. L. 1972. Prenatal stress feminizes and demasculinizes the behavior of males. Science 175:82–84.

Ward, I. L., and J. Weisz. 1980. Maternal stress alters plasma testosterone in fetal males. Science 207:328–329.

Yalcinkaya, T. M., P. K. Siiteri, J.-L. Vigne, P. Licht, S. Pavgi, L. G. Frank, and S. E. Glickman. 1993. A mechanism for virilization of female spotted hyenas in utero. Science 260:1929–1931.

Zielinski, W. J., J. G. Vandenbergh, and M. M. Montano. 1991. Effects of social stress and intrauterine position on sexual phenotypes in wild-type house mice (*Mus musculus*). Physiol. Behav. 49:117–123.

Zielinski, W. J., F. S. vom Saal, and J. G. Vandenbergh. 1992. The effect of intrauterine position on the survival, reproduction, and home range size of female house mice (*Mus musculus*) Behav. Ethol. Sociobiol. 30:185–191.

Part IV

CASE STUDIES OF MATERNAL EFFECTS

Undoubtedly the greatest insights to the inner workings of maternal effects are obtained from detailed experimental analysis of a single system. In this last part, four chapters do a particularly thorough job of exploring the mechanisms, causes, and consequences of maternal effects in a fly, a lizard, a turtle, and a plant. In chapter 16, David Denlinger summarizes two decades of work concerning the physiological and biochemical mechanisms associated with diapause control in the flesh fly *Sarcophage bullata*. In chapter 17, Barry Sinervo presents a compilation of his recent work addressing the evolutionary significance of maternal effects in a lizard. Sinervo's work is especially notable given that most of it was conducted in the wild, permitting a direct assessment of the consequences to fitness, and the time course of natural selection on maternally affected traits. In chapter 18, Willem Roosenburg and Peter Niewiarowski explore the adaptive significance of environmental sex determination (ESD) in a terrapin turtle. They also present a comprehensive review of ESD in reptiles. And finally, in chapter 19, Susan Mazer and Lorne Wolfe present a very detailed case study of the importance of density-mediated maternal effects on seed size in the wild radish *Raphanus sativus*. These studies exemplify the techniques and approaches required for evolutionary studies of maternal effects and serve as excellent guides for future work in this area.

16

Maternal Control of Fly Diapause

DAVID L. DENLINGER

V_{ery} few insects, even in the tropics, sustain continuous development throughout the year. Seasonal changes in temperature, rainfall, and food availability are powerful driving forces promoting seasonal development. For most insects the solution to this periodic, but predictable, environmental challenge is to enter diapause, a stage-specific developmental arrest characterized by suppressed metabolism. Seasonal changes in day length, supplemented by temperature cues, are the signals most widely exploited by insects for determining the correct time for diapause entry (reviewed by Saunders 1982; Tauber et al. 1986; Danks 1987; Zaslavski 1988). Depending on the species, diapause may be expressed in the embryo, larva, pupa, or adult. The environmental cues used to program diapause are perceived by the brain, which in turn controls the hormonal milieu that directs the insect's developmental fate (reviewed by Denlinger 1985).

In most cases, the perception of environmental cues and the execution of the diapause program are played out within the life of a single individual, but this is not always so. More and more examples of maternal control of diapause are being discovered (reviewed by Mousseau and Dingle 1991). In such cases, the mother exerts control over the diapause fate of her progeny. This scenario appears to be especially common among flies and parasitic wasps, but it is also well documented for some moths (the commercial silkmoth, *Bombyx mori*) and representatives of a few other insect orders. An arrest during embryonic development is the form of diapause most commonly controlled by the mother.

Despite the fascinating questions posed by maternal control of diapause, very few investigations have probed the mechanisms involved in this form of diapause regulation, with the notable exception of *B. mori* (reviewed by Yamashita 1996). The photoperiod received by the female of *B. mori* during her embryonic life determines whether she will produce diapause hormone as an adult. It is the action of this neurohormone on the female's ovaries that directs the eggs she lays to enter embryonic diapause. This is the only case, thus far, where the regulator of a maternal effect has been identified.

A different, and somewhat more complex, form of maternal control is exhibited in the

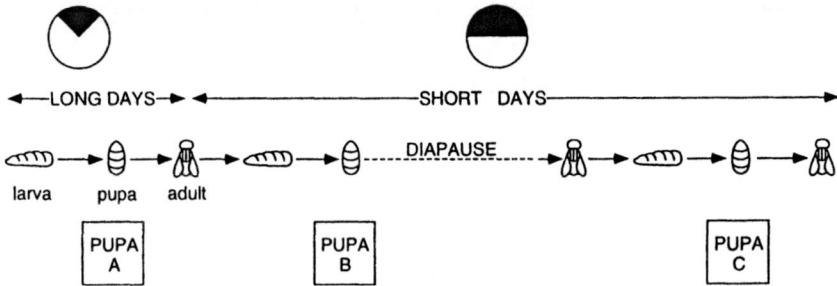

Figure 16.1 The influence of day length on the pupal developmental fate of the flesh fly *Sarcophaga bullata*. Pupa A does not enter diapause because it was reared as an embryo and larva at long day lengths. Pupa B can respond to short day length by entering diapause because its mother was reared at long day length. Pupa C can *not* respond to short day length by entering diapause because its mother was reared at short day length. It is this influence of the mother's photoperiodic history on the diapause fate of her progeny that we refer to as the maternal effect.

flesh fly *Sarcophaga bullata*. In this case, the mother determines not the diapause fate of the next immediate stage, the embryo, but instead the diapause fate of the pupa. If the mother has experienced pupal diapause (or the short day length that causes her to enter diapause) she completely prevents the expression of pupal diapause in her progeny (figure 16.1). By contrast, when there is no diapause in the maternal generation diapause is affected by the environmental conditions experienced by the progeny. It is this maternal effect operating in flesh flies that is the focus of this review. Experiments with *S. bullata* have identified the environmental regulators of the maternal effect and have probed the physiological and molecular processes involved in transferring this information from the mother to her offspring, thus making the maternal control of diapause in this flesh fly one of the best-understood maternal effects.

This review summarizes our current understanding of the following questions:

1. Does the diapause history of both the mother and father influence the expression of diapause in the progeny? *No, only the mother's experience is involved.*
2. Is this maternal effect caused by the experience of diapause, or is it the short days used to program the diapause? *Maternal short day exposure, not diapause itself, is essential.*
3. How is the capacity for diapause restored in subsequent generations? *The rearing of one generation at long day lengths is sufficient.*
4. Can environmental or chemical stresses alter expression of the maternal effect? *Thus far, only deuterium oxide has proven effective.*
5. Does maternal age influence expression of the maternal effect? *Yes, the effect is diminished in older females.*
6. What is the adaptive significance of the maternal effect? *It enables progeny of overwintering flies to develop in early spring without diapause, even though day length is still short.*
7. When is the photoperiodic information from the female's brain transferred to her ovary? *Sometime between pupariation and the third day of adult life.*

8. Is the nervous connection between the adult brain and ovary essential for expression of the maternal effect? *No.*
9. Is the maternal effect influenced by the mother's central nervous system? *Yes, the central nervous system contains a factor that can influence expression of the maternal effect.*
10. Do any of the known insect hormones or neuromodulators influence expression of the maternal effect? *Yes, GABA and octopamine.*
11. Can the maternal effect be associated with unique patterns of gene expression? *Yes, differences in ovarian proteins and mRNA are both apparent.*

16.1 Transgenerational Control

We discovered the maternal effect controlling pupal diapause in the flesh fly *S. bullata* during the course of experiments designed to genetically select a strain with a higher diapause incidence. By starting with individuals that had entered diapause in response to short day length, we anticipated increasing the diapause incidence in the next generations. But that did not occur. Instead, none of the offspring entered diapause, even when they were reared under strong diapause-inducing conditions of short day length (12 hours light:12 hours dark) and low temperature (20°C; Henrich and Denlinger 1982). This striking observation had previously eluded us and others who worked on flesh fly diapause because diapause experiments consistently utilized parent flies that had been reared under long day length (i.e., nondiapause conditions).

To determine whether both the mother and father contribute to this transgenerational effect, reciprocal cross matings were carried out between parents that had experienced pupal diapause and those that had not. Such crosses indicated that only the mother's diapause history is of consequence (table 16.1). The father's diapause history has no influence on the developmental fate of the progeny. As indicated by the data in table 16.1 and numerous more recent experiments, the maternal effect is quite pronounced. At most, just a few individuals, usually fewer than 1–2%, will enter pupal diapause if their mother had experienced diapause as a pupa.

The possibility remained that it wasn't the mother's diapause per se that caused the maternal effect. The mother's pupal diapause is elicited in response to the short day length she received as a late embryo and young larva, and possibly this is the signal responsible for expression of the maternal effect. We tested this idea by rearing mothers under short day lengths but then averting diapause by transferring her to a higher temperature (25°C) just before she pupariated. This manipulation produced a mother with a short-day history but no diapause experience. Such females also expressed the maternal effect; thus, it is not the experience of having been in diapause that causes the maternal effect to be expressed but the mother's exposure to short day length during her embryonic and larval life. Of course, under natural conditions short day length and the experience of pupal diapause usually coincide.

When successive generations of flies are reared under short day length, they consistently fail to express diapause in each generation (figure 16.2). But, if the progeny of females exposed to short day length are reared for one generation under long day length, the next generation of flies is again capable of responding to short day length by entering diapause. Thus, the capacity to express diapause can be restored by rearing a single generation at long day length.

In association with diapause, mature larvae typically wander longer searching for a

Table 16.1 Pupal diapause incidence at 12:12-hour light:dark cycle, 20°C, among progeny of *Sarcophaga bullata* with a diapause history (D) and with no diapause history (N).

Cross (♀ × ♂)	Replication	No. of females	No. of larvae	Diapause incidence (%)
N × N	a	5	265	70.6
	b	6	148	46.0
D × D	a	3	148	0
	b	2	110	0
N × D	a	4	231	80.9
	b	5	217	54.4
D × N	a	4	231	0
	b	5	229	0

From Henrich and Denlinger (1982).

pupariation site, they deposit additional hydrocarbons on the interior surface of their puparium, and as adults they produce fewer progeny than nondiapausing individuals reared under long day lengths. In this regard, fli 3 expressing the maternal effect have interesting characteristics. Though they have received the same short-day signals as diapausing individuals, they are similar to nondiapausing individuals in not being programmed for diapause. In such flies the duration of the wandering period, quantities of hydrocarbons on the puparium, and fecundity rates are between the extremes observed in the other two groups of flies (Rockey et al. 1991). Thus, the maternal effect switches the developmental program to nondiapause, but the progeny retain some characteristics of diapause. Like the expression of diapause, it is the diapause history of the female, rather than of the male, that influences the fecundity rate.

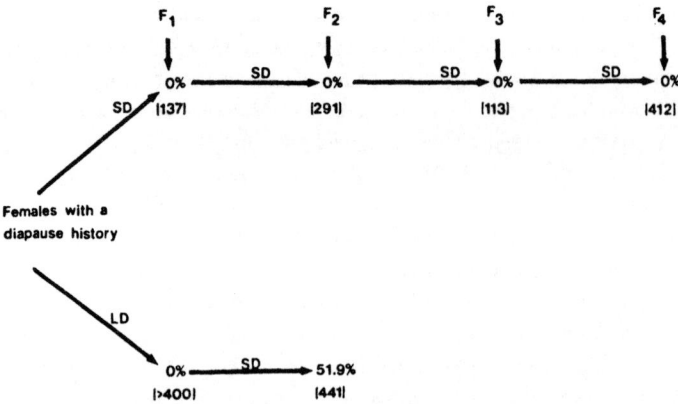

Figure 16.2 Diapause incidence in descendants of females that had experienced diapause as pupae. One line (SD) was continuously reared under short day length through the F_4 generation, and as a result of the maternal effect, no diapause was expressed. By contrast, when progeny from the same group of females were exposed to long day length for one generation (LD) and then exposed to short day length in the following generation, the capacity for diapause was restored. From Henrich and Denlinger (1982).

How common this type of maternal effect may be among other flesh flies remains unknown. Most of our experiments utilized a population of *S. bullata* originating in Lexington, Massachusetts, but similar experiments with populations of this same species from Ohio and Michigan demonstrated a strong maternal effect operating in these populations as well. Expression of the maternal effect was also evident in our long-established laboratory colony of *S. crassipalpis*, collected originally in Illinois, but in this case the effect was much more modest.

We suspect that this trait may gradually be lost in long-established laboratory colonies. Over 15 years have elapsed since we first observed the maternal effect in *S. bullata*, and initially we routinely observed 0% diapause incidence in response to the maternal effect. The maternal effect is still quite evident in this colony, but it is now common to observe 1–2% diapause among progeny of short-day females. Our stock colony is routinely maintained in the laboratory under long day length at 25°C and is thus not subjected to selection that would maintain either diapause or the diapause maternal effect. In contrast, field populations are exposed annually to strong selective pressure. The cost to wild flies for not entering diapause or not developing at the proper time of year is death, a toll not exacted on our laboratory colony. We assume the relaxation of selective pressure for expression of the maternal effect accounts for its gradual erosion under laboratory conditions.

16.2 Lack of Plasticity in Expression of the Maternal Effect

Expression of the maternal effect in *S. bullata* is rigidly fixed. Attempts to alter its expression by subjecting the mother to a range of environmental stresses have failed (Webb and Denlinger, in press). Heat shocks of 35°C and 40°C with durations of 30 minutes to 2 hours administered to first instar larvae, pharate adults, and newly emerged adults did not influence expression of the maternal effect, nor did exposing wandering larvae, pupae, and pharate adults to 4°C for 1–10 days and to −10°C for 30 minutes to 1 hour, or food deprivation during larval development or during adult life. Also, females of different sizes (pupal weights from 30 to 90 mg) expressed the maternal effect equally well. These results suggest that expression of the maternal effect is an "all-or-none" response. Females either block expression of diapause in all their progeny or fully permit their progeny to respond to the environmental cues that program diapause.

If the fly has not been programmed for diapause by short day length during its embryonic and larval development and if its mother has not been raised under long day length, it has absolutely no capacity to enter pupal diapause, even if it is reared under the strongest diapause-inducing conditions. However, those flies that have a capacity to enter diapause can have a range of responses from 0 to 100% diapause (Denlinger 1971, 1972a). At day lengths longer than 13.5 hours diapause will not be expressed, and even at day lengths shorter than 13.5 hours the response can range from a very low incidence of diapause (at high temperatures) to 100% (at low temperatures). Small pupae enter diapause less readily than large pupae, and females enter diapause less readily than males. Until shortly before the actual entry into diapause, the decision can still be reversed. For example, a heat shock at the time of pupariation (Denlinger 1976) or physical shaking of newly formed pupae (Denlinger et al. 1981) can avert diapause. The diapause program can thus be altered by a number of manipulations right up to the actual entry into diapause, and indeed even during the first few days of diapause the pupa can still easily switch to nondiapause (Denlinger 1985).

But, once the decision to enter diapause has been made and the pupa has firmly entered diapause, the program regulating the duration of diapause is rigidly fixed. No distinctions can be made between a diapause generated in response to strong diapause-inducing conditions (e.g., short days and low temperatures) and a diapause derived from weak diapause-inducing conditions (e.g., short days and high temperatures). Diapause expressed in pupae originating from a batch having a low diapause incidence has the same characteristics as diapause in a batch of pupae having a high diapause incidence. Thus, several aspects of diapause regulation are rigid responses: expression of the maternal effect, entry into diapause at the individual level, and the program controlling how long the diapause will persist. Flexibility in the system is evident only for the population response to diapause entry. Although it is an all-or-none decision at the individual level, what proportion of the individuals within a population will enter diapause is flexible, and it is at this level that the fine tuning of the population response is achieved.

16.3 Altering Expression of the Maternal Effect with Deuterium Oxide

Deuterium oxide (D_2O, or heavy water), a nonradioactive isotope of water, exerts a interesting effect on biological systems. The substitution of deuterium for hydrogen increases the activation energy of the bond and thereby decreases the rates of reactions. This has consequences for the transport of ions across cell membranes and for membrane polarization. The increased viscosity of D_2O also slows the rate of diffusion. One of the most conspicuous physiological effects of D_2O is a lengthening of the circadian period, an effect that has interesting implications for insect diapause. In both *Ostrinia nubilalis* (Beck 1980) and *Sarcophaga crassipalpis* (Rockey and Denlinger 1983), the diapause incidence is reduced dramatically when the photosensitive stages are exposed to D_2O.

As observed in *S. crassipalpis*, the addition of 20% D_2O to the larval diet averts diapause in *S. bullata* under conditions that normally would result in diapause (Webb and Denlinger 1997). Curiously, it exerts the opposite effect on females programmed to express the maternal effect. Rather than averting diapause, addition of D_2O to the larval diet enables some progeny from short-day females to enter diapause (i.e., it reverses the maternal effect). The diapause fate of the subsequent generation is also affected. Progeny from females reared on the 20% D_2O diet have a high capacity for diapause. This implies that D_2O exerts an effect similar to that of rearing the mothers under long day length. Though the female actually is exposed to short day length, the D_2O presumably causes her to interpret the short days as long days. Hence, her progeny do not receive the maternal factor that precludes the expression of pupal diapause. But, the effect of D_2O on progeny of females with a short-day history is less easy to interpret. Normally, the diapause-suppressing maternal effect is strongly expressed in such flies, but when larvae from the short-day flies are reared on the D_2O diet, a significant portion enter pupal diapause. In this case, short days cannot simply be misinterpreted as long days. If that were the case, one would predict less diapause, not the higher incidence of diapause that is observed. This result implies that D_2O somehow interferes with expression of the maternal effect within the progeny. From this, we can conclude that the information transferred from the mother to her progeny requires some additional processing within the progeny, and it is the processing of this information that is derailed when the progeny consume D_2O.

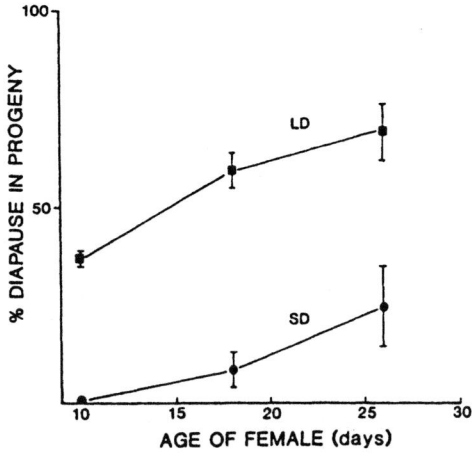

Figure 16.3 Incidence of pupal diapause in progeny produced by females of *Sarcophaga bullata* at different ages. LD refers to females that had been reared under long day length during their embryonic and larval development (hence, they did not express the maternal effect), and SD refers to females reared under short day length during this same photosensitive period (hence, they express the maternal effect). Each point represents mean (±SE) of three to five replicates of 40 females. Range in number of progeny was 3107–7711 for first broods, 2049–3072 for second broods, and 977–1128 for third broods. From Rockey and Denlinger (1986).

Since the chemical properties of water are so crucial for all life processes, it is likely that the effects observed with D_2O can be attributed to a variety of causes. For this reason, D_2O has limited potential for probing specific reactions. Yet, in this case it has proven useful in establishing the fact that expression of the maternal effect can be altered and that the information transferred to the progeny requires further processing within the body of the progeny.

16.4 Influence of Maternal Age

An influence of maternal age on the diapause fate of the female's progeny has been documented frequently. Examples among dipterans (e.g., Ring 1967; Vinogradova and Zinovjeva 1972; Saunders 1987) and hymenopterans (e.g., Simmonds 1948; Saunders 1962; McNeil and Rabb 1973) are especially well known, but similar reports exist for other insect orders as well. In most cases, older females produce a higher proportion of diapausing offspring. This is also true for the flesh fly *S. bullata* (Rockey and Denlinger 1986). When reared under a daily 12:12-hour light:dark cycle at 25°C, the diapause incidence in progeny of long-day flies (i.e., not expressing the diapause maternal effect) was 37% in the first brood but increased to nearly 70% in the third brood (figure 16.3). A similar increase in the incidence of diapause was observed in progeny of short-day flies (i.e., those expressing the maternal effect). Though none of the first brood entered diapause, 24% of the third brood did so. Manipulations such as switching the females to opposite photoregimes during the first brood (Rockey and Denlinger 1986) or denying the female a protein meal for the first 10 days (Webb and

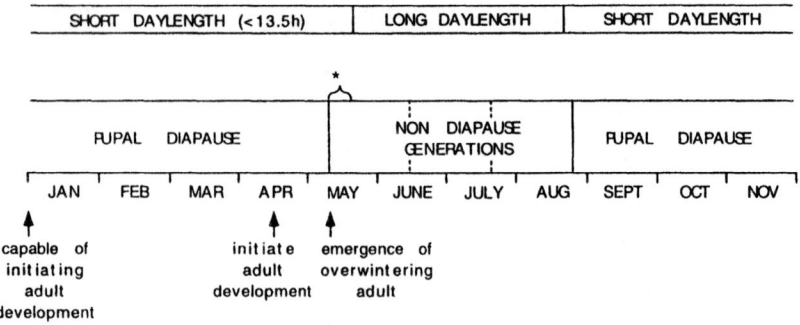

Figure 16.4 Seasonal development in the flesh fly *Sarcophaga bullata*, at 40°N. The adaptive significance of the maternal effect is evident in early May (see interval noted by star). The overwintering females emerge in early spring when the day length is still shorter than the critical day length (13.5 hours). They can exploit this early portion of the year because the maternal effect prevents the progeny from an untimely entry into diapause in response to the short day lengths that prevail at this time. After completing several additional generations during the long days of summer, the flies can again respond to the short days in August by entering an overwintering diapause. Though the diapausing pupae are fully capable of responding to warm temperatures needed for diapause termination as early as January, the prevailing temperatures of winter delay the onset of adult development until mid-April.

Denlinger, in press) have no influence on the diapause incidence observed in later broods. Protein deprivation delays production of the first brood, thus yielding an older fly but one that is producing her first brood. Despite this delay her progeny yields the same diapause incidence as observed in typical first broods. This suggests that the observed influence of maternal age is not a function of the female's absolute age, but rather the consequence of brood number. Females producing their first brood late in life produce a brood having the characteristics of a first brood. Brood number rather than female age is the critical factor.

16.5 Adaptative Significance

The adaptive significance of the diapause maternal effect in *S. bullata* is readily apparent (figure 16.4). Under natural conditions, this maternal effect serves to suppress diapause in the first generation of flies produced in the spring. In central Illinois (Denlinger 1972b) and central Ohio (Chen et al., 1991) the adults that have overwintered as diapausing pupae emerge in early May. At this time of year, the prevailing day length is interpreted as being short (day lengths shorter than 13.5 hours of light/day are interpreted as being short for these flies). Indeed, experiments with flies reared under long day length and taken into the field at this time of year readily demonstrate that their progeny will respond to the prevailing day lengths by entering pupal diapause (Denlinger 1972b). To do so at this favorable time of year would be an obvious mistake. The maternal effect plays the important role of preventing this untimely entry into diapause and enables the flies to thereby exploit early spring.

In central Ohio, *S. bullata* completes two or three uninterrupted generations before entering an overwintering pupal diapause in late summer and early autumn (Chen et al. 1991). The first generation that is produced does not enter diapause due to the maternal effect. Though only one generation at long day length is needed to restore the fly's capacity to express diapause, several nondiapausing generations are completed before the flies again encounter short days in late August. By this time, the flies are fully competent to respond to short day length by entering diapause. In this flesh fly example, the interpretation of the day length signal is rigid. The information is encoded as being either a short day or a long day. It is a threshold response and the fly does not distinguish between gradations of short or long. Alternatively, some insects can detect the direction of change in day length (Tauber et al. 1986), and if the insect has this capacity, it could differentiate between short but increasing day lengths in the spring and short but decreasing day lengths in the autumn. But, for an insect to make this distinction, it would presumably need a fairly long photosensitive period to accurately assess directional change. The photosensitive period in flesh flies is very brief, only four days in *S. crassipalpis* (Denlinger 1971), and such a short sensitive period would certainly be inadequate for making an accurate determination of directional change. The maternal effect in flesh flies thus enables the fly to preserve a successful mechanism for evaluating day length as a predictor of winter and yet not make the wrong developmental decision during the short days of spring.

The maternal effect in *S. bullata* is thus somewhat akin to the "interval timer" described for the aphid *Megoura viciae* (Lees 1960). In both cases, a certain interval must elapse before a particular developmental response can be expressed. The first spring generations of both flesh flies and aphids are unable to respond to short day length. With aphids, however, it is not diapause expression but the decision to produce parthenogenic or sexual offspring that is affected.

The significance of the effect of maternal age in influencing diapause is perhaps less clear than the significance of the maternal effect. Broods produced by an older female are, of course, produced later in the season than broods she produces as a young female, and hence the increased propensity for diapause may very well have some adaptive significance for channeling her offspring into diapause at the appropriate time. The impact that maternal age exerts on the diapause decision, however, pales in contrast to the influence of the maternal effect and the role of photoperiod. Maternal age exerts a rather subtle effect and at best plays a rather minor role in modulating the diapause response.

16.6 Transfer of Information from Mother to Progeny

The photoperiodic information received by the female of *S. bullata* during her embryonic and larval development is somehow transferred to her offspring, where it determines the offspring's capacity for expressing diapause when that offspring reaches the pupal stage. The mother's brain is the logical repository of the photoperiodic information she has received (Giebultowicz and Denlinger 1986), but at some point this information must be transferred to the egg. Our first experiments were designed to define when this information transfer occurs.

Eggs are fertilized internally at the time of ovulation (at 25°C this occurs 5 days after adult eclosion), and then, unlike most insects, the embryos are retained within the uterus until embryogenesis is completed 5 days later. The mother then gives birth to active, first instar larvae. The embryos are simply housed within the uterus. No nutritive material is transferred from mother to embryo, and indeed the success of culturing such embryos in an artificial uterus (Denlinger 1971) demonstrates that the mother provides no nutritive resources during this period. To test the possibility that the photoperiodic information might be transferred from mother to embryo at some point during embryonic development, embryos were removed from the mothers at five daily intervals following ovulation and cultured in vitro (Henrich and Denlinger 1982). The results indicated clearly that the diapause information was already transferred by the time of fertilization: progeny from females with no diapause history readily entered diapause while progeny from females that had been in diapause failed to enter diapause. By neck ligating females during oogenesis, we could also eliminate the last 3 days of oogenesis as being of consequence for the information transfer. Embryos developing from females that were neck ligated on the third day of adult life had already received the maternal information. Females ligated earlier than the third day did not successfully complete egg maturation, but within the limits of this experimental technique we were able to establish that the photoperiodic information from the female's brain is already transferred to her oocytes by the third day of adult life.

In our next approach we examined the mother during her larval life in an attempt to determine how early the information might be transferred from her brain to the ovary. One testable hypothesis is that the photoperiodic information is transferred from the larval brain to the larval ovary prior to pupariation. This possibility could be examined by transplanting ovaries between the two different types of females (Rockey et al. 1989). Ovaries transplanted into recipient larvae successfully differentiate in their hosts and produce fertile offspring. The diapause fate of such progeny was consistently identical with the fate of the host's natural progeny. Thus, we conclude that the information critical for the maternal effect is not yet transferred to the ovary by the end of larval life. This experiment thus narrows the window of transfer to sometime between pupariation and the third day of adult life.

To test the possibility that the maternal effect is transferred during adulthood from the central nervous system to the ovary through a nervous conduit, we severed the medial abdominal nerve (nerve XVII) in females on the day of adult eclosion. Although survivorship from this type of surgery was low, it was sufficient to establish that nerve transection does not alter the diapause fate of the female's progeny (Rockey et al. 1989). Likewise, hemolymph transfusions between larvae, pupae, and adults of both types of females were ineffective in altering the response. The one manipulation that did yield a positive response was injection of a central nervous system extract. When females programmed for the maternal effect were injected as third instar larvae with a central nervous system extract from long-day females, 38.9% of the females ($n = 18$) produced some diapausing progeny, while none of the saline-injected controls did so. Yet, the overall incidence of diapause among the progeny remained low (5.3%, $n = 1104$). This tantalizing result, however, suggests the presence of a central nervous system factor that may be involved in the response.

Attempts to mimic or alter the maternal effect with known insect hormones 20-hydroxyecdysone and juvenile hormone or with estrogens or androgens have failed (Rockey et al. 1989), but we have had modest success in altering the response with γ-aminobutyric acid (GABA) and one of its antagonists, picrotoxin (Webb and Denlinger in press). An injection of GABA into pupae or pharate adults reduced the incidence of diapause in progeny of females with a long-day history (diapause producers), while an injection of picrotoxin reversed the maternal effect (i.e., allowed the expression of diapause) in females with a short-day history. Though huge quantities of GABA are needed to elicit this response, the fact that GABA and its antagonist elicit opposite responses suggests the possibility that a GABA-mediated response within the mother influences the diapause potential of her progeny. Injections of octopamine and pilocarpine were also modestly effective in countering the maternal effect. Octopamine is a well-known insect neurotransmitter frequently associated with hormone release, and pilocarpine is best known as a cholinergic agonist, but how these agents may be involved in expression of the maternal effect is unclear. Several GABA agonists (muscimol and baclofen), GABA antagonists (SRR-95331, 5-aminovaleric acid, and bicuculline), dopamine, serotonin, physostigmine, dibutyryl cAMP, 8-bromo-cGMP, and reserpine all proved ineffective in altering the maternal effect.

We thus conclude that an unidentified factor extracted from the central nervous system, GABA, and octopamine are all possibly involved in regulating the maternal effect. And, the critical transfer of the information goes from the mother's brain to her ovary sometime between pupariation and the third day of her adult life. Much remains unknown about the physiological mechanisms involved.

16.7 Toward a Molecular Understanding

The obvious limitations imposed by the use of classic approaches of ligation, ovarian transplants, and testing the effects of tissue extracts, known hormones, and pharmacological agents prodded us to seek alternative methods for identifying the maternal determinant. To supplement these classic approaches we have begun searching for differences in gene expression in the ovaries of flies expressing the maternal effect and those that do not (Zhang 1994; Denlinger et al. 1995). Two-dimensional electrophoresis indicates differences in proteins synthesized by ovaries of females with long-day and short-day histories, and with the use of differential display we have detected a number of differences in mRNA populations between these two types of flies. Several bands that showed differential expression were excised from a differential display gel, reamplified, and cloned. Northern blot analysis indicates that at least one of the clones we have isolated is expressed only in ovaries of flies with a short-day history (i.e., flies expressing the maternal effect). The 585–base pair segment that we have sequenced indicates that the closest homology is with mouse putative primordial protein, a protein of unknown function. None of the other clones thus far isolated have proven to be specific for flies with either a long-day or short-day history. While this work is still at an early stage, we anticipate that this approach will enable us to gain new access to this challenging problem.

Questions Raised

Though the adaptive significance of the maternal effect operating in the populations of *S. bullata* observed in Massachusetts, Ohio, and Michigan seems apparent, it is equally obvious that such a maternal effect should not operate in populations farther north, where summers are much shorter. The northern boundary for this species is unknown, but clearly such a response should not be operating in localities where the fly could complete only one generation a year. The three populations evaluated for the maternal effect are all from roughly the same latitude, and it is thus not surprising to find few differences in their expression of the maternal effect. Additional sampling of populations from both farther north and farther south would help define how this trait changes in relation to geography. The maternal effect observed in *S. bullata* depends on photoperiod, but related flesh flies in the Old World tropics enter a pupal diapause regulated instead by temperature (Denlinger 1974). Does diapause of such a different type also depend on a maternal effect or has the maternal effect evolved more recently as the flies, which are thought to be of tropical origin, invaded the temperate regions?

At the mechanistic level, this and other maternal effects also offer a fascinating challenge. How is photoperiodic information that is garnered by one generation transferred to the pupal stage of the next? A maternal determinant that can so dramatically affect the developmental fate of a female's progeny has interesting regulatory properties for gene expression. Its characteristics suggest that a simple developmental switch is used to turn "on" or "off" this important developmental decision. A switch controlling a developmental decision of such a magnitude is indeed worthy of pursuit. At this point only a few small pieces of this intriguing puzzle have been tentatively identified.

Acknowledgment   Research from my laboratory discussed in this review was supported in part by the USDA Competitive Grants Program, most recently Grant 94-37302-0502.

References

Beck, S. D. 1980. Insect Photoperiodism. 2nd ed. Academic Press, New York.
Chen, C.-P., D. L. Denlinger, and R. E. Lee Jr. 1991. Seasonal variation in generation time, diapause and cold hardiness in a central Ohio population of the flesh fly, *Sarcophaga bullata*. Ecol. Entomol. 16:155–162.
Danks, H. V. 1987. Insect Dormancy: An Ecological Perspective. Biological Survey of Canada, Ottawa.
Denlinger, D. L. 1971. Embryonic determination of pupal diapause in the flesh fly, *Sarcohaga crassipalpis*. J. Insect Physiol. 17:1815–1822.
Denlinger, D. L. 1972a. Induction and termination of pupal diapause in *Sarcophaga* (Diptera: Sarcophagidae). Biol. Bull. 142:11–24.
Denlinger, D. L. 1972b. Seasonal phenology of diapause in the flesh fly *Sarcophaga bullata*. Ann. Entomol. Soc. Am. 65:410–414.
Denlinger, D. L. 1974. Diapause potential in tropical flesh flies. Nature 252:223–224.
Denlinger, D. L. 1976. Preventing insect diapause with hormones and cholera toxin. Life Sci. 19:1485–1490.
Denlinger, D. L. 1981. The physiology of pupal diapause in flesh flies. Pp. 131–160. In *Current Topics in Insect Endocrinology and Nutrition*, G. Bhaskaran, S. Friedman, and J. G. Rodriguez, eds. Plenum Press, New York.
Denlinger, D. L. 1985. Hormonal control of diapause. Pp. 353–411 in G. A. Kerkut and L. I. Gilbert (eds.), Comprehensive Insect Physiology, Biochemistry, and Pharmacology. Vol. 8. Pergamon Press, Oxford.
Denlinger, D. L., K. H. Joplin, R. D. Flannagan, S. P. Tammariello, M.-L. Zhang, G. D. Yocum, and K.-Y. Lee, 1995. Diapause-specific gene expression. Pp. 289–297 in A. Suzuki, H. Kataoka, and S. Matsumoto (eds.), Molecular Mechanisms of Insect Metamorphosis and Diapause. Industrial Publishing and Consulting, Tokyo.
Giebultowicz, J. M., and D. L. Denlinger. 1986. Role of the brain and ring gland in regulation of pupal diapause in the flesh fly, *Sarcohaga crassipalpis*. J. Insect Physiol. 32:161–166.
Henrich, V. C., and D. L. Denlinger. 1982. A maternal effect that eliminates pupal diapause in progeny of the flesh fly, *Sarcophaga bullata*. J. Insect Physiol. 28:881–884.

Lees, A. D. 1960. The role of photoperiod and temperature in the determination of the parthenogenetic and sexual forms of the aphid *Megoura viciae* Buckton. II. The operation of the "interval timer" in young clones. J. Insect Physiol. 4:154–175.

McNeil, J. N., and R. L. Rabb. 1973. Physical and physiological factors in diapause initiation of two hyperparasites of the tobacco hornworm, *Manduca sexta*. J. Insect Physiol. 19:2107–2118.

Mousseau, T. A., and H. Dingle. 1991. Maternal effects in insect life histories. Annu. Rev. Entomol. 36:511–534.

Ring, R. A. 1967. Maternal induction of diapause in the larva of *Lucilia caesar* L. (Diptera: Calliphoridae). J. Exp. Biol. 46:123–126.

Rockey, S. J., and D. L. Denlinger. 1983. Deuterium oxide alters pupal diapause response in the flesh fly, *Sarcophaga crassipalpis*. Physiol. Entomol. 8:445–449.

Rockey, S. J., and D. L. Denlinger. 1986. Influence of maternal age on incidence of pupal diapause in the flesh fly, *Sarcophaga bullata*. Physiol. Entomol. 11:199–203.

Rockey, S. J., B. B. Miller, and D. L. Denlinger. 1989. A diapause maternal effect in the flesh fly, *Sarcophaga bullata*: Transfer of information from mother to progeny. J. Insect Physiol. 35:553–558.

Rockey, S. J., J. A. Yoder, and D. L. Denlinger. 1991. Reproductive and developmental consequences of a diapause maternal effect in the flesh fly, *Sarcophaga bullata*. Physiol. Ent. 16:447–483.

Saunders, D. S. 1962. The effect of the age of female *Nasonia vitripennis* (Walker) (Hymenoptera, Pteromalidae) upon the incidence of larval diapause. J. Insect Physiol. 8:309–318.

Saunders, D. S. 1982. Insect Clocks. 2nd ed. Pergamon Press, Oxford.

Saunders, D. S. 1987. Maternal influence on the incidence and duration of larval diapause in *Calliphora vicina*. Physiol. Entomol. 12:331–338.

Simmonds, F. J. 1948. The influence of maternal physiology on the incidence of diapause. Philos. Trans. R. Soc. Lond. (ser. B) 233:385–414.

Tauber, M. J., C. A. Tauber, and S. Masaki. 1986. Seasonal Adaptations of Insects. Oxford University Press, New York.

Vinogradova, E. B., and K. B. Zinovjeva. 1972. Maternal induction of larval diapause in the blowfly, *Calliphora vicina*. J. Insect Physiol. 18:2401–2409.

Webb, M.-L. Z., and D. L. Denlinger. 1997. Deuterium oxide prevents expression of a diapause maternal effect in the flesh fly, *Sarcophaga bullata*, and alters development and fecundity. Eur. J. Entomol. 94:177–182.

Webb, M.-L. Z., and D. L. Denlinger. In press. GABA and picrotoxin alter expression of a maternal effect that influences pupal diapause in the flesh fly, *Sarcophaga bullata*. Physiol. Entomol.

Yamashita, O. 1996. Diapause hormone of the silkworm, *Bombyx mori*: Structure, gene expression and function. J. Insect Physiol. 42:669–679.

Zaslavski, V. A. 1988. Insect Development, Photoperiodic and Temperature Control. Springer-Verlag, Berlin.

Zhang, M.-L. 1994. A maternal effect that influences pupal diapause in progeny of the flesh fly, *Sarcophaga bullata* Parker (Diptera: Sarcophagidae). Ph.D. dissertation, Ohio State University, Columbus.

17

Adaptation of Maternal Effects in the Wild

Path Analysis of Natural Variation and Experimental Tests of Causation

BARRY SINERVO

The process of adaptation fundamentally involves Darwin's principle of evolution by natural selection in which variation in heritable traits is shaped by natural selection, and individuals that leave the most surviving progeny in the next generation have the highest fitness. The mean and variance of the heritable trait change across generations, resulting in adaptation to the selective environment. It is axiomatic to refer to many maternal effects undergoing the process of adaptation. Maternal effects such as egg size are thought to confer a strong fitness advantage to the offspring (Lack 1954; Williams 1966; Smith and Fretwell 1974; McGinley et al. 1987), and because egg size can be genetically transmitted to offspring (e.g., Sinervo and Doughty 1996), the reproductive trait of egg size should respond to the force of natural selection. However, a maternal effect such as egg size could affect the reproductive phenotype of the offspring as a nongenetic or "environmental effect" (Falconer 1965; Riska et al. 1985; Kirkpatrick and Lande 1989). I first discuss some of the obstacles in adaptational analyses of maternal effects in natural populations, and then I describe experimental manipulations of a suite of maternal effects in lizards that overcome many of these obstacles.

The possibility of maternal traits having both genetic and nongenetic effects on the reproductive phenotype of offspring greatly complicates adaptational analyses of maternal effects. Heritable genetic variation that forms the basis of maternal effects remains confounded with heritable nongenetic variation unless elaborate mating designs are used to estimate the magnitude of the purely additive genetic component (e.g., half-sibling designs; Falconer 1981). Such problems are particularly vexing in natural populations where matings cannot be controlled. Experimental and statistical protocols must be developed to disassociate these two "heritable" effects on offspring phenotype if we are to understand microevolutionary forces that act on maternal effects during the process of adaptation.

If nongenetic maternal effects remain confounded with genetic sources of variation, then estimates of response to selection (e.g., R for the simple formulation of response to selection, $R = h^2 s$) in natural populations can be under- or overestimated, depending on the degree to which maternal effects confound estimation of heritability. Conversely, if measures of natural selection acting on maternal effects in the parental generation do not take into account the delayed effects that reach the offspring and grandoffspring generations as cross-generational maternal effects, then response to selection might be underestimated (e.g., through the term s, the selection differential in the equation $R = h^2 s$). An additional complication arises in the interpretation of heritability, selection, and response to selection across generations when it is noted that most maternal traits are phenotypically plastic depending on the environment within a single generation (see by Mazer, Schmitt, and Fox and Mousseau, this volume). Differences in the environment between the dams and their female offspring might give rise to a change in the maternal trait across generations that confounds interpretation of response to selection in a heritable trait.

Laboratory estimates of additive genetic variation are attractive because half-sibling designs (Falconer 1981) can be used to estimate maternal effects in the absence of such environmental effects. Paternal half-sibling designs provide "clean" estimates of additive genetic variation in traits, but such laboratory estimates of heritability are of limited utility if the goal is to understand response to natural selection in natural populations. In addition, it is impossible to use elaborate paternal half-sibling mating designs in free-ranging animals because matings cannot be controlled. Moreover, most maternal reproductive traits are not expressed in the paternal phenotype (e.g., sex-limited expression), and standard quantitative genetic estimates of the sire's breeding value for female traits (as transmitted from sires to a large number of daughters in a commercial animal breeding experiment) are not feasible in pedigrees based on free-ranging animals. Finally, the maternal effect that is estimated from a half-sibling design pools maternal effect variation from a number of sources, with no specific maternal trait being identified (Riska 1991).

Parent-offspring regression of maternal traits (mother-daughter) provide the best estimates of additive genetic variation in nature because such estimates of heritability are not confounded with dominance deviations as are estimates from correlations between full siblings (Falconer 1981). However, maternal effects confound parent-offspring estimates of additive genetic variation. Allometric engineering of offspring size (Sinervo and Huey 1990; Sinervo 1993; Sinervo and Doughty 1996) can be used to experimentally assess specific maternal effects that arise from yolk volume per se. Such experimental manipulations of yolk volume can be viewed as a controlled perturbation of the effect that yolk volume has on offspring phenotype (e.g., offspring egg size and clutch size). In addition, the experimental manipulation of yolk volume also provides an experimental assessment of the strength of natural selection on maternal traits that might arise from offspring survival selection as a function of egg size.

The cautions described above for adaptational analyses of single-trait maternal effects like egg size and offspring survival apply equally well to multivariate suites of maternal traits such as the life history trade-offs that involve egg size and clutch size. The egg size or clutch size expressed by a female at maturity may result in additional life history trade-offs in which current effort (e.g., egg size per offspring or total effort per clutch) may affect future reproductive success (Reznick 1985). Multiple regression and the allied technique, path analysis, provide methods for elucidating the complex inheritance pattern of maternal effects (Riska et al. 1985) as well as their selective consequences (Crespi and Bookstein 1989; Mitchell 1993). These analytical methods are particularly powerful when performed on data from animals that have been subjected to experimental manipulations that alter a suite of life history traits

(Sinervo and DeNardo 1996). In the case of lizards, clutch size and egg size can be manipulated by a variety of hormonal and surgical protocols (Sinervo and Licht 1991a,b; Sinervo and DeNardo 1996) to experimentally alter reproductive investment of mature females and study the survival costs to these females.

We can also induce a controlled perturbation of specific maternal trait in a single generation and measure the effect of such perturbations on free-ranging animals. In addition, we can measure the selective consequences of maternal traits for both parents and offspring that are reflected as life history trade-offs, such as offspring quantity and quality trade-off and the trade-off between current effort versus future reproductive success. Detection of an effect of egg size or clutch size under such controlled conditions would validate the magnitude and direction of phenotypic and genetic correlations that are found in traditional quantitative genetic parent-offspring designs. Such maternal effects could be both heritable and acted on by natural selection (e.g., adaptive value). In this chapter, I combine path analytic methods and experiments to test the fitness consequences of two maternally affected traits: clutch size and egg size. I complement purely statistical path analyses of cause and effect with manipulations of maternal traits (e.g., egg size and clutch size). Experimental manipulations directly test the causal assumptions that underlie the statistically- based path analytic methods.

17.1 Materials and Methods

Side-blotched lizards (*Uta stansburiana*) were studied in a semi-isolated population that is located on a 250-meter-long outcrop that is adjacent to Billy Wright Road, Merced County, California (~2 km east of Los Baños Creek). Data reanalyzed and synthesized in this chapter are from previous studies of natural selection on offspring size (Sinervo et al. 1992; Sinervo and Doughty 1996), heritability (Sinervo and Doughty 1996), sexual selection (Sinervo and Lively 1996), the proximate physiological control of clutch size and egg size (Sinervo and Licht 1991a,b), costs of reproduction (Sinervo and DeNardo 1996), and natal dispersal (Doughty and Sinervo 1994; Doughty et al. 1994). We collected data on offspring survival from 1989 to 1997 and data on survival of the female parent as a function of reproductive allocation from 1991 to 1997. Here, we reanalyze data on offspring survival from the 1989–1990 hatchling releases.

Near-term gravid females from the study site were obtained from March to September of 1989 and 1990. We captured all lizards on this outcrop in early March (~120–150 females) and determined their reproductive state by abdominal palpation. Females received a unique toe clip that permits individual identification for life. Periodic palpation of free-ranging animals allows ovulation to be predicted to within 4 days (Sinervo and Licht 1991b), and females can be brought into the lab when their ovulated eggs are being shelled. Even though females were fed ad libitum in the laboratory, most if not all energy in the eggs was obtained in the wild because of the presence of the shell (Packard and Packard 1988). Females were maintained in ovipositoria with a moist peat moss/sand substrate for oviposition, provided a thermal gradient with an incandescent bulb, and provided with full-spectrum Vita-lites and Phillips BL40 UV.

Egg mass (nearest 0.01 g) and clutch size were obtained within 12 hours of laying. We used postlaying mass and snout-vent length of the female as indices of maternal size. The female parents were returned to their territories after oviposition, and they invariably resumed residence on the territory used prior to the brief stay ($\sim$10 days) in the lab required to obtain eggs. We monitored survival of the female parent to production of subsequent clutches to estimate survival costs of reproduction that might arise from allocation to reproduction. We thereby estimated lifetime reproductive success (Clutton-Brock 1988) and complement such correlative studies of the effect of female investment on future reproductive success with manipulations of investment (see below).

On discovering freshly laid eggs, we removed a portion of the yolk from half of the eggs in a clutch. Yolk was aspirated using a sterile syringe (25 gauge). Experimentally miniaturized eggs produce hatchlings miniaturized in proportion to the amount of yolk removed (Sinervo 1990). The remaining control eggs in the clutch were sham manipulated by inserting a syringe but withdrawing no yolk. We also produced gigantized hatchlings by bringing several females into the lab during early vitellogenesis and ablating a subset of follicles on one or both ovaries (Sinervo and Licht 1991a,b). Because the manipulation is performed after energy to produce the clutch has been acquired, these females ovulate and lay a smaller clutch of enlarged eggs.

Eggs from all females were individually incubated in moist vermiculite (-200 kPa) at 28°C, and the vermiculite was changed weekly. Incubation conditions in the laboratory control for maternal effects arising from oviposition site choice that would otherwise arise in natural populations. On hatching, hatchlings received a unique toe clip (which identifies them for life). Hatchlings were released at "nest sites" randomly regarding sibship (hatchlings were randomized regarding the female parent's territory). Number of hatchlings released at nest sites and the locations of nest sites were chosen based on the natural density of female territories (Doughty and Sinervo 1994). A total of 384 and 558 hatchlings were released in the 1989 and 1990 cohorts, respectively (partitioning of hatchlings by treatment can be found in Sinervo et al. 1992).

All progeny matured within 1 year. Maturation is spread out over 2 months during March and April. Reproductive condition of the female offspring was carefully monitored by abdominal palpation. When oviducal eggs were detected, we brought the female progeny into the laboratory to lay their eggs as described above for the female parent. We returned female progeny to their territory immediately after laying their eggs and collected them again on the second clutch of the reproductive season (approximately 1 month later). A subset of these females were collected on their third clutch. These procedures have been followed for all females from 1989 to 1996. We process $\sim$60–200 mature female progeny per year.

From 1991 to 1997 we have been manipulating the reproductive investment (e.g., clutch size and total clutch mass) of field active females to complement such correlative studies (Sinervo and DeNardo 1996). Clutch size is reduced by ablating a subset of the follicles on the ovary of the female (Sinervo and Licht 1991a,b; Sinervo and DeNardo 1996). Clutch size is enhanced by surgical implanting an Elvax pellet that contains follicle-stimulating hormone (Sinervo and DeNardo 1996). Finally, the amount of energy allocated to male versus female offspring is altered by surgically implanting Silastic capsules containing the hormone corticosterone (Sinervo and DeNardo 1996). Effects of these manipulations on egg size, clutch size, and total clutch mass are summarized in figure 17.1.

A variety of controls for the effects of surgical manipulation of females and for the miniaturization and gigantization procedure are described in detail elsewhere (Sinervo 1990; Sinervo and Huey 1990; Sinervo and Licht 1991a,b; Sinervo et al. 1992). Sham surgical manipulation of the female parent per se does not affect her egg size or clutch size or her survival

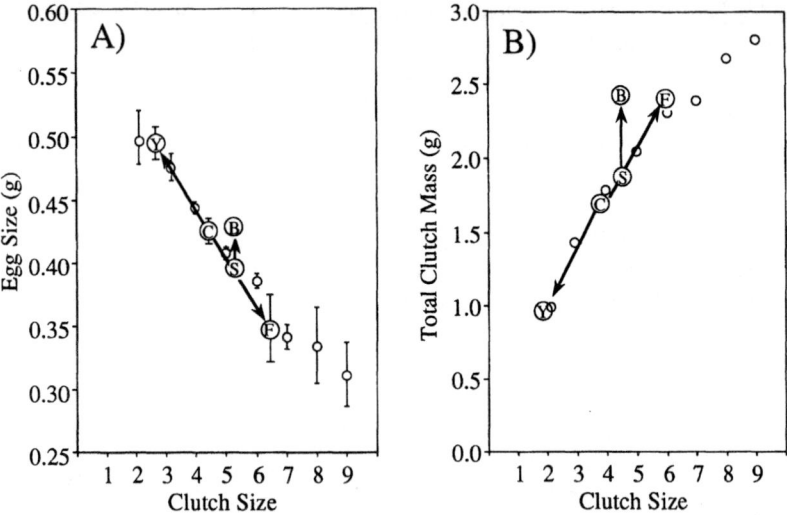

Figure 17.1 Summary of published data on experimental manipulations relative to natural covariation in mean (±SE) egg size (g) versus clutch size (panel A), and total clutch mass (g) versus clutch size (panel B) among female *Uta stansburiana* from populations (O) from three populations from inner coast range (adapted from Sinervo and Licht 1991a,b; Sinervo 1994). The results from two complementary experimental manipulations of clutch size are also plotted. Females receiving exogenous gonadotropin ovine follicle-stimulating hormone (F) produced larger clutches of small eggs relative to either sham or control females (C). Females with surgically reduced clutch size achieved by follicle "yolkectomy" (Y) produced a smaller clutch of large eggs. The third manipulation, exogenous corticosterone (B), did not affect clutch size per se but increased the amount of energy invested in eggs relative to females receiving a saline implant (S). Moreover, this energy was selectively channeled into female offspring (see Sinervo and DeNardo 1996).

in the wild. Moreover, eggs that are gigantized (by follicle ablation in female parent) and then miniaturized back to "normal size" (by removing yolk at oviposition) have the same viability to hatching and offspring survival in the wild as sham-manipulated eggs. The "double manipulations" (miniaturized giant eggs) and the sham manipulations indicate that manipulations of offspring size mimic the effects of natural differences in yolk volume among females. For example, selection gradients that describe the decreased offspring survival from naturally small eggs are comparable in magnitude to the selection gradients assessed with experimentally augmented variation in egg size (Sinervo et al. 1992).

Statistical Analysis of Natural Selection and Response to Selection

Natural selection on egg size was measured on a combined sample of natural and experimentally manipulated variation. Preliminary path analyses suggested that inferences from experimental manipulation and natural variation were consistent (e.g., see Sinervo et al. 1992). Experimental variation provides robust estimates of selection coefficients (Sinervo et al. 1992). However, estimates of selection differentials and selection gradients that act on experimental and natural variation must be standardized to natural variation per se if the experimental selection differentials are to be used in estimating the univariate response to selection of egg size.

Estimates of selection that are made from experimental variation isolate the effect of egg size on fitness and thus provide a clean "univariate" selection differential for egg size. In a similar fashion, multiple regression or path analysis provides a partial regression coefficient that statistically isolates the direct effects of egg size from the indirect effects that arise from selection on correlated traits. Experimental estimates of selection on egg size are not confounded with selection on other traits and thus are not susceptible to the criticisms of multivariate analyses of selection in which unmeasured traits might distort the true estimates of the strength of selection (Mitchell-Olds and Shaw 1987).

Univariate estimates of selection are expressed as the selection differential. The selection differential can be estimated from

$$s = \text{Cov}(W, X), \tag{17.1}$$

where W is relative fitness (raw measures of fitness standardized to mean fitness) and X is the trait value (Arnold and Wade 1984). For the univariate case, s can also be derived from the regression of relative fitness on the value of the trait. The slope of the fitness regression is given by (Arnold and Wade 1984):

$$\beta = \text{Cov}(W, X)/\text{Var}(X) \tag{17.2}$$

Values of β derived for experimental and natural variation must be reexpressed in terms of natural variation [e.g., $\text{Var}(X)$]. To accomplish this, w, mean fitness of natural variation, is used to compute W, relative fitness, and $\text{Var}(X)$ is computed from natural variation in egg size and not from experimental variation. Statistical inference regarding the significance of selection gradients (e.g., an individual survives, $W = 1$, or and individual is dead, $W = 0$) was based on jackknife estimates of the standard error of the regression (Mitchell-Olds and Shaw 1987; Sinervo et al. 1992). The estimate of the selection differential for natural variation based on experimentally derived selection gradients is given by

$$s = \beta_{\text{exp}} \times \text{Var}(X), \tag{17.3}$$

where β_{exp} is the experimental selection gradient standardized by mean fitness of natural variation, which is multipled by $\text{Var}(X)$, the natural variance in egg size.

Natural selection on egg size was partitioned into episodes of selection according to the following life history stanzas that are delimited by dispersal events (Doughty and Sinervo 1994):

1. fecundity selection acting on females, measured from the regression of clutch size on egg size;
2. viability selection acting on offspring from hatching to 1 month of age (June–August);
3. viability selection acting on offspring from 1 month of age to onset of the overwintering period in the fall (Septempter–November censuses);
4. survival from the fall to emergence from overwintering when maturation commences (March);
5. survival from onset of maturation to production of the first clutch (April);
6. survival from production of the first clutch to cessation of reproduction (August);

7. costs of reproduction" that arise from selection on natural variation in reproductive effort of adult females; and

8. experimental assessment of costs of reproduction using surgical and hormonal manipulations of clutch size, and egg size of the female parent.

Selection differentials were computed for episodes 1-7, and the "net" selection differential acting on offspring size across the lifetime of side-blotched lizards was computed as the sum of these episodic selection differentials. Below and in section 17.3 I consider selection differentials for the effects of experimental manipulation of reproductive investment on survival costs of reproduction. Separate selection differentials were computed for male and female offspring. In most episodes, selection on egg size was directional. However, for survival selection on male offspring during the first month of life, significant stabilizing selection was detected and selection estimates include an additional quadratic term (γ). Non-parametric fitness regression was used to visualize the fitness surfaces (Schluter 1995). The appropriateness of a quadratic regression model (e.g., quadratic vs. directional selection) was based on such exploratory data analysis. The reproductive success of individuals that survive to produce a clutch in their second year (~ 10–20% of mature animals or $<5\%$ of newly hatched animals) was ignored, as its contribution to total variance in fitness is of trivial magnitude compared to the magnitude of selection that occurs on egg size during the first year of life, which includes selection on costs of reproduction during the first year of life.

Selection on offspring is not limited only to early life history stages. The impact of reproductive investment at maturation on future reproductive success or costs of reproduction must also be assessed. However, natural variation in female investment (e.g., total clutch mass, number of clutches produced, etc.) is confounded by many environmental effects. One of the most important effects in lizards arises from the social environment. Competition and agonistic interactions among females (B. Sinervo, personal observations) affect clutch size and allocation to offspring (Sinervo and DeNardo 1996). Such effects appear to be correlated with a simple and readily quantified variable—female density. The "local density" that an individual female experiences can be estimated by counting the number of neighboring females that make incursions onto a female's territory. Details on estimating territory size and the number of neighbors (i.e., density), and the justification of the following path analytic model are found in Sinervo and DeNardo (1996). The effect of density on a female during her first clutch is spatially autocorrelated with the density of females on her second and third clutches. Females are sedentary, and the density of neighbors that a female experiences on her first clutch is correlated with the density of females on subsequent clutches. Thus, there is a strong a priori expectation that female density will have a correlative effect on current investment (e.g., in the first clutch) and survival to subsequent clutches. Density can be considered a "common environmental cause" (Li 1975) that affects both current effort and future success. The correlation set up by density can be removed by regression analysis, which yields residuals for both current investment and future success that are free of the common cause of density. Such residuals would be expected to be indicative of any underlying costs of reproduction. Path analytic estimates of natural selection on current investment can be compared to estimates that are based on experimental manipulations of effort. Follicle ablation reduces clutch size and total clutch mass. Administering follicle-stimulating hormone or corticosterone enhances total clutch mass, but the physiological pathway is somewhat different for the two hormones (Sinervo and DeNardo 1996), and the effect on allocation (e.g., clutch size vs. offspring size) also differs (see figure 17.1).

Response to natural selection in any given year is estimated from

$$R = h^2 s, \tag{17.4}$$

where h^2 is derived from a path analysis of a daughter's egg size on the dam's egg size and values for s, the selection differentials, are from the analyses of selection described by equations (17.1) and (17.3) (Falconer 1981). The path analytic model for estimating heritable variation in the egg size of daughters includes terms for

1. egg size of the dam: a correlation between daughter and dam egg size is indicative of additive genetic variation;
2. experimental manipulation of yolk volume: if the effect of yolk volume is significant, it would suggest that the additive effect estimated for egg size of the dam (term 1) is confounded with yolk volume per se;
3. the difference between an individual egg and the average egg size for the clutch: if significant, this would imply that allocation differences among eggs in the same clutch impact the phenotype of the offspring;
4. clutch size;
5. female body size: post-laying mass, which is largely correlated with female age;
6. lay date; and
7. additional maternal effects are controlled by incubating eggs in a common laboratory environment, and by releasing hatchlings randomly regarding the territory of the female parent.

We compared the evolutionary response to natural selection as predicted from equation (17.4) with the observed change in egg size across generations. A reasonable fit between predicted and observed change in egg size would imply that most salient maternal effects involved in life history adaptation (e.g., natural selection) have been identified.

17.2 Results

Estimates of the selection differentials for all episodes are given in table 17.1. Eleven of the 12 possible selection differentials that comprise selection up to 1 month of age were significant. Thus, fecundity and survival selection during the earliest phases of the life history are under very strong selection. Selection on egg size was significant for only one of eight possible cases in the interval between 1 month of age and maturation. Surprisingly, selection on egg size affected survival of progeny during adult phases, and two of the four possible selection differentials were significant.

Variation in the egg size of daughters was significantly correlated with variation in the dam's egg size, and the heritability of egg size is $h^2 = 0.61 \pm 0.29$ (Figure 17.2 Li 1975; Sinervo and Doughty 1996) for the pooled 1989 and 1990 cohorts (figure 17.2A). Furthermore, we demonstrate a correlation between egg size of the female parent and clutch size of their daughters which is consistent with an underlying negative genetic correlation between egg size and clutch size. This genetic correlation imposes a selective constraint on the evolution of egg size. Although it is generally advantageous to produce large offspring, (e.g., a positive selection differential for viability selection as a function of egg size was found for most episodes) (table 17.1), a female that produces large eggs tends to produce fewer offspring and has lower fecundity.

Table 17.1 Selection differentials acting on egg size for episodes of selection during the lifetime of male and female progeny from side-blotched lizards from the first versus later clutch hatchlings.

Episode year	Clutch	Sex	Mean fitness	Directional selection gradient	Selection differential	Stabilizing selection gradient
Fecundity						
1989	First	Both	6.29	−3.66	−0.0111	
	Later	Both	5.79	−1.61	−0.0026	
1990	First	Both	5.57	−2.50	−0.0071	
	Later	Both	4.61	−2.11	−0.0131	
Survival (0–1 month)						
1989	First	Female	0.656	**3.14**	**0.0034**	
		Male	0.575	**22.49**	**0.0246**	**36.24**
	Later	Female	0.630	0.77	0.0020	
		Male	0.481	**16.58**	**0.0320**	**−16.18**
1990	First	Female	0.700	**1.37**	**0.0043**	
		Male	0.580	**−1.65**	**−0.0052**	
	Later	Female	0.412	**3.96**	**0.0251**	
		Male	0.500	**4.34**	**0.0248**	**5.05**
Survival (1 month to maturation)						
1989	First	Female	0.381	9.04	0.0092	
		Male	0.435	6.46	0.0066	
	Later	Female	0.588	−0.08	−0.0002	
		Male	0.231	3.20	0.0063	
1990	First	Female	0.429	1.62	0.0058	
		Male	0.217	−0.77	−0.0017	
	Later	Female	0.431	**3.94**	**0.0250**	
		Male	0.421	−0.73	−0.0040	
Survival (maturation to first clutch)						
1989–1990	First	Female	0.524	**−2.89**	**−0.0126**	
	Later	Female	0.610	0.30	0.0013	
Survival (first clutch to later clutches)						
1989–1990	First	Female	0.444	−0.87	−0.0029	
	Later	Female	0.545	**2.48**	**0.0182**	

Selection gradients and selection differentials in boldface are significant at $P < 0.05$.

It is noteworthy that experimental manipulation of yolk volume did not affect egg size, clutch size, or size at maturity of the progeny (not shown). Our experimental design thus allows us to descriminate between cross-generational genetic sources of variation on egg size (e.g., heritable and genetic) from experimentally induced effects of egg size that arise from a controlled manipulation of yolk volume (see Sinervo and Doughty 1996). Other salient maternal effects such as incubation temperature and hydric environment of eggs are tightly controlled by incubating eggs in a common laboratory environment. Maternal effects in the wild such as territory quality are eliminated by randomizing sib-ships across the study site. Other maternal effects such as lay date, yolk constituents, and oviductal environment may

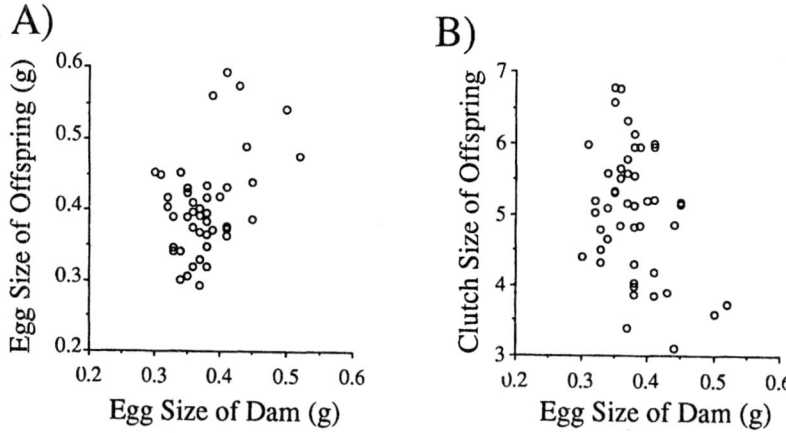

Figure 17.2 A, significant positive correlation ($P < 0.05$) between a dam's egg size and the daughter's egg size for free-ranging female side-blotched lizards is indicative of heritable variation for egg size. B, significant negative correlation ($P < 0.05$) between a dam's egg size and the daughter's clutch size is indicative of a negative genetic correlation between egg size and clutch size. Data are pooled from 1989 and 1990 cohorts, and all progeny mature in 1 year. Many salient maternal effects were controlled by incubating eggs in a controlled laboratory environment (see section 17.1). In addition, yolk was removed from half of the eggs in each clutch. Because experimental yolk removal ($P > 0.05$, not shown) has no effect on reproductive traits of progeny when they mature, the maternal effect of yolk size per se does not confound estimates of additive genetic variation and covariation (note that egg size has profound effects on offspring survival; see table 17.1). Data are from Sinervo and Doughty (1996). Only data for first clutch eggs are shown, but estimates of heritability that include first and second clutch eggs in a repeated-measures design indicate significant heritability in egg size ($h^2 = 0.61 \pm 0.29$).

confound our estimates of additive genetic variation in egg size. However, path analysis of such uncontrolled effects suggests that they do not affect the parent-offspring correlation (Sinervo and Doughty 1996). Additional manipulations of these maternal traits could also be made to verify the results from path analysis (Sinervo, unpublished observations).

Results of the path analysis of costs of reproduction for 1 year (1992) are given in figure 17.3. In 1992, density appears to have had a positive effect on current effort and a positive effect on survival to subsequent clutches (e.g., number of clutches produced). This results in a net positive environmentally driven correlation between current effort and survival to subsequent clutches that, once removed, yields residuals between current effort and future success that are negative and reflect costs of reproduction. The selection differential for such a cost of reproduction is -0.31. It is instructive to compare this estimate of selection derived from natural variation to the value computed from an experimental reduction in clutch size that is achieved by ablating a subset of follicles from a female's clutch. The selection differential acting on costs of reproduction as assessed by follicle ablation experiments is -0.22.

1992

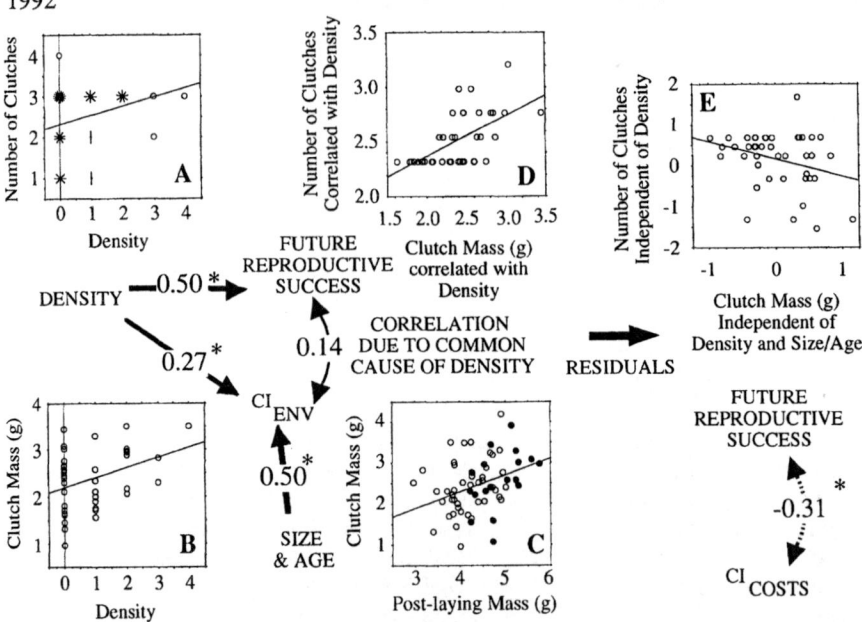

Figure 17.3 The local density of females on an individual female side-blotched lizard's territory affects future reproductive success (number of clutches produced during the reproductive season; A) and current investment (CI) on the first clutch of the reproductive season (total clutch mass, g; B). Current investment on the first clutch is also correlated with female size (and age; C; open circles are 1-year-olds; solid circles are ≥2-years-olds). D, rather than assume that the environmental effects correlated with density (e.g., aggregation) affect current investment and future reproductive success through separate causal pathways (e.g., B, Density→ CI $_{ENV}$, and A, Density→ future reproductive success), we assumed that the factors that enhance current effort act through the joint cause Density to determine future success and current investment. Because density is positively correlated with both current effort and future reproductive success, a positive correlation between current effort and success would be set up by factors associated with density. For example, females may aggregate at high density at locations that are particularly rich in resources. E, residuals from this regression analysis reflect the correlation ($P < 0.05$) between current investment and future reproductive success (CI $_{COSTS}$ → future reproductive success) that is independent of the environmental effects of density or phenotypic effects of size and age. Significance of path coefficients: *$P < 0.05$.

17.3 Discussion

Significant selection differentials acting on offspring size were computed for 14 of 24 episodes of selection. Survival selection on egg size is very strong during the first month of life and can also be of large magnitude during adult reproduction. The largest single-episode selection differential was for fecundity selection in 1990 ($s = -0.0263$). However, such fecundity selection acting on offspring size was counteracted by single-episode survival selection of comparable magnitude (e.g., $s = 0.025$ for survival of female offspring from hatching to 1 month during 1990). The magnitude of the selection differential acting on adult survival was also large in some cases (e.g., $s = 0.018$ for later clutch offspring from first reproduction to survival to the end of the reproductive season). Thus, the selective impact of the egg size is typically of large magnitude early in ontogeny, but continued selection occurred during adult phases of the offspring's life history. The relative magnitude of selection differentials can be referenced to the standard deviation in egg size to assess the biological impact of a given selection differential on variance in egg size. A selection differential of 0.020 would be roughly equivalent to a magnitude of 0.5–1.0 units of standard deviation in the natural variation in egg size.

Seasonal Shifts in Optimal Egg Size

Seasonal shifts in egg size late in the reproductive season are common in many organisms and presumably reflect an adaptive response by the female to the deteriorating environment that late-born offspring face (Hutchinson 1951; Lack 1954; Kerfoot 1974; Ferguson and Bohlen 1978; Nussbaum 1981; Ferguson and Snell 1986; Glazier 1992). It is worth comparing the magnitude of natural selection acting on first versus later clutches in side-blotched lizards. In this population, females lay eggs in early-season clutches that are significantly smaller than late-season eggs. The "net selection differential" (sum of selection from all episodes) acting on the first clutch (1989 and 1990 combined) is $s = -0.0044$ (table 17.1). In essence, females tracked the optimum egg size, as the magnitude of the selection differential was quite small relative to the variation in egg size observed among females. A selection differential of -0.0044 is less than 0.1 of a standard deviation in the natural range of egg size and would have resulted in only a trivial response to selection. However, it is important to realize that the net selection differential resulted from a number of strong selection differentials that varied in direction (e.g., fecundity versus survival selection). In contrast to the first clutch, the net selection differential for the second clutch was large and positive ($s = 0.047$) and an order of magnitude larger than that observed for first-clutch eggs (table 17.1).

The experimental manipulations indicate that the pattern of producing larger eggs on the second, third, and fourth clutches is quite adaptive in side-blotched lizards; however, females were not laying optimal-sized eggs in later clutches, and the observed egg size falls more than 2 standard deviations below the optimal egg size (table 17.1). This discrepancy between the "optimality" of first-clutch offspring and the lack of optimality of later-clutch offspring may result from stringent constraints on egg size arising from the diameter of the pelvic girdle of vertebrates (Congdon and Gibbons 1987; Sinervo and Licht 1991a; Luetenegger 1976, 1979). Female lizards that were induced to lay eggs as large as the optimum for later clutches (e.g., >0.50 g) incurred a higher frequency of burst eggs at laying, and the females also became egg-bound (unable to pass the eggs in the oviduct through the pelvic girdle) at a higher frequency.

If functional constraints on maximum offspring size were the only factor influencing the survival of the female parent, we would expect females laying the smallest eggs or perhaps intermediate-sized to have the highest survival. Indeed, size of eggs produced during an offspring's first clutch was subject to strong optimizing selection (see Sinervo 1994). Female offspring producing both large and small eggs in their first clutch had low survival after laying (Sinervo 1994) whereas female offspring that laid intermediate sized eggs had high survival. The lower survival of females laying small eggs may arise from correlated selection on other reproductive traits or perhaps from a lower level of vigor that is indicative of poor health.

Experimental data are unavailable for the 1989 and 1990 cohorts to verify which of these potential causes of adult mortality was in operation. However, results obtained from subsequent years (1991–1994; Sinervo and DeNardo 1996) indicated that females that laid the smallest eggs tend to lay the largest clutches, produce the largest total clutch mass, and have low survival in some years compared to females that lay smaller total clutch mass. The observed patterns of selection among years are complex, but it appears that these effects are causally related to the costs of reproduction arising from reproductive investment in a clutch (Landwer 1994; Sinervo and DeNardo 1996).

Are Changes in Egg Size among Years an Evolutionary Response to Natural Selection or an Induced Effect of the Abiotic or Biotic Environment?

It is also of interest to compare predicted response to natural selection on the heritable trait egg size with the observed change in egg size between generations. The net selection differential was 0.028 and 0.017 for the 1989 and 1990 cohorts respectively. Results from these 2 years indicate that egg size was under relatively strong and consistent selection. Although survival selection tended to favor large female offspring, survival selection favored males that hatched from small eggs in one case (e.g., 1 month survival of the 1990 first clutch) and tended to be stabilizing in the three other cases (1 month survival for the other three cohorts). Results from subsequent years (1991–1996) confirm these observations (B. Sinervo, unpublished data). Moreover, results from 1989 and 1990 suggest that egg size should have increased across generations in response to a net positive selection differential. The observed response, $R_{observed}$, to selection in 1989 was 0.045 and in 1990 was 0.0035 (e.g., difference in mean egg size between years; figure 17.4). Predicted reponses to selection based on $R = h^2 s$ are 0.019 and 0.010 for 1989 and 1990, respectively (where h^2 is from Sinervo and Doughty 1996, and s is from table 17.1).

There is a considerable discrepancy between observed and predicted responses to selection, which suggests that other abiotic or biotic factors might have influenced the difference in offspring size among years (e.g., generations). The years 1989–1991 were the last 3 years of a long-term 7-year drought, and thus the climate was remarkably constant from year to year. Even though abnormally low levels of rainfall were seen in all 3 years, little variation in rainfall occurred among years. Thus, it is unlikely that the drought per se can explain the discrepancy between observed and predicted response to selection. A biotic environmental factor that might influence egg size, which also varied among years, was lizard density (Sinervo and DeNardo 1996). Lizard density more than doubled between 1989 and 1990, and this may have induced an increase in egg size between years. Indeed, individual female side-blotched lizards that experience a low density (e.g., females that maintain nonoverlapping territories) allocate energy differently to reproduction than females that experience a high density (e.g., one or more females that overlap their territories; Sinervo and DeNardo 1996). Thus, individual females may alter offspring size in response to the local social environment, and an overall change in density between years would result in a change in egg size for purely environmental reasons. The plasticity in egg size and clutch size (Sinervo and DeNardo 1996) seen in a high-density year (1990) compared to a low-density year may be quite adaptive. If females can accurately predict the more intense selection that occurs under high density, they might provision their offspring with additional energy.

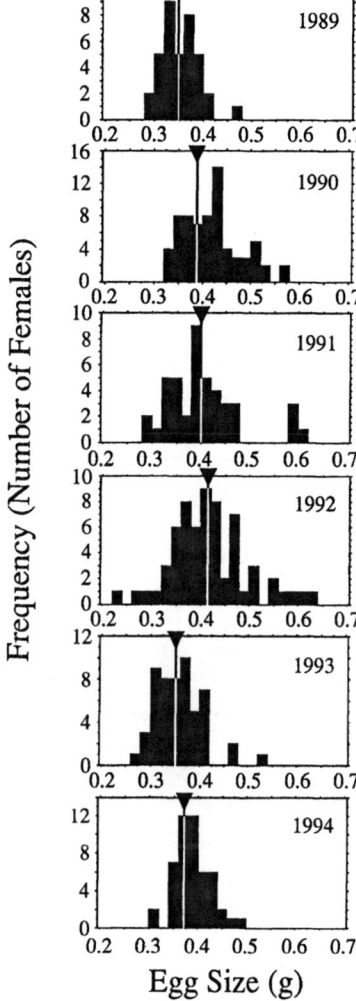

Figure 17.4 Distribution of mean egg size in first clutch eggs for individual side-blotched lizard females from a population located near Los Baños, California. The difference between 1989 and 1990 is significant ($P < 0.01$), as is the difference between 1992 and 1993 ($P < 0.001$). Data are from Sinervo and DeNardo (1996). Arrows indicate the mean egg size.

We have isolated a putative mechanism for plastic adjustments of egg size, clutch size, and total investment in reproduction. The stress steroid corticosterone (Johnson et al. 1992) changes across the reproductive cycle of female side-blotched lizards (Wilson and Wingfield 1992). Experimentally manipulating corticosterone produces concomitant changes in the relative energy that female side-blotched lizards invest in male versus female offspring (Sinervo and DeNardo 1996). We hypothesize that females under high density have higher levels of corticosterone than females under low density. Thus, adaptive plasticity in egg size (e.g., induced by the mother's environment) can readily confound estimates of adaptive response to natural selection (e.g., genetic change in heritable variation that controls egg size). Presumably, the large increase in observed egg size from 1989 to 1990 (0.045; figure 17.4) is due to both a genetic response that is calculated to be 0.017 and a putative plastic response of ~0.028. Regardless of the source of environmentally induced variation in egg size, the mismatch between observed and predicted response to selection can be used as an indicator of environmental effects at work in nature.

Costs of Reproduction and Further Selection on Maternal Iinvestment

Sinervo and DeNardo (1996) used path analysis to remove the effects of female density on allocation to reproduction. If an individual remains in the same location throughout life, there is a strong a priori expectation that local density conditions experienced by an individual during first reproduction will also have correlated effect on future reproductive success because of spatial and temporal autocorrelation (Sinervo and DeNardo 1996). Removing correlated effects of the environment on current and future reproductive success using path analysis (Li 1975) yields "residuals" between current investment and future reproductive success that are free from the confounding effects of environment (e.g., density). Perhaps such residuals might explain selection on adult female survival as a function of current investment that would reflect a cost of reproduction (Partridge and Harvey 1985; Reznick 1985, 1992; Partridge 1992).

Experimental manipulations of reproductive effort that involve surgical and hormonal treatments applied to free-ranging females were used to confirm the strength of selection detected in the path analysis of natural variation.

A large drop in average egg size between 1992 and 1993 is consistent with a genetic response to selection that acts against females who expend too much effort in large offspring. The magnitude of the selection differential computed from experimental variation, $s = 0.078$ translates into a predicted response to selection of $R = h^2s = 0.047$. The magnitude of predicted change from this single episode of selection, while large, is consistent with the large observed change in egg size between 1992 and 1993 of 0.060 g (figure 17.4). Natural selection was strongly correlated with a rise in the number of predatory snakes on side-blotched lizards that occurred between 1992 and 1993 (Sinervo and Doughty 1996). Presumably, the greater number of snakes placed females that produce large offspring at a greater risk owing to the increased foraging required to produce such offspring. In this case strong selection on maternal investment is hypothesized to act as females yolk up the eggs for their first clutch, a critical 1-month episode during which mortality risk was presumably elevated compared to previous years (e.g., 1989–1992) owing to the presence of greater numbers of predatory snakes.

Toward an Understanding of the Adaptive Value of Maternal Investment in Nature

While the potential sources of "inducing" environmental effects and "selective" environmental agents appear to be many, experimental manipulations of phenotypic traits (e.g., egg size, current effort) provide unambiguous answers concerning the magnitude of selective effects in natural populations. In this regard, these manipulations are useful for testing hypotheses generated from a path analysis of natural variation. Selection on the trait egg size at different stages of the life history can be very strong. Survival selection acts on newly hatched individuals, and the selective effects of maternal investment can act on adult phases of the life history and influence survival of progeny as adults. Selection on parents can act via the offspring number-quality trade-off (e.g., clutch size vs. egg size; figure 17.1) or via the trade-off between current versus future reproductive success. With such strong selection, it is difficult to maintain additive genetic variation in egg size if the genetic factors underlying this maternal effect have a quantitative genetic basis (Mousseau and Roff 1987).

The suite of traits described above for female side-blotched lizards, which include egg size, clutch size, offspring survival, adult survival, and reproduction, appear to be associated with a simple heritable throat color polymorphism in female offspring (B. Sinervo unpublished data) and associated with alternative mating strategies in male offspring (Sinervo and Lively 1996). Sinervo and Lively (1996) demonstrate that strong frequency-dependent selection acting on males could maintain frequencies of three male morphs in an indefinite cycle of morph frequency that resembles the rock-paper-scissors game where paper beats rock, scissors beat paper, and rock beats scissors (Maynard Smith 1982; Sinervo and Lively 1996). As in the rock-paper-scissors game, the wide-ranging, "ultradominant" strategy of orange males is defeated by the "sneaker" strategy of yellow males, which is in turn defeated by the mate-guarding strategy of blue males; the orange strategy defeats the blue strategy to complete the dynamic.

If the male and female morphs are due to the same set of genes, then frequency-dependent selection acting on males alone could maintain substantial genetic variation in both alternative male and female strategies while at the same time prohibiting a stable equilibrium in morph frequency. Results described earlier indicate that females respond to local density by adjusting offspring size. Preliminary results indicate that females adjust the size of their male vs. female offspring as a function of their own morphotype (e.g., orange vs. yellow female morph) and the density of female morphs in their vicinity. As discussed in section 17.3, a putative mechanism for such socially mediated plasticity is the steroid stress hormone corticosterone.

Frequency-dependent adjustments of offspring size or mate choice could have interesting but as yet poorly understood effects on the oscillations of the male rock-paper-scissors game. Indeed, models of maternal effects (Kirkpatrick and Lande 1989) and sexual selection (Lande 1981) predict oscillations in traits. Theoretical results suggest that such maternal plasticity might act as an additional "heritable nongenetic component," which might in turn might accentuate or attenuate the genetic response to natural and sexual selection (Kirkpatrick and Lande 1989; Wade, this volume). Sinervo and Lively (1996) predict a 16-year cycle in morph frequency based on a simple genetic model and the observed 5-year cycle is far faster, suggesting that current evolutionarily stable strategy (ESS) models and predictions for the maintenance of the alternative male morphs might be improved by developing comparable ESS models of female morphs and synthesizing such plastic ESS strategies with quantitative genetic models of mate choice and sexual selection. Females may have the capacity to choose mates with "good genes" (O'Donald 1983; Pomiankowski 1988; Iwasa et al. 1991) and vary energy placed in propagules from such matings depending on the current social environment (e.g., morph frequency).

Acknowledgments I thank P. Statdler and R. Schrimp for generously permitting work on their land, and P. Doughty, T. Frankino, K. Zamudio, and K. Allred for assistance with gravid females in the lab and field. Tony Frankino provided constructive criticism on the manuscript. This research was supported by NSF Grants BSR 89-19600 and DEB 93-07999 to B.S.

References

Arnold, S. J., and M. J. Wade. 1984. On the measurement of natural and sexual selection: Theory. Evolution 38:709–719.

Clutton-Brock, T. H. 1988. Reproductive Success: Studies of Individual Variation in Contrasting Breeding Systems. University of Chicago Press, Chicago.

Congdon, J. D., and J. W. Gibbons. 1987. Morphological constraint on egg size: A challenge to optimal egg size theory? Proc. Natl. Acad. Sci. USA 84:4145–4147.

Crespi, B. J., and F. L. Bookstein. 1989. A path-analytic model for the measurement of selection on morphology. Evolution 43:18–28.

Doughty, P., and B. Sinervo. 1994. The effects of habitat, time of hatching, and body size on dispersal in *Uta stansburiana*. J Herpetol. 28:485–490.

Doughty, P., G. Burghardt, and B. Sinervo. 1994. Dispersal in the side-blotched lizard: A test of the mate-defense dispersal theory. Anim. Behav. 47:227–229.

Falconer, D. S. 1965. Maternal effects and selection response. Pp. 763–774 in S. J. Geerts (ed.), Genetics Today. Pergamon Press, Oxford.

Falconer, D. S. 1981. Introduction to Quantitative Genetics. Longman, New York.

Ferguson, G. W., and C. H. Bohlen. 1978. Demographic analysis: A tool for the study of natural selection of behavioral traits. Pp. 227–243 in N. Greenberg and P. D. MacLean (eds.), Behavior and Neurology of Lizards. DHEW, Washington DC.

Ferguson, G. W., and H. L. Snell. 1986. Endogenous control of seasonal change of egg, hatchling, and clutch size of the lizard *Sceloporus undulatus garmani*. Herpetologica 42:185–191.

Glazier, D. S. 1992. Effects of food, genotype, and maternal size and age on offspring investment in *Daphnia magna*. Ecology 73:910–926.

Hutchinson, G. E. 1951. Copepodology for the ornithologist. Ecology 32:571–577.

Iwasa, Y., A. Pomiankowski, and S. Nee. 1991. The evolution of costly mate preferences. II. p. 550 The "handicap" principle. Evolution 45:1431–1442.

Johnson, E. O., T. C. Kamilaris, G. P. Chrousos, and P. W. Gold. 1992. Mechanisms of stress: A dynamic overview of hormonal and behavioral homeostasis. Neurosci. Biobehav. Rev. 16:115–130.

Kerfoot, W. C. 1974. Egg-size cycle of a cladoceran. Ecology 55:1259–1270.

Kirkpatrick, M., and R. Lande. 1989. Selection, inheritance, and evolution of maternal characters. Evolution 43:485–503.

Lack, D. 1954. The Natural Regulation of Animal Numbers. Clarendon Press, Oxford.

Lande, R. 1981. Models of speciation by sexual selection on polygenic traits. Proc. Natl. Acad. Sci. USA 78:3721–3725.

Landwer, A. J. 1994. Manipulation of egg production reveals costs of reproduction in the tree lizard (*Urosaurus ornatus*). Oeocologia 100:243–249.

Li, C. C. 1975. Path Analysis—A Primer. Boxwood Press, Pacific Grove, Calif.

Luetenegger, W. 1976. Allometry of neonatal size in eutherian mammals. Nature 263:229–230.

Luetenegger, W. 1979. Evolution of litter size in primates. Am. Nat. 114:525–531.

Maynard Smith, J. 1982. Evolution and the Theory of Games. Cambridge University Press, Cambridge.

McGinley, M. A., D. H. Temme, and M. A. Geber. 1987. Parental investment in offspring in variable environments: Theoretical and empirical considerations. Am. Nat. 130:370–398.

Mitchell, R. J. 1993. Path analysis: Pollination. Pp. 211–231 in S. M. Scheiner and J. Gurevitch (eds.), Design and Analysis of Ecological Experiments. Chapman and Hall, London.

Mitchell-Olds, T., and R. G. Shaw. 1987. Regression analysis of natural selection: Statistical and biological interpretation. Evolution 41:1149–1161.

Mousseau, T. A., and D. A. Roff. 1987. Natural selection and the heritability of fitness components. Heredity 59:181–197.

Nussbaum, R. A. 1981. Seasonal shifts in clutch size and egg size in the side-blotched lizard, *Uta stansburiana* Baird and Girard. Oecologia 49:8–13.

O'Donald, P. 1983. Sexual selection by female choice. Pp. 53–66 in P. Bateson (ed.), Mate Choice. Cambridge University Press, Cambridge.

Packard, G. C., and M. J. Packard. 1988. The physiological ecology of reptilian eggs and embryos. Pp. 523–605 in C. Gans and R. B. Huey (eds.), Biology of the Reptilia. Liss, New York.

Partridge, L. 1992. Measuring reproductive costs. Trends Ecol. Evol. 7:99–100.

Partridge, L., and P. H. Harvey. 1985. Costs of rerproduction. Nature 316:20–21.

Pomiankowski, A. 1988. The evolution of female mate preference for male genetic quality. Pp. 136–184 in P. Harvey and L. Partridge (eds.), Oxford Surveys in Evolutionary Biology. Oxford University Press, New York.

Reznick, D. 1985. Costs of reproduction: An evaluation of the empirical evidence. Oikos 44:257–267.

Reznick, D. 1992. Measuring costs of reproduction. Trends Ecol. Evol. 7:42–45.

Riska, B. 1991. Maternal effects in evolutionary biology. Pp. 719–724 in E. Dudley (ed.), The Unity of Evolutionary Biology. Dioscorides Press, Portland, Ore.

Riska, B., J. J. Rutledge, and W. R. Atchley. 1985. Covariance between direct and maternal genetic effects in mice, with a model of persistent environmental influences. Genet. Res. (Cambridge) 45:287–297.

Schluter, D. 1995. Adaptive radiation in sticklebacks: Trade-offs in feeding performance and growth. Ecology 76:82–90.

Sinervo, B. 1990. The evolution of maternal investment in lizards: An experimental and comparative analysis of egg size and its effects on offspring performance. Evolution 44:279–294.

Sinervo, B. 1993. The effect of offspring size on physiology and life history: Manipulation of size using allometric engineering. Bioscience 43:210–218.

Sinervo, B. 1994. Experimental tests of allocation paradigms. Pp. 73–90 in E. R. Pianka and L. J. Vitt (eds.), Lizard Ecology. Princeton University Press, Princeton, N.J.

Sinervo, B., and D. F. DeNardo. 1996. Costs of reproduction in the wild: Path analysis of natural selection and experimental tests of causation. Evolution 50:1299–1313.

Sinervo, B., and P. Doughty. 1996. Interactive effects of offspring size and timing of reproduction on offspring reproduction: Experimental, maternal, and quantitative genetic aspects. Evolution 50:1314–1327.

Sinervo, B., and R. B. Huey. 1990. Allometric engineering: An experimental test of the causes of interpopulational differences in locomotor performance. Science 248:1106–1109.

Sinervo, B., and P. Licht. 1991a. Proximate constraints on the evolution of egg size, egg number, and total clutch mass in lizards. Science 252:1300–1302.

Sinervo, B., and P. Licht. 1991b. The physiological and hormonal control of clutch size, egg size, and egg shape in *Uta stansburiana*: Constraints on the evolution of lizard life histories. J. Exp. Zool. 257:252–264.

Sinervo, B., and C. M. Lively. 1996. The rock-paper-scissors game and the evolution of alternative male reproductive strategies. Nature 380:240–243.

Sinervo, B., P. Doughty, R. B. Huey, and K. Zamudio. 1992. Allometric engineering: A causal analysis of natural selection on offspring size. Science 258:1927–1930.

Smith, C. C., and S. D. Fretwell. 1974. The optimal balance between size and number of off-spring. Am. Nat. 108:499–506.

Williams, G. C. 1966. Adaptation and Natural Selection. Princeton University Press, Princeton, N.J.

Wilson, B. S., and J. C. Wingfield. 1992. Correlation between female reproductive condition and plasma corticosterone in the lizard *Uta stansburiana*. Copeia 1992:691–697.

18

Maternal Effects and the Maintenance of Environmental Sex Determination

WILLEM M. ROOSENBURG & PETER NIEWIAROWSKI

One of the more perplexing issues in evolutionary biology has been the adaptive significance of environmental sex determination (ESD). Two forms of ESD can be identified based on the nature of the environmental factor that influences sex determination. The first, biotic ESD, influences sexual development by either the density of conspecifics or the quality of hosts. The apparent adaptive advantage of biotic ESD is intuitive; however, the maternal effect of female choice may play an important role determining the sex of the offspring. Far less obvious is the adaptive significance of abiotic ESD; in this form the factors influencing the sex of offspring include temperature, pH, and photoperiod. Temperature is most frequently cited as the environmental factor influencing sex in vertebrates with ESD. Despite considerable efforts to understand why ESD is widespread among reptiles and apparently has persisted for long periods of evolutionary time in certain lineages, empirical data supporting an adaptive basis for this type of sex determination is sparse. A common theme arising from long-term studies of ESD is that conditions during incubation influence both growth and sex. Thus, maternal effects on growth that are gender specific may interact with female nest-site choice as a mechanism to enhance the fitness benefit of ESD.

In this chapter we discuss two potential maternal effects that may play an adaptive role in ESD of long-lived reptiles: (1) female nest-site choice and (2) maternal hormonal contribution to developing embryos. Then, using a feasible ($r \geq 0$) demography approach, we explore how these maternal effects can influence population-level processes in the diamondback terrapin, *Malaclemys terrapin*, a turtle with ESD. Our findings suggest that maternal effects can be important in maintaining ESD but also can play a major role in influencing life history traits.

18.1 Introduction

ESD is a condition in which the sex of an individual is determined by environmental conditions experienced by that individual (Charnov and Bull 1977; reviewed in Bull 1980, 1983;

Korpelainen 1990; Ewert and Nelson 1991; Janzen and Paukstis 1991a). ESD differs from genotypic sex determination (GSD), in which gender is determined through Mendelian inheritance of specific sex chromosomes, for example, XY or ZW systems found in many organisms. The apparent ease with which ESD can result in skewed sex ratios has led to the assumption that there must be strong selection maintaining ESD (Bull and Charnov 1988). The Charnov-Bull model (1977) describes how ESD may be adaptive and predicts that the fitness of the individual is correlated with the gender and the environmental conditions that occur during sex determination; that is, males from certain environments have greater fitness than females from the same environment and vice versa.

The two general categories of ESD are biotic and abiotic ESD. With biotic ESD, the component of the environment that influences sex is related to the size or number of conspecifics that are present. Examples include sequential hermaphrodism (protandry or protogyny; Charnov 1982) or cases in which the number of conspecifics or host condition influences gender, as is observed in a variety of invertebrates with ESD (reviewed in Bull 1983; Korpelainen 1990). In biotic ESD, the form of the mating system increases the relative advantage of one sex over the other. Two general conditions arise in biotic ESD depending on which gender is likely to receive the greatest increase in fitness due to the current environmental conditions. In the first case, when increasing female body size or condition improves fitness through fecundity advantage, decreasing host quality or increasing conspecific density results in a higher proportion of males (e.g., Mermithid nematode worms; reviewed in Korpelainen 1990). In the other case, males are produced only in the presence of females (e.g., the echiurid worm *Bonellia viridis;* Bull 1983). The selective advantage of biotic ESD is that it maximizes the relative fitness of a particular sex in response to the number and gender of other conspecifics, thus ensuring appropriate sex ratios.

In organisms with abiotic ESD, gender determination is a function of the abiotic environment. Temperature, pH, and photoperiod have been identified as environmental factors that influence sex (reviewed in Bull 1980, 1983; Korpelainen 1990; Ewert and Nelson 1991; Janzen and Paukstis 1991a). In abiotic ESD, the fitness consequences of incubation effects on gender do not appear to be linked to population sex ratio, and thus the adaptive significance of abiotic ESD has, for the most part, remained elusive. Several studies have suggested an adaptive basis for ESD by linking environmental conditions, gender, and fitness during sex determination. Conover and colleagues demonstrated in the Atlantic silverside, *Menidia menidia*, that cool temperatures early in the season produce females that benefit from a longer growing season, and their larger body size results in more eggs (Conover 1984; Conover and Heins 1987). In crocodilians and turtles, incubation temperature affects growth independent of sex, and thus for sexually dimorphic species it appears that rapid growth is matched to the faster growing sex (Joanen et al. 1987; Spotila et al. 1994; Allsteadt and Lang 1995; Rhen and Lang 1995). In some cases incubation conditions displaced from threshold levels result in growth rates correlated with the fitness of a particular gender (Allsteadt and Lang 1995; Rhen and Lang 1995). These studies suggest that the larger sex benefited the most from rapid growth, through either the fecundity advantage for females or male-male competition. Curiously, increased locomotor performance occurs in hatchlings from intermediate temperatures; this increased performance may decrease hatchling survivorship by increasing predation vulnerability (Janzen 1995). The puzzling nature of abiotic ESD has been compounded by the fact that it occurs predominately in long-lived organisms in which the correlated effects of environment and sex on fitness are difficult to determine (Bull and Bulmer 1989).

Abiotic ESD occurs in four vertebrate groups: some fish, 80% of turtle species, 14% of lizard species, and all crocodilians studied to date (Janzen and Paukstis 1991a). In these organisms, temperature during early ontogeny determines the gender of the developing individual (reviewed in Bull 1980, 1983; Korpelainen 1990; Ewert and Nelson 1991; Janzen and Paukstis 1991a). In reptiles, sex determination usually occurs during the middle third of the incubation period when the gonadal primordia begin to differentiate (Yntema 1979; Bull and Vogt 1981; Pieau and Dorizzi 1981). Several patterns of environmental sex determination have been described in reptiles, the nomenclature of which varies among sources. Perhaps the best designation for the patterns is the nomenclature of Ewert et al. (1994): FMF, females at cool and warm temperatures, is characteristic of all groups of reptiles shown to have ESD; MF, males at cool temperatures and females at warm temperatures, is found only in turtles; and FM, females at cool temperatures and males at warm temperatures, formerly was found in lizards and some crocodilians, but recently most of these groups have been classified as FMF (Viets et al. 1993, 1994; Lang and Andrews 1994). Since its discovery, a large number of researchers have investigated both the mechanism and adaptive significance of ESD. Because of the potential for ESD to generate skewed sex ratios, it is assumed that selection maintains ESD.

Several researchers have suggested that ESD is a plesiomorphic condition (Ewert and Nelson 1991; Janzen and Paukstis 1991b; Burke 1993) and may have no adaptive value (Bull 1980; Mrosovsky 1980). An alternative to this hypothesis is that the evolution of ESD may be correlated with other traits that affect fitness and thus antagonistic pleiotropy prevents the evolution of GSD. Thus, a fruitful direction for research may be to identify traits that allow organisms with ESD to maximize the fitness of their offspring. One manner that organisms with ESD might be able to increase the fitness of their offspring is through maternal effects that covary with offspring gender and its respective fitness. We discuss two possible maternal effects that may be adaptive in organisms with ESD: (1) the use of nest-site choice as a mechanism to manipulate the sex of the female's progeny and (2) maternal manipulation of hormone levels in the eggs to ensure that embryos incubating near the pivotal temperature become the appropriate sex. We elaborate on both of these mechanisms and suggest how they may interact with offspring condition, another possible maternal effect, to develop an adaptive basis for ESD. We combine our discussion of these hypotheses with heuristic modeling that explores how maternal effects might impact population dynamics.

18.2 Maternal Manipulation of Sex Ratio

One of the assumptions of the Charnov-Bull hypothesis is that females have no control over the gender of their offspring (no patch choice; Charnov and Bull 1977). However, with the discovery of abiotic ESD in reptiles, it was suggested that the Charnov-Bull assumption of no maternal patch choice may be relaxed as an adaptive advantage for ESD (Bull 1980). Early speculation on this issue suggested that maternal manipulation of sex ratio could serve as a mechanism to maintain balanced sex ratios in populations. For most vertebrates with abiotic ESD, this argument was considered implausible because they are long-lived and late maturing. Additionally, nest-site decisions would have to predict the sex ratio of the population several years into the future when those offspring mature. It is unlikely that mothers are able to predict sex ratio several years into the future based on the current sex ratio.

Bull (1983) suggested that species with ESD might be able to manipulate the sex ratio of their offspring if mothers could pass to their offspring fitness benefits that covary with gender. The ability of females to manipulate the sex ratio of their offspring as a function of her condition was first described for mammals with GSD (Trivers and Willard 1973). The application of maternal condition-dependent sex ratio decisions has been modified to apply to organisms with ESD and has been referred to as Triversian selection (Mrosovsky 1994) or the maternal condition-dependent choice hypothesis (MCDC) (Roosenburg 1996). We review this model and elaborate on how the model can apply to the current data on organisms with abiotic ESD.

Triversian selection (Mrosovsky 1994) and the MCDC hypothesis (Roosenburg 1996) invoke a maternal condition-dependent sex ratio decision (in the sense of Trivers and Willard 1973). An underlying assumption of this model is that environmental factors affect male and female fitness differentially, similar to the Charnov-Bull model (Charnov and Bull 1977). However, an additional fitness difference between male and female offspring is dependent on maternal allocation to her offspring, whereby allocation decisions should influence female behavior (Mrosovsky 1994; Roosenburg 1996). Thus, two potential maternal effects can evolve according to this scenario. First, females may make allocation decisions concerning progeny that determines the most appropriate sex for those offspring (e.g., egg size or some other qualitative difference among offspring). Second, females must manipulate sex ratio to produce the appropriate sex based on the allocation decision (e.g., oviposition or nest site choice). Female allocation can involve resources that determine offspring size or any other physiological condition that imparts a fitness difference between males and females. Furthermore, sex ratio manipulation can occur either behaviorally through oviposition site choice or physiologically through hormonal manipulation of the offspring. Females with the ability to accurately manipulate the sex of their offspring and to associate allocation decisions and offspring sex should be favored by selection. Thus, these maternal effects can be adaptations if they exist in natural populations.

Nest-Site Choice

Maternal control of oviposition site in organisms with both biotic and abiotic ESD can have a direct effect on the gender of the offspring, and differential maternal allocation of resources among offspring may serve as a basis for making oviposition decisions (Trivers and Willard 1973; Mrosovsky 1994; Roosenburg 1996). Perhaps one of the reasons that the fitness consequences of abiotic ESD are difficult to untangle is that the potential impact of maternal effects such as propagule size and oviposition site choice cannot be detected easily in laboratory-based studies.

For most species with abiotic ESD, all of the eggs in a single clutch become either male or female (Vogt and Bull 1984; Bull 1985; Janzen 1994; Roosenburg and Place 1995). Thus, sex ratio decisions can be made with some probability that offspring of a single reproductive bout all become either male or female. However, a female's ability to accurately predict the sex ratio of her offspring is complicated in most species with ESD because the temperature-sensitive period occurs during the middle third of the incubation period, more than 14 days after oviposition (Yntema 1979; Bull and Vogt 1981; Pieau and Dorizzi 1981). Nonetheless, the ability of females to discriminate among different nesting sites with different thermal properties has been documented for lizards (Bull et al. 1988) and turtles (Roosenburg 1996).

For female nest-site choice to evolve as a mechanism to maintain ESD, the following two assumptions must be met: (1) the impact of the maternal condition on offspring quality must affect the fitness of one sex more than the other (in the sense of Trivers and Willard 1973); (2) females must be able to assess future offspring quality, through either her condition or the quality of environmental conditions during vitellogenesis. If these two assumptions hold, then females that can selectively produce the sex that is most likely to benefit from the differences in offspring quality will have a fitness advantage (Roosenburg 1996).

One case that supports the MCDC hypothesis and Triversian selection model is the diamondback terrapin, *Malaclemys terrapin* (Roosenburg 1996). The diamondback is a sexually dimorphic species with an MF pattern of ESD. Male terrapins mature at approximately 300 g and females mature at approximately 1100 g. Commensurate with the size dimorphism are differences in growth and age of maturity; females mature between 8 and 13 years of age, whereas males mature between 4 and 7 years (Roosenburg 1991). In this terrapin population, egg size ranges from 6 to 12 g among clutches, but egg size varies little within clutches (Roosenburg and Kelley 1996). Egg size among clutches varies independent of clutch size or female size (Roosenburg and Dunham 1997). Finally, egg size affects the growth of juvenile female terrapins. At 3 years of age, female terrapins developing from larger eggs are 30–40% larger than those developing from smaller eggs (Roosenburg and Kelley 1996). Compared to the terrapin growth rates in the field, this difference could result in a several year difference in age at maturity between female offspring from large versus small eggs (W. M. Roosenburg, unpublished data). Interestingly, a similar effect of egg size on growth was not observed for male terrapins. Finally, in the terrapin system females place smaller eggs in cooler nest sites and larger eggs in warmer nest sites, a pattern consistent with the prediction of matching gender with the fitness advantage of larger egg size (Roosenburg 1996). These findings suggest that female nest-site choice using egg size as a cue for discrimination may provide an adaptive basis for ESD and contribute to its evolutionary persistence.

Studies have reported findings that also can be considered consistent with Triversian selection or the MCDC hypothesis in other species with ESD. Rhen and Lang (1994) detected strong clutch effects and clutch × incubation temperature effects on the growth rate of juvenile snapping turtles that were independent of egg size. Some clutches exhibited faster growth rates when incubated at male-producing temperatures and others grew at similar rates regardless of incubation temperature. Their data suggested that some clutches might be predisposed to be either better males or females independent of egg size. These data do not imply that female nest-site choice is occurring in snapping turtles; however, they do suggest that a basis for the discrimination among clutches by female snapping turtles exists. Although detailed field work attempting to detect nest-site choice in snapping turtles is lacking, certainly female ability to match incubation sites and clutch predisposition regarding gender would be consistent with the model of female choice. Furthermore, several studies have revealed that egg size is a primary factor contributing to hatchling size in turtles, and these size effects persist into early growth in desert tortoises (Spotila et al. 1994) and snapping turtles (Brooks et al. 1991; Bobyn and Brooks 1994). The combination of sexual dimorphism and the persistent effects of egg size could function as exaptations that promote the evolution of female ability to discriminate among nesting sites regarding gender. More information on nest-site choice and egg-size effects is needed to determine whether MCDC is a common phenomenon or occurs in isolated cases in species with ESD.

One study explicitly investigated the possibility that female nest-site choice might serve as a mechanism to manipulate sex ratio in a lizard with ESD (Bull et al. 1988). In this study, temperatures among different nest sites had a normal distribution centered around intermediate temperatures. Had sex ratio comprised an important criterion for nest N site choice, a bimodal distribution would have seemed more likely. Other studies of squamates with GSD have found that offspring performance increased at intermediate incubation temperatures (Burger 1989, 1990, 1991), suggesting that intermediate temperatures result in offspring of higher fitness. However, Shine et al. (1995) studied a lizard with GSD in which offspring growth and performance were related to incubation temperature and gender: males from cool incubation temperatures grew and performed better than females, and vice versa for females from warm incubation temperatures. In this system, a bimodal distribution of nest temperatures may be expected due to the effect of incubation temperature on offspring fitness, as in the case of MCDC. Nonetheless, bimodal distributions in the wild may be difficult to detect because, first, nest-site choice in oviparous organisms must ensure survival of the offspring, and the fitness effects regarding gender are likely to be secondary. Second, depending on the range of environments available and the specific requirements for incubation, the resolution required to detect nest-site differences regarding sex ratio may be overwhelmed by the accuracy with which nest-site discrimination can be measured. Furthermore, detecting these subtle differences in the field would be complicated by fluctuation in nest-site temperatures. Future field studies should incorporate measurement of nest temperatures, size or quality of the propagules, the hatching success, and the resulting sex ratio of that reproductive bout.

Studies of turtle hatchling performance have been made in species with both ESD and GSD. In snapping turtles (having ESD), offspring swimming and running speed was greatest when incubated at intermediate temperatures (Janzen 1995) and in wetter substrates (Miller et al. 1987); however, performance was negatively correlated with hatchling survival (Janzen 1995). Other studies have suggested that larger snapping turtle hatchlings have greater survivorship, as measured by the ability of hatchlings to traverse the distance between the nesting site and the adjoining marsh (Janzen 1993a; F. J. Janzen, J. V. Tucker, and G. L. Paukstis, personal communication). In this study, hatchling survivorship appeared to be correlated with locomotor performance. Similarly, in *Apalone mutica* (a turtle with GSD), hatchling size and performance were greatest for individuals incubated at higher temperatures (Janzen 1993b). Clearly, further studies of the relevant environmental factors that affect fitness need to be performed before a thorough understanding of nest-site choice can be obtained. For example, it would be interesting to compare the selective environments of snapping turtles and *Apalone* to determine whether increased performance in *Apalone* has a similar effect on survivorship as in snapping turtles. These studies indicate the complexity of environmental effects during incubation and other different life cycle stages and suggest that ESD represents another variable that potentially complicates maternal decisions as females attempt to maximize their fitness.

Maternal Hormonal Manipulation of Offspring Gender

Females of species with abiotic ESD might be able to manipulate the gender of their offspring through the differential allocation of hormone or hormonal precursors to follicles or eggs. Researchers attempting to identify the molecular mechanism of ESD have demonstrated that hormones applied to eggs incubated under laboratory conditions can reverse temperature effects on gender (reviewed in Crews 1994; Crews et al. 1994; Pieau 1996). These studies have shown that eggs treated with 17-β estradiol or estrone can result in females incubated at male-producing temperatures. A variety of aromatase inhibitors also have successfully biased sex ratios toward males when eggs were incubated near the pivotal temperature (Rhen and Lang 1995; Pieau 1996). (Aromatase is the primary enzyme responsible for the transformation of androgens into estrogen. Aromatase levels typically increase at female-producing temperatures during the temperature-sensitive stage in organisms with ESD and are directly implicated as part of the sex determination pathway [Jeyasuria et al. 1994; Pieau 1996]).

Furthermore, studies have indicated that there is a dosage effect when hormones are used to reverse sex (Crews et al. 1991, 1994; Wibbels et al. 1991) and a synergistic effect between incubation temperature and hormonal treatment; that is, the closer the eggs are incubated to a pivotal temperature, the lower the hormone dosage necessary to reverse sex, suggesting that maternal hormonal manipulation may be effective only at incubation conditions near the threshold of sex change (Wibbels et al. 1991). Recent work also has demonstrated that yolk contains estrogen and androgens that may be converted by aromatase, suggesting that differential maternal allocation of steroids could serve as a mechanism to manipulate sex ratio, at least in American alligators (Conley et al. 1996). Thus, females may be able to bias the sex ratio of their eggs by differentially allocating among clutches of eggs estrogen or other hormones that may play a role in sex determination. Unfortunately, hormonal manipulation of sex determination was unsuccessful when red-eared slider eggs were treated with high levels of estrogen benzoate at oviposition, suggesting that differing estrogen levels at oviposition may not effect sex determination in these turtles (Gutzke and Chyimy 1988). As we learn more about the molecular mechanism of ESD (reviewed in Pieau 1996), we may find that differential allocation by the mother that effects the production of aromatase or the level of precursors in the embryo may cause subtle changes in the threshold temperature of a clutch. Hormonal manipulation may be particularly effective for species in which offspring fitness is highest at intermediate incubation temperatures close to threshold temperatures, yet egg size or clutch effects might benefit one gender more than the other. Slight differences in hormone levels of reptile eggs have been observed in crocodilians and turtles (J. W. Lang, personal communication); however, it remains unknown whether hormone levels can actually result in gender differences.

Finally, one experimental factor that plagues most of these studies is that incubation in the lab occurs at constant temperatures, yet nests in the field experience diel temperature cycles (Jeyasuria et al. 1994; Plummer et al. 1994). Clearly, more work is needed to thoroughly understand the possibility of hormonal manipulation of sex ratio as a maternal effect, particularly during incubation conditions that accurately reflect the conditions that occur in natural nests.

18.3 Why Should a Female Produce Males?

Fisherian selection argues that frequency dependence will result in a 50:50 sex ratio in populations of sexually reproducing organisms (Fisher 1930). However, biased sex ratios can be maintained in populations with polygamous mating systems. Turtles species with ESD are polygamous, and sex ratios vary widely from male biased to female biased (Gibbons 1990). The diamondback terrapin population described above has a 3:1 female-biased sex ratio

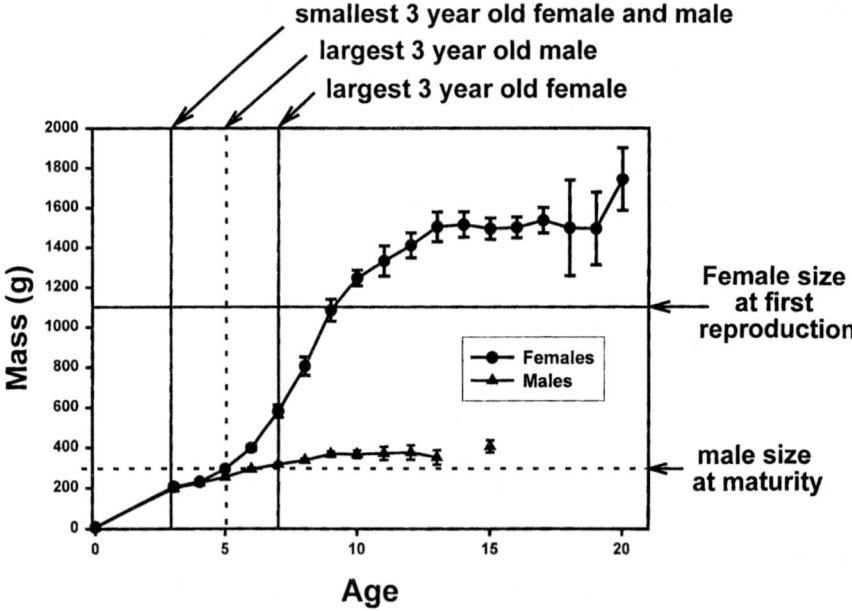

Figure 18.1 Age-mass relationship for male and female terrapins from the Patuxent River population. The two horizontal lines on the graph indicate the minimum size at maturity for males (300 g) and the minimum size at first reproduction for females (1100 g). The vertical lines intersect the growth curve at the minimum and maximum size of 3-year-old turtles from a laboratory growth experiment (Roosenburg and Kelley 1996). Differences in time to maturity were interpolated based on the intersection of the lines with the age axis. Female sizes in the experiment spanned 4 years in size classes while males spanned only 2 years. A considerable amount of the variation (60%) in female size was attributable to variation in egg size; the variation in male size could not be related to egg size.

(Roosenburg et al. 1997). In this section, we put aside the argument of Fisherian selection for sex ratio and model how age at first reproduction for females, as influenced by egg size, should influence when mothers should produce later maturing female offspring, assuming earlier reproducing male offspring have higher fitness.

Differences in age of maturity between male and female terrapins generate a demographic paradox (figure 18.1). Because males mature much earlier than females, male offspring would always have higher fitness simply by reaching maturity at an earlier age (Cole 1954) if the only difference between male and female life tables is the age at maturity. To understand when it would be beneficial for female terrapins to select nesting sites that had a higher probability of producing female offspring, we examined the demographic parameters that would result in female offspring having greater fitness than male offspring. In our investigation we did not consider adult sex ratio as a driving force in nest-site choices to produce female or male offspring.

Based on the findings of Roosenburg (1996), we asked three questions concerning how the interaction between nest-site choice and egg size could maintain ESD. First, what egg size is necessary for males and females to have a similar age of first reproduction, if any?

Second, what is the magnitude of the effect of egg size on the age of first reproduction? Finally, what is the relationship between age at first reproduction and adult survivorship or how might decreasing survivorship influence selection on egg size, as a maternal effect? Our exercise was intended as a heuristic attempt to understand how the maternal effects of egg size can serve as a basis for nest-site choice and thus maintain populations with ESD. However, our findings can be applied also to the evolution of egg size as a life-history trait in general. Maternal effects on egg size can accelerate or impede the rate at which evolution might change the trait of egg size (reviewed in Arnold 1994; Bernardo 1996) and the manner in which selection might act on egg size may change over the lifetime of the individual, particularly for long-lived organisms.

We addressed the first question, what size egg would be required for female offspring to mature at the same age as male offspring, by using the regression equation from Roosenburg and Kelley (1996). This equation estimates the linear relationship between egg mass and mass at 3 years (the point at which males from the experiment were becoming sexually mature):

$$\ln(\text{final mass}) = \ln[\text{egg mass}(1.18)] + 3.17 \qquad (18.1)$$

Using equation (18.1) we estimated that the minimum egg size of 25.8 g would be needed to produce a female that would mature at the same age as a male. This is more than 2.5 times the average egg mass and almost twice the mass of the largest viable egg ever observed in the Patuxent population. Although 20 g eggs have been observed in Patuxent terrapins, these eggs are deformed in shape and none has produced a viable hatchling (W. M. Roosenburg, personal observation). However, the occurrence of large eggs suggests that egg size is not restricted by morphological constraints on the female's ability to produce enlarged eggs (Congdon and Gibbons 1985, 1987). Additionally, if we assume that there is a linear trade-off between egg size and clutch size, this finding would result in a 2.5-fold reduction in the number of eggs being produced. Such a trade-off has not been detected across the observed range of egg sizes in the terrapin population (Roosenburg and Dunham 1997). Nonetheless, increasing egg size to 25.8 g is likely to lead to a decrease in total fecundity that would have to be offset by increases in either adult or juvenile survivorship. Current estimates of turtle juvenile survivorship are low (Congdon et al., 1993, 1994); however, there is experimental evidence that suggests increasing hatchling size increases juvenile survivorship (Janzen 1993a; Janzen et al., personal communication). Currently, in the terrapin study the number of hatchlings marked that were later recaptured as adults is small (<10) and an accurate test of hatchling size on fitness cannot be accomplished.

To answer our second question, what is the magnitude of the effect of egg size on growth, we used the relationship between age and mass of wild-caught males and females from the Patuxent River terrapin population. Figure 18.1 is the age-mass relationship between male and female terrapins, illustrating (1) the differences in age and size of maturity between males and females, (2) the differences in growth, and (3) the sexual-size dimorphism in terrapins. To determine the effect of size on time to maturity, we drew vertical lines to intersect the growth curves at the size of the largest and smallest 3-year-old male and female terrapins from a growth experiment (figure 18.1; Roosenburg and Kelley 1996). We estimated that the male sizes from the growth experiment reflected the average size of wild individuals between 3 and 5 years of age. The range of female sizes in the growth experiment was similar to that of wild individuals between 3 and 7 years of age (figure 18.1). Sixty percent of the differences in size of the 3-year-old females was attributable to the effect of egg size, while no egg size effect was detected for males (Roosenburg and Kelley 1996). Thus, the maternal effect of egg size can cause up to a 4-year difference in age at maturity for female offspring but not for males. The size range of 3-year-old females from the growth experiment was consistent with the range of 3-year-old wild-caught female terrapins (Roosenburg 1996), suggesting that some of the natural variation in size of juvenile female terrapins can be attributable to variation in egg size. The minimum age of first reproduction determined through inguinal palpitation of female terrapins is 8 years. However, the average 8-year-old female is too small to reproduce and the average age at which females reach the minimum size at maturity is 10 years. Thus, by accelerating female age of first reproduction, the maternal effect of egg size can have a considerable effect on the terrapin's life history.

Our third question is why females do not produce eggs that would result in female offspring capable of reaching maturity at the same time as male offspring (4 vs. 8 years, respectively). Females might be able to produce eggs large enough to mature at the same age as males by increasing the maternal investment, thus accelerating growth to result in an earlier age of first reproduction. We varied egg size to estimate the demographic consequences of adjusting egg and clutch size so that females could attain maturity between 3 and 7 years of age (figure 4 of Roosenburg and Kelley 1996; assuming linear individual growth rates). We assumed that there is a linear trade-off between number and size of eggs, such that increasing egg size resulted in a decrease in clutch size. We then calculated how adult survival might be affected by decreasing the age of first reproduction through increasing egg size, while simultaneously adjusting our simulation for the change in offspring number due to the linear trade-off between offspring size and number.

We estimated a life table for the diamondback terrapin based on observations from the Patuxent River terrapin population. We used the Euler equation

$$1 = \int l(x)m(x)e^{-rx} \qquad (18.2)$$

to determine the intrinsic rate of growth, r, of the population; $l(x)$ is the survivorship to time x, $m(x)$ is the reproductive output at time x, and e is the base of the natural logarithm. We considered positive values for r as feasible solutions that are sufficient to maintain populations. Negative solutions indicate that the population is declining and therefore are considered nonfeasible. We varied age at maturity (between 3 and 7 years) as a linear function of egg size and assumed that total clutch mass was fixed. Assuming a stationary population ($r = 0$), we then modeled the demographic consequences of increasing maternal provisioning, adjusting the age at first reproduction, and then solving for adult survivorship. All other aspects of hatchling demographics were held constant (i.e., juvenile survival and reproductive rates after age at first reproduction).

Several interesting conclusions can be drawn from this modeling exercise. First, decreasing the age of first reproduction can result in decreasing adult survivorship, independent of a trade-off between offspring size and number (figure 18.2). Second, the trade-off between offspring size and its effect on age of maturity results in two combinations of egg and clutch size that have lower adult survivorship when $r = 0$ (figure 18.3A). Thus, if decreasing adult survivorship is necessary to maintain a stable population, then there are two possible manners this can be accomplished: (1) by increasing egg size and decreasing age of first reproduction and clutch size or (2) by decreasing egg size and increasing age of first reproduction and clutch size (figure 18.3B–D). The latter case appears to match that observed in the ter-

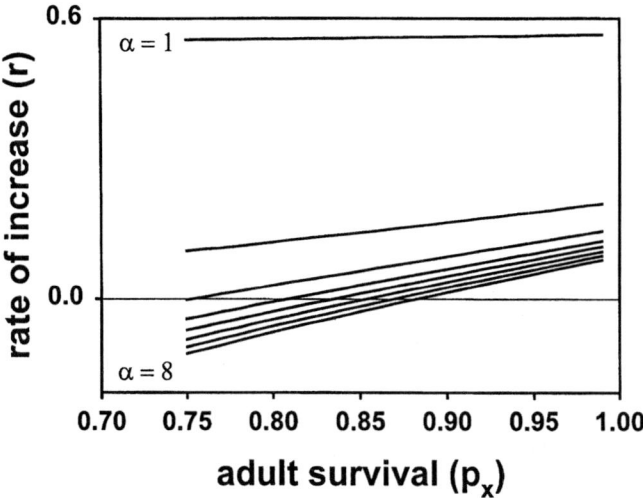

Figure 18.2 The relationship between adult survivorship and the intrinsic rate of increase in the population for different ages of first reproduction. Each line represents the relationship for a different age of first reproduction (α). Adult survivorship (p_x) is age-specific adult survivorship after the onset of maturity, which was held constant for all age classes after first reproduction. Increasing the age at first reproduction requires compensation in other areas in the life table, in this case increasing adult survivorship.

rapin population (i.e., minimum age at maturity of 7 years; figure 18.3). Minimum adult survival rates producing a stationary population at minimum age of maturity of 7 years are lower than for a minimum age at maturity of 3 years (figure 18.3A). In other words, the maximum level of adult mortality that can sustain a stable population is higher at the observed age at first reproduction than for any earlier age at maturity.

Two aspects of this result are interesting. First, while the observed age at maturity appears to be favored demographically, differences in allowable adult mortality among the different potential ages at first reproduction are not large. Second, increasing maternal investment in eggs sufficient to produce a female hatchling that matures at 3 years old incurs a smaller cost in terms of minimum adult survival than more intermediate changes in maternal investment (ages 4, 5, and 6). Because populations of long-lived organisms such as turtles can be extremely sensitive to changes in adult survivorship (Congdon et al. 1993, 1994), it may be that the necessary increase in survivorship to produce larger eggs and smaller clutches cannot be accomplished.

None of these analyses incorporate effects or trade-offs other than that between clutch size and offspring size. For example, we have no idea whether sex ratio considerations may affect maternal investment directly or through an interaction with female nest-site choice. Furthermore, several assumptions involved in the analyses are somewhat arbitrary. For example, we assumed that the function relating egg size to age at maturity is linear. Finally, the role of nest-site selection and ESD could dilute selection on variation in maternal investment by reducing the predictability of producing a male or female hatchling. It is presently unclear how nest-site selection, ESD, and maternal investment interact in the context of

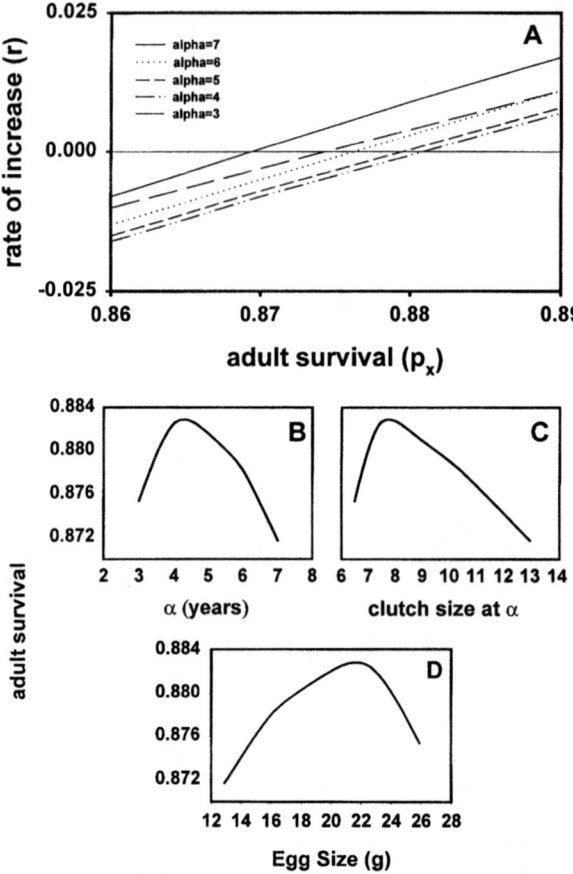

Figure 18.3 A, the relationship between adult survivorship and intrinsic rate of increase incorporating the effects of egg size on age at first reproduction, assuming a direct trade-off between egg size and number. Notice that the lines for 7 and 3 years of age can support the lowest survivorship. B–D use the age-specific survivorship (p_x) from A to illustrate the relationship between age of first reproduction (α, B), clutch size (C), and egg size (D) taken from the solution for A. Notice that the lowest adult survivorship can support one of two strategies: (1) small egg size, late maturing, and large clutch size; and (2) large egg size, early maturing, and small clutch size.

demographic consequences for the trade-off between egg size and clutch size modeled above.

18.4 Conclusions

Understanding the adaptive significance and the maintenance of abiotic ESD will always be complicated by several factors. First, the long-lived nature of most species with abiotic ESD makes it difficult to take a multigenerational approach to study ESD. Additionally, the periods in which selection could strongly favor ESD may be episodic. In the interim, ESD may be a neutral trait. Second, virtually all of the studies that have investigated the adaptive significance of ESD have relied on constant incubation conditions. A variety of studies at this point have demonstrated that incubation conditions in natural environments fluctuate and show diel cycling. Studies are needed to determine how fluctuating and natural incubation conditions affect hatchling growth and other fitness-related traits, particularly regarding maternal effects. Furthermore, the impact of hormonal sex reversal on eggs incubated at cycling temperatures remains unstudied. Third, many species with ESD have small clutches, making it difficult to use a split-clutch experimental design necessary to thoroughly study maternal effects. Perhaps sea turtles would be one of the best species because they frequently produce clutches of more than 100 eggs. Unfortunately, the protected status of sea turtles makes it difficult to manipulate large numbers of hatchlings.

Our study also has explored the manner in which egg size might evolve in response to a changing demographic environment. Although, our work was inspired by trying to understand maternal effects and their impact on populations with ESD, our conclusions apply to the evolution of propagule size as a trait acted on directly by selection or through maternal control of propagule size. Our modeling exercise demonstrates how the demographic environment can have counterintuitive consequences for selection on egg size. Our findings suggest that selection due to demographic factors may force egg size away from optimum sizes as predicted by optimality theory (Smith and Fretwel 1974; Brockelman 1975). Perhaps one interesting technique that can be applied to understanding how maternal effects and other factors influence egg size is allometric engineering experiments that manipulate offspring size (Sinervo and Huey 1990; Sinervo and Licht 1991).

Clearly, maternal effects can play a major role in the maintenance of ESD. However, despite increasing evidence on clutch-specific gender effects, the mechanisms by which maternal effects contribute to fitness differences among clutches relative to their gender remain poorly understood and require further study, particularly in natural settings. We suspect that ESD arose without the ability of maternal nest-site choice and that nest-site choice will not be a universal explanation for the presence of ESD among the diversity of taxa in which it occurs. However, the adaptive manner in which the maternal effects of nest-site choice and propagule quality can interact provide a context in which ESD can be maintained.

Acknowledgments   We are grateful to T.A. Mousseau for inviting us to participate in the maternal effects symposium. We also thank Dr. and Mrs N. Dodge and Mr. and Mrs. L. McCormick-Goodhart for permission to use their property as a study site. We are grateful to C. Fox, F. Janzen, and K. Kelley for helpful comments on the manuscript. We also thank the numerous field assistants and watermen who assisted in the terrapin demographic study for the last 10 years, particularly T. Crelly and P. Daly. The demographic study has been funded by grants to W. M. R. from the Maryland Department of Natural Resources, Sigma Xi, and the Shell Foundation.

References

Allsteadt, J., and J. W. Lang. 1995. Sexual dimorphism in the genital morphology of young American alligators, *Alligator mississippiensis*. Herpetologica 51:314–325.
Arnold, S. J. 1994. Multivariate inheritance and evolution: A review of concepts. Pp. 17–48 in C. R. B. Boake (ed.), Quantitative Genetic Studies of Behavioral Evolution. University of Chicago Press, Chicago.

Bernardo, J. 1996. Maternal effects in animal ecology. Am. Zool. 83–105.

Bobyn, M. L., and R. J. Brooks. 1994. Interclutch and interpopulation variation in the effects of incubation conditions on sex, survival, and growth of hatchling snapping turtles (*Chelydra serpentina*). J. Zool. Lond. 233:233–257.

Brockelman, W. Y. 1975. Competition, the fitness of offspring, and optimal clutch size. Am. Nat. 109:677–699.

Brooks, R. J., M. L. Bobyn, D. A. Galbraith, J. A. Layfield, and E. G. Nancekivell. 1991. Maternal and environmental influences on growth and survival of embryonic and hatchling snapping turtles (*Chelydra serpentina*). Can. J. Zool. 69:2667–2676.

Bull, J. J. 1980. Sex determination in reptiles. Q. Rev. Biol. 55:3–21.

Bull, J. J. 1983. The Evolution of Sex Determining Mechanisms. Benjamin/Cummings, Menlo Park, Calif.

Bull, J. J. 1985. Sex ratio and nest temperatures in turtles: Comparing field and laboratory data. Ecology 66:1115–1122.

Bull, J. J., and M. G. Bulmer. 1989. Longevity enhances selection of environmental sex determination. Heredity 63:315–320.

Bull, J. J., and E. L. Charnov. 1988. How fundamental are Fisherian sex ratios? Oxford Sur. Ecol. Biol. 5:96–135.

Bull, J. J., and R. C. Vogt. 1981. Temperature sensitive periods of sex determination in emydid turtles. J. Exp. Zool. 218:435–440.

Bull, J. J., W. H. N. Gutzke, and M. G. Bulmer. 1988. Nest-site choice in a captive lizard with temperature-dependent sex determination. J. Evol. Biol. 2:177–184.

Burger, J. 1989. Incubation temperature has long-term effects on behavior of young pine snakes (*Pituophis melanoleucus*). Behav. Ecol. Sociobiol. 24:201–207.

Burger, J. 1990. Effects of incubation temperature on behavior of hatchling pine snakes: Implications for reptilian distributions. Behav. Ecol. Sociobiol. 28:297–303.

Burger, J. 1991. Effects of incubation temperature on behavior of young black racers and (*Coluber constrictor*) and kingsnakes (*Lampropeltis getulus*). J. Herpetol. 24:158–163.

Burke, R. L. 1993. Adaptive value of sex detrmination mode and hatchling sex ratio bias in reptiles. Copeia 1993:854–859.

Charnov, E. L. 1982. The Theory of Sex Allocation. Princeton University Press, Princeton, N.J.

Charnov, E. L., and J. Bull. 1977. When is sex environmentally determined? Nature 266:828–830.

Cole, L. 1954. The population level consequences life history phenomena. Q. Rev. Biol 29:103–137.

Congdon, J. D., and J. W. Gibbons. 1985. Egg components and reproductive characteristics of turtles: Relationships to body size. Herpetologica 41:194–205.

Congdon, J. D., and J. W. Gibbons. 1987. Morphological constraint on egg size: A challenge to optimal egg size theory. Proc. Natl. Acad. Sci. USA 84:4145–4147.

Congdon, J. D., A. E. Dunham, and R. C. van Loben Sels. 1993. Delayed sexual maturity and demographics of Blanding's turtle (*Emydoidea blandingii*): Implications for conservation and management of long-lived organisms. Conserv. Biol. 7:826–833.

Congdon, J. D., A. E. Dunham, and R. C. van Loben Sels. 1994. Demographics of common snapping turtles (*Chelydra serpentina*): Implications for conservation of long-lived organisms. Am. Zool. 34:397–408.

Conley, A. J., Lang, P. Elf, J. Corbin, S. Dubowsky, and A. Fivizzani. 1996. Developmental changes in yolk steroids in the alligator, a reptile with temperature-dependent sex deter-miantion. [Abstract]. Western Regional Conference of Comparative Endocrinology, Berkeley, March.

Conover, D. O. 1984. Adaptive significance of temperature-dependent sex determination in a fish. Am. Nat. 123:297–313.

Conover, D. O., and S. W. Heins. 1987. Adaptive variation in environmental and genetic sex determination in a fish. Nature 326:496–498.

Crews, D. 1994. Temperature, steroids and sex determination. J. Endocrinol. 142:1–8.

Crews, D., J. J. Bull, and T. Wibbels. 1991. Estrogen and sex reversal in turtles: A dose-dependent phenomenon. Gen. Comp. Endocrinol. 81:357–364.

Crews, D., J. M. Bergeron, D. Flores, J. J. Bull, J. K. Skipper, A. Tousignant, and T. Wibbels. 1994. Temperature-dependent sex determination in reptiles: Proximate mechanisms, ultimate outcomes, and practical applications. Dev. Genet. 15:297–312.

Ewert, M. A., and C. E. Nelson 1991. Sex determination in turtles: Diverse patterns and some possible adaptive values. Copeia 1991:50–69.

Ewert, M. A., D. R. Jackson, and C. E. Nelson. 1994. Patterns of temperature-dependent sex determination in turtles. J. Exp. Zool. 270:3–15.

Fisher, R. A. 1930. The Genetical Theory of Natural Selection. Dover, New York.

Gibbons, J. W. 1990. Se ratios and their significance among turtle populations. Pp. 171–182 in J. W. Gibbons (ed.), Life History and Ecology of the Slider Turtle. Smithsonian Press, Washington, D.C.

Gutzke, W. H. N., and D. B. Chimry. 1988. Sensitive periods during embryogeny for hormonally induced sex determination in turtles. Gen. Comp. Endocrinol. 71: 265–267.

Janzen, F. J. 1993a. An experimental analysis of natural selection on body size of hatchling turtles. Ecology 74:332–341.

Janzen, F. J. 1993b. The influence of incubation temperature and family on eggs, embryos, and hatchlings of the smooth softshell turtle (*Apalone mutica*). Physiol. Zool. 66:349–373.

Janzen, F. J. 1994. Vegetational cover predicts the sex ratio of hatchling turtles in natural nests. Ecology 70:1593–1599.

Janzen, F. J. 1995. Experimental evidence for the evolutionary significance of temperature-dependent sex determination. Evolution 49:864–873.

Janzen, F. J., and G. L. Paukstis. 1991a. Environmental sex determination in reptiles: Ecology, evolution, and experimental design. Q. Rev. Biol. 66:149–179.

Janzen, F. J., and G. L. Paukstis. 1991b. A preliminary test of the adaptive significance of environmental sex determination in reptiles. Evolution 45:435–440.

Jeyasuria, P., W. M. Roosenburg, and A. R. Place. 1994. The role of P-450 aromatase in sex determination in the diamondback terrapin, *Malaclemys terrapin*. J. Exp. Zool. 270:95–111.

Joanen, T., L. McNease, and M. W. J. Ferguson. 1987. The effects of egg incubation temperature on post-hatchling growth of American alligators. Pp. 533–537 in G. J. W. Webb, S. C. Manolis, and P. J. Whitehead (eds.), Wildlife Management: Crocodiles and Alligators. Surrey Beatty, New South Wales, Australia.

Korpelainen, H. 1990. Sex ratios and conditions required for environmental sex determination in animals. Biol. Rev. 65:157–184.

Lang, J. W., and H. V. Andrews. 1994. Temperature-dependent sex determination in crocodilians. J. Exp. Zool. 270:28–44.

Miller, K., G. C. Packard, and M. J. Packard. 1987. Hydric conditions during incubation influence locomotor performance of hatchling snapping turtles. J. Exp. Biol. 127:401–412.

Mrosovsky, N. 1980. Thermal biology of sea turtles. Am. Zool. 20:531–547.

Mrosovsky, N. 1994. Sex ratios of sea turtles. J. Exp. Zool. 270:16–27.

Pieau, C. 1996. Temperature variation and sex determination in reptiles. Bioessays 18:19–26.

Pieau, C., and M. Dorizzi. 1981. Determination of the temperature sensitive stages for sexual differentiation of the gonads in embryos of the turtle *Emys orbicularis.* J. Morphol. 170:373–382.

Plummer, M. V., C. E. Shadrix, and R. C. Cox. 1994. Thermal limits of incubation in embryos of softshell turtles (*Apalone mutica*). Chelonian Conserv. Biol. 1:141–144.

Rhen, T., and J. W. Lang. 1994. Temperature-dependent sex determination in the snapping turtle: Manipulation of the embryonic sex steroid environment. Gen. Comp. Endocrinol. 96:234–254.

Rhen, T., and J. W. Lang. 1995. Phenotypic plasticity for growth in the common snapping turtle: Effects of incubation temperature, clutch, and their interactions. Am. Nat. 146:726–747.

Roosenburg, W. M. 1991. The diamondback terrapin: Population dynamics, habitat requirements and opportunities for conservation. Pp. 227–233 in J. A. Mihursky and A. Chaney (eds.), New Perspectives in the Chesapeake System: A Research and Management Partnership, Proceedings of a Conference. Pub. No 137. Chesapeake Research Consortium, Solomons, Md.

Roosenburg, W. M. 1996. Maternal condition and nest site choice: An alternative for the maintenance of environmental sex determination. Am. Zool. 36:157–168.

Roosenburg, W. M., and A. E. Dunham. 1997. Allocation of reproductive output: Egg- and clutch-size variation in the diamondback terrapin. Copeia 1997:290–297.

Roosenburg, W. M., and K. C. Kelley. 1996. The effect of egg size and incubation temperature on growth in the turtle, *Malaclemys terrapin.* J. Herpetol. 30:198–204.

Roosenburg, W. M., and A. R. Place. 1995. Nest predation and hatchling sex ratio in the diamondback terrapin: Implications for management and conservation. Pp. 65–70 in S. Nelson (ed.), Towards a Sustainable Coastal Watershed: The Chesapeake Experiment, Proceedings of a Conference. Pub. No. 149. Chesapeake Research Consortium, Solomons, Md.

Shine, R., M. J. Elphick, and P. S. Harlow. 1995. Sisters like it hot. Nature 378:451–452.

Sinervo, B., and R. B. Huey 1990. Allometric engineering: An experimental test of the causes of interpopulational differences in performance. Science 248:1106–1109.

Sinervo, B., and P. Licht. 1991. Hormonal and physiological control of clutch size, egg size, and egg shape in the side-blotched lizard (*Uta stansburiana*): Constraints on the evolution of lizard life histories. J. Exp. Zool. 257:252–264.

Smith, C. C., and S. D. Fretwell. 1974. The optimal balance between size and number of offspring. Am. Nat. 108:499–506.

Spotila, J. R., L. C. Zimmerman, C. A. Binckley, J. S. Grumbles, D. C. Rostal, E. C. Beyer, K. M. Phillips, and S. J. Kemp. 1994. Effects of incubation conditions on sex determination, hatching success, and growth of hatchling desert tortoises, *Gopherus agassizii.* Herpetol. Monogr. 8:116–123.

Trivers, R. L., and D. E. Willard. 1973. Natural selection of parental ability to vary the sex ratio of offspring. Science 179:90–92.

Viets, B. E., A. Tousignant, M. A. Ewert, L. C. E. Nelson, and D. Crews. 1993. Temperature-dependent sex determination in the leopard gecko, *Eublepharis macularis.* J. Exp. Zool. 265:679–683.

Viets, B. E., M. A. Ewert, L. G. Talent, and C. E. Nelson. 1994. Sex-determining mechanisms in squamate reptiles. J. Exp. Zool. 270:45–56.

Vogt, R. C., and J. J. Bull. 1984. Ecology of hatchling sex ratio in map turtles. Ecology 65:582–587.

Wibbels, T., J. J. Bull, and D. Crews. 1991. Synergism between temperature and estradiol: A common pathway in turtle sex determination? J. Exp. Zool. 260:130–134.

Yntema, C. L. 1979. Temperature levels and periods of sex determination during incubation of eggs of *Chelydra serpentina.* J. Morphol. 159:17–27.

19

Density-Mediated Maternal Effects on Seed Size in Wild Radish

Genetic Variation and Its Evolutionary Implications

SUSAN J. MAZER & LORNE M. WOLFE

19.1 Introduction

The quality of the environment in which plants develop can have an immediate and sometimes a persistent effect on their appearance, on their reproductive success, and on the quality and performance of their offspring. This fact has been observed by many ecologists, evolutionists, and breeders, although the perspective from which it has been examined depends strongly on the characters and processes of interest. For example, population ecological studies often aim to determine or to measure direct environmental effects on individual growth and reproduction, which can have immediate effects on the demography and distribution of the next generation, and on the outcome of competitive interactions in natural popualtions (Van Andel and Vera 1977; Marshall et al. 1986; Stratton 1989; Miao et al. 1991a,b). Ecologists have also examined environmental effects on progeny quality because of their imporance in determining life history attributes of the subsequent generation (Gutterman and Evenari 1972; Gutterman 1980, 1992; Cresswell and Grime 1981; Parrish and Bazzaz 1985; Fenner 1986; Garbutt and Bazzaz 1987; Miao et al. 1991b; Wulff and Bazzaz 1992; Hume 1994). From these studies, we know that the quality of a maternal plant's environment often has a strong effect on the nutrient content of its seeds, the probability that they will germinate, and the growth rate and competitive ability of the seedlings derived from them.

In contrast, evolutionary studies are primarily concerned with the consequences of environmental variation for long-term evolutionary change. In these studies, environmentally induced variation is examined as a factor that influences the expression of heritable variation, determines the outcome of competitive interactions among genotypes, contributes to the maintenance of genetic variation, influences the ability of natural selection to mold phenotypic responses to the environment, and influences patterns of gene dispersal and population genetic structure (Alexander and Wulff 1985; Stratton 1989; Aarssen and Burton 1990; Mazer and Schick 1991a,b; Mazer and Wolfe 1992; Schmitt et al. 1992; Curtis et al. 1994; Delesalle and Blum 1994; Wolfe 1995; Young and Schmitt 1995; Lacey 1996).

Breeders and agronomists have approached the phenomenon from both perspectives. Their interest in determining environmental effects on traits that are of economic importance (e.g., germination rate, seed yield) have motivated studies aiming to determine the conditions associated with improved plant performance or seed quality (Austin 1966a,b; Sawhney and Quick 1985; Drew and Brocklehurst 1990). When their goal has been to predict the response to artificial selection, they have examined environmental sources of phenotypic variation and maternal effects on offspring quality, both of which may mask the expression of additive genetic variation, diminishing the response to selection (Durrant 1962; Hill 1967; Gutterman et al. 1975).

Depending on the question of interest, similar research conducted independently by these groups of investigators can generally be divided into studies of phenotypic plasticity and studies of maternal environmental effects on offspring. Phenotypic plasticity is identified when the phenotype expressed by a genotype regarding any morphological or physiological trait depends on the environment in which the genotype is raised or sampled (Bradshaw 1972; Erskine and Khan 1977; Deschamp and Cooke 1985; Schlicting 1986, 1989; Schlicting and Levin 1986; Sultan 1987; Devlin 1988; Stearns 1989; Mazer and Schick 1991a,b; Mazer 1992; Gavrilets and Scheiner 1993; Scheiner 1993, and references therein; Bernardo 1996a). Maternal environmental effects on offspring phenotype represent a special case of phenotypic plasticity and may be defined as environmentally induced influences transmitted from an off-spring-bearing individual to its progeny, which are potentially detectable in any trait expressed during the life cycle of the progeny (Cook 1975; Corey et al. 1976; Schaal 1984; Alexander and Wulff 1985; Antonovics and Schmitt 1986; Roach and Wulff 1987; Mazer 1989, 1992; Mazer and Schick 1991a,b; Gutterman 1992; Schmitt et al. 1992; Plattenkamp and Shaw 1993; Wolfe 1993; Lacey 1996). While maternal environmental effects on offspring phenotype may be induced by the same kinds of cues that are responsible for phenotypic plasticity, they differ from the latter by being cross-generational.

Detecting Maternal Environmental Effects and Evaluating Their Evolutionary Significance

Although environmental effects on morphological and life history traits are commonly observed within and across generations, their evolutionary significance is poorly understood. It is clear that environmental variance (expressed as plasticity or as maternal effects on offspring phenotype) can obscure the expression of consistent phenotypic differences among genotypes. This observation contributes to the view that environmentally induced phenotypic changes constrain the rate of evolution of traits subject to them (Falconer 1989; Kirkpatrick and Lande 1989). An alternative perspective is that the nature of phenotypic plasticity in fitness-related traits is a genetically determined trait that may be open to natural selection (Schlicting and Levin 1986; Schmitt et al. 1992; Wulff et al. 1994). If so, then natural selection among genotypes that express (or fail to express) particular environment-specific phenotypes for any particular trait may influence not only the evolution of the sensitivity of such traits to environmental change, but also the evolution of the trait's mean phenotypic value (Case et al. 1996).

Measuring the degree to which an offspring's phenotype is influenced by its maternal parent's environment, and evaluating the evolutionary significance of such maternal environmental effects, can be a tricky matter. First, in plants, maternal environmental effects on offspring phenotype are not easily defined as a property of, nor can their magnitude be easily measured for, individuals. Since a given seed-bearing individual does not usually produce seeds in more than one environment (independent of changes in age; e.g., Wolfe 1992), the expression of maternal environmental effects on offspring phenotype can be reliably measured only at the genotype or population levels. Second, as is the case for phenotypic plasticity, there may exist variation among genotypes in the direction of or the degree to which maternal environmental effects on progeny phenotype are expressed. So, depending on the trait, the genotypes, and the population examined, maternal environmental effects may or may not contribute significantly to total phenotypic variation in a trait, and the expression of such effects may or may not be open to natural selection.

To measure maternal environmental effects on progeny phenotype in plants requires several steps. The primary step is to cultivate to adulthood genetically homogeneous or highly related seeds, or vegetatively produced clones, across a range of environments. After collecting the seeds produced by these adults in each environment, the phenotype of traits expressed by the seeds must be recorded (e.g., seed mass, germination rate, seedling growth rates, and subsequently expressed fitness components). Differences among environments regarding mean progeny phenotype may be the result of two non-mutually exclusive phenomena. First, if the parental generation was not wholly represented by identical genotypes in all environments, and if different genotypes were favored by natural selection in different environments, then the phenotypic differences among the progeny produced in the different environments may have a genetic basis. Second, the differences among progeny may be purely environmentally induced such that the environment in which a seed-bearing parental plant is cultivated determines initial progeny size and/or subsequently expressed characters.

It is difficult, however, to rule out environment- and genotype-specific mortality as sources of phenotypic differences among progeny produced in distinct environments. Even if total survivorship rates are similar in all environments (but not 100%), the genotypes that *are* culled may differ among environments, resulting in genetic differences among the survivors in different environments. That is, if adults or their developing gametes or progeny are subject to environment-specific selection regimes before reproducing or during seed development, this could result in groups of progeny that differ genetically and phenotypically among environments (Case et al. 1996; Mazer and Gorchov 1996). If environment-specific natural selection can be eliminated as a potential cause of differences observed between progeny produced in different environments, then the phenotypic differences among them can be attributed to maternal environmental effects.

Because maternal environmental effects may be viewed as a special case of phenotypic plasticity, the two phenomena may be the focus of some shared and basic evolutionary questions. For example, is there genetic variation in the expression of maternal environmental effects or phenotypic plasticity that may be open to natural selection? What patterns of environmentally induced phenotypic change may be favored by natural selection? In the case study described in this chapter, we focus on a trait that is perhaps the easiest one in which to detect and to measure maternal environmental effects in plants: seed size. We ask whether genotypes from a natural population differ regarding the direction or magnitude of maternal environmental effects on seed size, and whether there are fitness differences among genotypes that exhibit different kinds of maternal environmental effects on seed size.

Evidence for the Adaptive Nature of Maternal Environmental
Effects on Offspring Phenotype: A Case Study of Wild Radish

To examine the question of whether maternal effects are adaptive, it is useful to consider what kind of evidence would be necessary to answer in the affirmative. As for any trait whose adaptive significance is being evaluated, one must be able to detect and measure alternate phenotypes of the trait, observe the relationship between phenotype and fitness, and determine which (if any) of the phenotypes expressed in nature has the highest expected fitness within or among the range of environments encountered. In the study described here, we consider whether the expression of density-mediated maternal environmental effects on seed size is adaptive.

In wild radish (*Raphanus sativus* L., Brassicaceae), under a wide range of environmental conditions, relatively large seeds develop into seedlings with higher survivorship, faster growth, and higher lifetime reproductive success than smaller seeds (Stanton 1984, 1985; Mazer 1987a; see section 19.4). Positive effects of seed size on germination and/or growth have also been observed in a number of other wild and cultivated species, particularly under competitive conditions (Erickson 1946; Black 1956, 1957; Harper and Obeid 1967; Twamley 1967; Cideciyan and Malloch 1982; Weis 1982; Pet and Garretsen 1983; Dolan 1984; Hendrix 1984; Marshall 1986; Wulff 1986a,b). These observations tempt one to predict that maternal plants or genotypes that respond to increased resource availability by producing larger seeds will enjoy a fitness advantage relative to those that show no such maternal environmental effect on offspring size. In other words, maternal environmental effects on seed size would appear to be "adaptive" when they involve the production of relatively large progeny (all other fitness components remaining equal). It is possible, however, that genotypes that respond to one set of environmental conditions by producing relatively large offspring (compared to genotypes that do not so respond) react to an alternate environment by producing relatively small progeny (or by expressing relatively low values for other fitness components). In this case, such "responsive" genotypes may not possess the most "adaptive" genotype when considered over all the conditions that they are likely to encounter.

The study described here was designed with these observations in mind and with three primary objectives. First, we aimed to detect genetic variation in the magnitude and/or the nature of maternal environmental effects on seed size that can be induced by the population density at which maternal plants are raised. We sought both to identify genotypes that either do or do not produce progeny whose size is density dependent, and to determine whether all density-sensitive genotypes respond in the same direction. Second, we compared responsive (R) and nonresponsive (NR) genotypes regarding seed size and several other fitness components. Third, by comparing the relative performance of R and NR groups within and across population densities, we evaluated the potential role of natural selection in molding the response of maternal plants to population density. We started with the view that to address the question of whether maternal environmental effects on seed size are adaptive in wild radish, the relative performance of alternate genotypes must be evaluated not in a single environment but across the range of environmental conditions they are likely to encounter in nature.

A breeding program was used to produce wild radish seed families of known parentage. From among 76 greenhouse-raised maternal plants (each producing one maternal seed family), we identified nine pairs of maternal seed families of which each pair consisted of two families sired by the same father but produced by different mothers, and of which the pair members differed in the expression of maternal environmental effects on seed size (mediated by sowing density). One type of seed family—the NR type—developed into adults that produced a constant seed size independent of the population density in which they were raised. The other, R family type developed into adults that produced seeds that significantly declined in mean individual seed mass as density increased. These two groups of plants were monitored under three population densities for a range of fitness components to address the following sets of questions:

1. What offspring (seed) size is favored by selection within density treatments? Do relatively large seeds enjoy fitness benefits over smaller seeds within all densities observed?
2. Given that there is genetically based variation in the expression of density-dependent seed size, do genotypes that express such maternal environmental effects on seed size (i.e., R families) always produce larger offspring than NR families?
3. Are there any fitness "costs" to adjusting offspring size? For example, do R maternal families exhibit any disadvantages regarding other fitness components?
4. Is the expression of maternal environmental effects on seed mass adaptive in wild radish? What is the relationship between phenotype (R vs. NR families) and fitness estimates within environments encountered by wild radish?

19.2 Study Organism and Source Material

Raphanus sativus L. is an annual, cosmopolitan, self-incompatible, weedy crucifer of disturbed habitats, and it has become naturalized and widely distributed throughout California. *R. sativus* and its congener *R. raphanistrum* have been the target of many studies of environmentally induced, age-related, additive and nonadditive genetic variation in quantitative morphological, life history, and fitness-related traits. In particular, individual seed mass, due to its ease of measurement, high phenotypic variation within populations, and influence on seedling survival and performance, has been particularly well studied as a target of selection (Stanton 1984, 1985; Mazer 1987a,b, 1989; Stanton et al. 1987; Nakamura and Stanton 1989; Mazer and Wolfe 1992). The floral biology of wild radish permits controlled breeding to produce distinct genetic lines. Since individuals typically produce several hundred flowers during their ~3-month greenhouse life span, one may replicate hand pollinations from several pollen donors within maternal plants (pollen recipients) to produce groups of seeds representing half- and full-sibships.

The source population for the seeds and genotypes used in this study is an old field near the Goleta Coastal Wetlands Preserve (Santa Barbara County, Calif.). This field is adjacent to the site of the fenced, irrigated experimental garden in which this study was performed. This proximity of the source population to the experimental garden helped to assure that the phenotypic variation observed among genotypes is typical for this population of wild radish, and not strongly influenced by observing the genotypes in a novel environment. Within naturalized populations of wild radish, population densities of seedlings and adult plants vary widely over small distances (within meters; S. J. Mazer and L. M. Wolfe, personal observation), so it is logical to manipulate sowing density in an effort to detect maternal environmental effects on progeny phenotype. The population densities created for this study fall well within the range of densities observed in natural populations (Mazer and Wolfe, personal observation).

Seeds Collected from 120 Field-Grown Individuals

Breeding Design

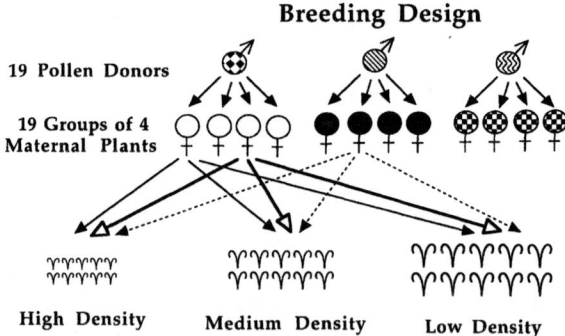

19 Pollen Donors

19 Groups of 4
Maternal Plants

High Density   Medium Density   Low Density

**A total of 2700 seeds representing 19 paternal families
and 76 maternal families were sown in the garden**

Figure 19.1 Breeding design used to raise an array of F_1 maternal seed families across a range of densities. For details, see section 19.3.

19.3 Materials and Methods

Breeding Design

Seeds from 120 naturally pollinated maternal plants were collected in August 1990 from the field population described in section 19.2. Several seeds per plant were sown individually in a sandy loam in 4-inch pots in a greenhouse at the University of California, Santa Barbara, in December 1990. As plants flowered, they were used in a nested breeding design performed in the greenhouse; each plant used was the progeny of a distinct field-collected maternal plant. We used 19 plants as pollen donors with four pollen recipients per male, for a total of 95 mating individuals (figure 19.1). We verified that all maternal plants were genetically compatible with the pollen donor assigned to them. (If flowers failed to develop into fruits following initial pollinations, the maternal plant was replaced with a compatible one from another field-collected maternal sibship.) Approximately 20 flowers on each maternal plant were hand-pollinated with pollen from its assigned donor (a total of >1,900 pollinations). Mature fruits were harvested in the greenhouse in March 1991. From each of the 76 maternal plants, approximately 36 F_1 seeds were arbitrarily chosen to contribute to the F_1 generation.

In June 1991, 2,700 seeds were sown in an irrigated experimental garden next to the field from which the parental seeds were originally collected. The experimental garden consisted of a total of nine plots, each represented by three low-, medium-, and high-density plots. Each plot consisted of a 10×30 grid of 300 seeds positioned randomly regarding family membership. Approximately 12 F_1 seeds from each greenhouse-grown maternal individual were assigned to each density. In the high-density plot, seeds were sown 5 cm apart (~400 plants/m^2); in the medium-density plot, seeds were sown 10 cm apart (~100 plants/m^2); in the low-density plot, seeds were sown 20 cm apart (~30 plants/m^2). Consequently, the absolute sizes of the high-density plots were smaller than those of the medium- and low-density plots. To minimize edge effects we planted a border of genetically variable wild radish seeds collected from the source population around the perimeter of each plot. Naturally occurring vegetation (*Avena* sp., *Bromus* sp., and *Convolvulus* sp.) was not removed from the garden plots; the primary difference among density treatments was the level of intraspecific competition.

Measurement of Reproductive Characters and Fitness Components

As each surviving plant began to flower, a flower in which the anthers had not yet dehisced was sampled to estimate pollen grain production per flower using an Elzone 180PC Particle Counter (for methodological details, see Devlin 1988; Mazer 1992). As plants began to senesce, ~10 dry fruits were sampled per individual. These fruits were used to estimate mean seed number per fruit and mean individual F_2 seed mass (estimated to 0.1 mg, based on ~30 seeds per plant).

Several traits reflecting the lifetime contribution to the next generation were recorded or estimated: lifetime flower and fruit production, lifetime seed production (i.e., lifetime fecundity, estimated as the product of fruit production and mean seed number per fruit), lifetime seed yield (lifetime seed production × mean individual seed mass), and lifetime pollen production (estimated as the product of flower production and mean pollen grain number per flower).

Identification of Families Exhibiting Maternal Environmental Effects on Seed Mass: Selection of responsive versus nonresponsive families for subsequent analyses

Of the 2,700 F_1 seeds sown in the garden, 2,052 survived to flower, and 1,801 produced at least one fruit. For 1,559 of the F_1 adults, the mean individual seed mass of their progeny was recorded. Using the mean individual seed mass produced by each F_1 adult, one-way analyses of variance (ANOVAs; with planting density as the independent variable) were conducted to detect those maternal sibships in which there were (or were not) effects of the F_1 planting density on the mean mass of the F_2 seeds. Maternal sibships in which the F_1 generation produced F_2 seeds whose mean individual seed mass did not differ among densities were identified as NR (nonresponsive) families. Maternal sibships in which the F_1 generation produced F_2 seeds whose mean individual seed mass did differ significantly among densities were identified as R (responsive) families. A total of nine of the 76 maternal plants, each pollinated by a different pollen donor, produced F_1 progeny that developed into adults whose mean F_2 seed mass depended on planting density (i.e., there were nine R maternal families).

To create a balanced design with which to determine whether the R and NR maternal siblingships differ in their contribution to future generations, we excluded from all subsequent analyses the paternal sibships that did not include any R maternal sibships. None of the remaining paternal sibships included more than one R maternal sibship. Each maternal sib-

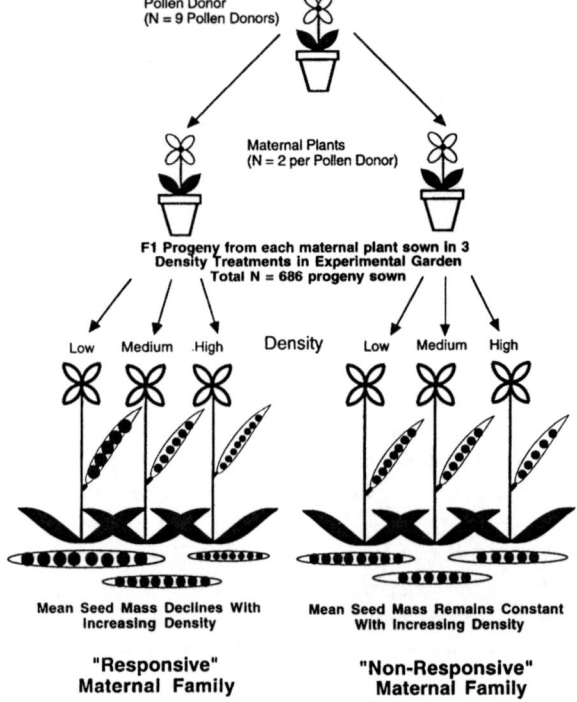

Figure 19.2 Subset of the breeding design (see figure 19.1) used to compare the performance of pairs of responsive and nonresponsive maternal families nested within pollen donors. This figure illustrates the family structure of one of the nine paternal families (and its descendants) analyzed in this study. For details, see section 19.3.

ship was paired with one randomly selected NR maternal sibship within each of the nine pollen donors (in a few cases, an NR maternal sibship with very low survivorship was deliberately excluded).

Figure 19.2 illustrates a representative pollen donor within which a pair of R and NR maternal families was nested. Each of the nine F_1 maternal siblingships that expressed density-dependent F_2 mean individual seed masses was therefore paired with a maternal sibship (nested within the same pollen donor) that produced seeds whose mass was not density dependent. The resulting nine R and nine NR sibships were represented by 686 sown seeds, of which 647 germinated, 561 survived to flower, 495 produced fruits, and 430 provided reliable estimates of mean F_2 individual seed mass (due to very low sample sizes or inviable seeds, 65 plants did not contribute estimates of this parameter). All flowering or fruiting individuals were used to estimate the mean values for fitness components and reproductive traits for each of the 18 sibships represented in each of the three densities.

Effect of Seed Mass on Expected Fitness within Each Density

Regression analyses of fitness estimates (estimated lifetime fecundity, estimated lifetime yield, and estimated lifetime pollen production) on the individual mass of sown F_1 seeds were conducted to evaluate the relationship between initial seed mass and subsequent performance within each density. Individuals that did not survive to flower were considered to have values of zero for all fitness measures. Correlations were estimated between the seed mass of F_1 seeds and F_2 mean seed mass to determine whether seed mass is effectively heritable in each of these densities. The density-specific slopes of these linear regressions were examined to determine whether the proportional change in fitness as a function of initial seed mass differed among densities. For example, if the slope of the regression of fecundity on seed mass is steeper at low density than at high density, this would suggest that the benefit of producing large seeds is, on average, particularly high when offspring are likely to experience low-density conditions.

Statistical Analyses

Once R and NR families were identified, ANOVAs (type III sums of squares; denominator mean squares calculated following Sokal and Rohlf 1995) were performed using data from each individual that survived to flower in each density treatment to detect significant differences between the means of fitness-related traits of the R and NR families while controlling for the effects of paternal genotype (a random effect) and plot (fixed effect). Individuals that did not survive to flower were excluded from the data set (rather than including them with phenotypic values of zero).

A second set of ANOVAs was conducted to detect significant differences between R and NR families regarding fitness components using a data set that comprised the mean of each of the 18 sibships expressed in each of the planting densities. Individuals that did not survive to flower were excluded from the data set; the R and NR families used did not differ significantly regarding survivorship, which was therefore not used to weight the fitness estimates. Due to the reduced degrees of freedom relative to the three-way ANOVAs described above, these data did not permit the simultaneous detection of plot and density effects. Statistically significant effects of the type of maternal sibship on the mean value of fitness components would indicate that R and NR sibships differ significantly in traits that influence their contributions to future generations.

19.4 Results

Effect of Seed Mass on Fitness Components within Each Density

In all density treatments, seed mass was positively and significantly correlated with all measured reproductive components (table 19.1). The proportional gain in each of these traits with increasing initial seed mass was greatest at low density and least at high density, indicating that the production of large seeds, particularly at low density, has significant fitness advantages.

Table 19.1 Summary of linear regressions describing the relationship between initial seed mass and subsequently expressed fitness components in low-, medium-, and high-density treatments.

Density		df	r	$p<$
Estimated lifetime fecundity vs. sown seed mass				
Low	$y = 12.37x + 26.65$	195	0.170	0.017
Medium	$y = 6.49x - 16.02$	185	0.202	0.006
High	$y = 3.28x - 13.91$	151	0.219	0.007
Estimated lifetime seed yield vs. sown seed mass				
Low	$y = 125.20x + 182.96$	194	0.167	0.019
Medium	$y = 65.14x - 220.66$	185	0.211	0.004
High	$y = 30.62x - 145.34$	151	0.220	0.007
Estimated lifetime pollen production vs. sown seed mass				
Low	$y = (8.55 \times 10^5)x + (7.4 \times 10^5)$	193	0.191	0.008
Medium	$y = (4.32 \times 10^5)x - (4.53 \times 10^5)$	183	0.210	0.005
High	$y = (2.27 \times 10^5)x - (4.72 \times 10^5)$	168	0.186	0.016

Seed mass is positively associated with estimated lifetime fitness components within all densities, but the absolute gain in fitness with an absolute increase in seed mass is usually highest at low density.

Plants derived from relatively large sown seeds did not produce significantly larger seeds than plants derived from small seeds (F_2 mean seed mass vs. F_1 sown seed mass; low density: $r = 0.085, p = 0.2882$, df $= 157$; medium density: $r = 0.154, p < 0.060$; df $= 151$; high density: $r = 0.088, p = 0.3655$, df $= 108$). Based on these parent-offspring regressions, the heritability of seed mass was close to zero in all three densities.

Identification of Families Exhibiting Maternal Environmental Effects on Seed Size

There were significant differences among the maternal sibships chosen for subsequent analyses regarding the effect of initial planting density on the size of seeds they produce. Table 19.2 summarizes the ANOVAs that distinguish between the R and NR maternal sibships nested within each of nine pollen donors. Within each pollen donor, one of the four maternal sibships expressed significant effects of planting density on mean individual seed mass at (or very near) the $p < 0.05$ level. Two maternal sibships (in paternal genotypes H and I) exhibited effects of density that were not quite statistically significant; however, the coefficients of variation (CV) among the means of the three densities in these two families (CV = 0.238 and 0.169, respectively) show that their sensitivity to density is similar to that expressed by the other R families. For each pair of maternal families, the norms of reaction for mean individual seed mass to planting density are illustrated in figure 19.3. Most R families produced the smallest seeds under the highest sowing density, although this was not universally true. For example, R sibships nested within Family D and Family G produced the smallest seeds at medium density. Similarly, most (but not all) families produced the largest seeds when sown at the lowest density.

Table 19.2 Summary of one-way ANOVAs conducted to detect effects of F_1 sowing density on F_2 mean individual seed mass.

Pollen donor ID	Responder (R) vs. nonresponder (NR)	F value (df)	p value	Coefficient of variation (among means of three densities)
A	NR	0.636 (2, 25)	0.54	0.061
A	R	3.588 (2, 20)	0.05	0.174
B	NR	0.325 (2, 21)	0.73	0.055
B	R	3.298 (2, 38)	0.05	0.139
C	NR	1.721 (2, 22)	0.20	0.109
C	R	3.623 (2, 21)	0.05	0.242
D	NR	0.437 (2, 19)	0.65	0.060
D	R	4.177 (2, 29)	0.03	0.162
E	NR	0.211 (2, 19)	0.81	0.041
E	R	4.733 (2, 13)	0.03	0.154
F	NR	0.746 (2, 19)	0.49	0.120
F	R	3.745 (2, 26)	0.04	0.143
G	NR	0.167 (2, 18)	0.85	0.053
G	R	5.979 (2, 18)	0.01	0.199
H	NR	0.096 (2, 22)	0.91	0.032
H	R	3.192 (2, 13)	0.08	0.238
I	NR	0.141 (2, 18)	0.87	0.027
I	R	3.422 (2, 15)	0.06	0.169

ANOVA results are reported for each of the 18 maternal siblingships used to identify responsive and non-responsive maternal seed families. The responsive maternal siblingships produced by pollen donors H and I show a marginally significant effect of density on F_2 mean individual seed mass, but are identified as responsive families because of their relatively high F values and coefficients of variation among density means. Boldface indicates significance at $p \leq 0.05$.

ANOVAs within Each Density: Effects of Plot, Paternal Genotype, and Type of Maternal Family (R vs. NR) on Reproductive and Fitness-Related Traits

There was no significant difference between the N and NR families within each density regarding the initial seed mass of the surviving plants (results not shown). Consequently, differences observed between the family types within densities regarding the fitness components of the F_1 adults could not have been caused by differences in the mass of the sown seeds from which they developed.

Within densities, individuals representing R families did not differ significantly from those representing NR families regarding any of the fitness-related traits reported here (figure 19.4). Nevertheless, at low (and often medium) densities, R families consistently produced larger plants with more flowers, higher estimated lifetime seed production and seed yield, and higher estimated lifetime pollen production than NR families. It appears that, in addition to producing seeds that are 10% heavier than those of NR families at low density, R families are also able to garner more resources than NR families for flower and fruit production (figure 19.4). Apparently, there is no evident fitness cost to the production of relatively large seeds by R families at low density. Although the difference in F_2 seed mass between R and NR families was not statistically significant, at low density R families tended to produce larger seeds and more of them than NR families. At high density, the difference in

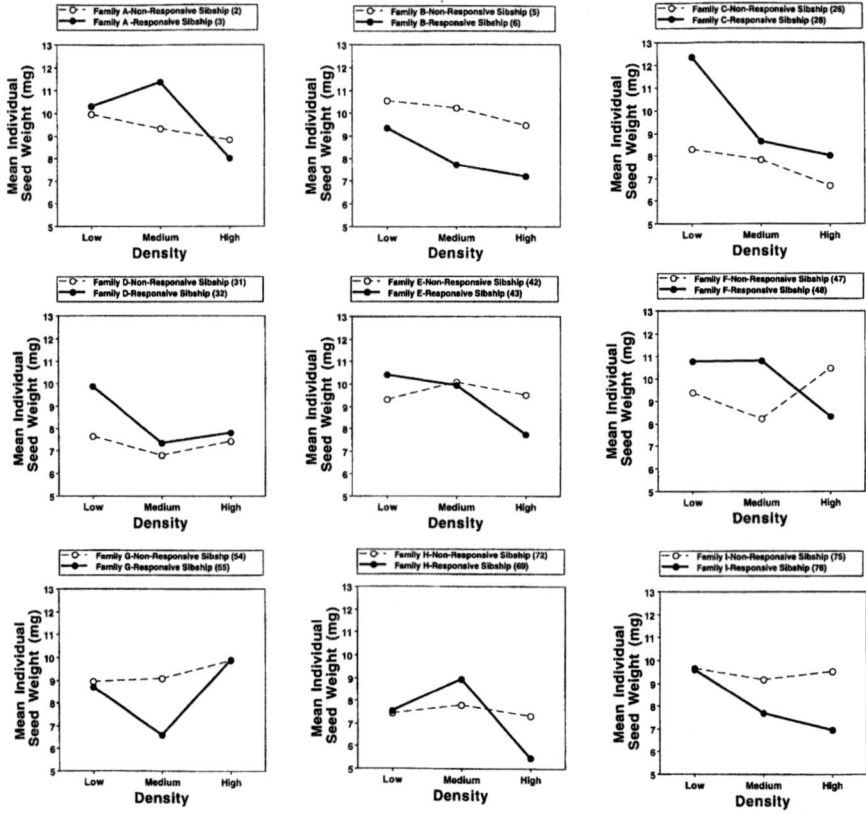

Figure 19.3 Norms of reaction for mean individual seed mass of F_2 generation seeds representing each pair of responsive versus nonresponsive maternal families nested within a distinct pollen donor. Solid lines represent norms of reaction of the responsive maternal families, in which F_2 mean individual seed mass differed significantly among density treatments. Dashed lines represent norms of reaction of the nonresponsive maternal families, in which F_2 mean individual seed mass was constant among density treatments. Numbers in parentheses are arbitrary identification numbers for each maternal sibship. Most responsive families produced the smallest seeds under the highest sowing density, although this was not universally true.

F_2 seed mass between R and NR families was as large as at low density but the relative seed masses of the two types were reversed, and there was no corresponding difference between them regarding other fitness-related traits (figure 19.4). While NR families produced seeds that were 10% heavier than those of R families at high density, they were unable to maintain higher reproductive output as well.

Figure 19.4 shows that the high level of seed production maintained by R families at low density affords them a clearer (23%) advantage over NR families when fitness is estimated as the product of seed number and seed mass (lifetime seed yield), although the ANOVA did not detect that this was statistically significant.

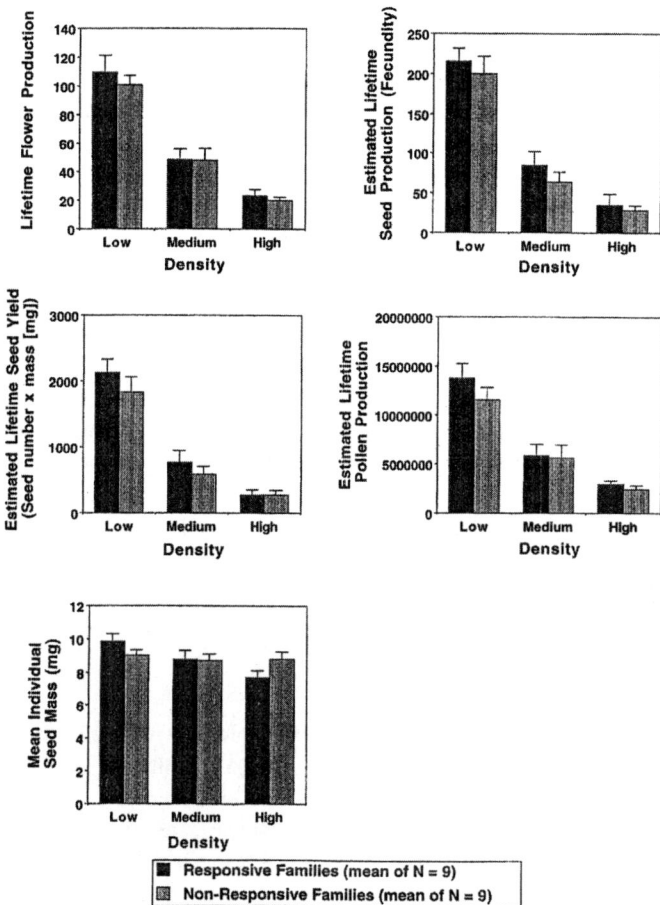

Figure 19.4 Density-specific means ($\pm$ SE) of reproductive components and estimates of lifetime fitness components for the nine responsive and nine nonresponsive families represented in low, medium, and high sowing densities. All families were present in all densities.

Two-Way ANOVAs: Effects of Density and Type of Maternal Sibship (R vs. NR)

The nearly significant density $\times$ sibship type interaction term ($p < 0.09$) influencing mean F_2 seed mass reflects the fact that in the low-density plots the R families produced slightly larger seeds than the NR families while in the high-density plots this pattern was reversed (figure 19.4). The absence of a significant density $\times$ sibship type interaction regarding any of the other fitness components indicates that the relative fitnesses of R and NR families were not strongly density dependent.

19.5 Discussion

Although maternal environmental effects on offspring phenotype are well known by population ecologists and by plant and animal evolutionary biologists (for examples of animals, see Mousseau and Dingle 1991; Reznick 1991; Bernardo 1996b), the degree to which some genotypes within populations may be highly canalized while others express strong maternal environmental effects has not been widely examined. In wild radish, we had previously observed that plants grown at high density (i.e., the treatment provided here) produce, across all seed families, smaller seeds than plants grown at low density (table 2 in Mazer and Wolfe 1992). So, we were not surprised to find this maternal effect on offspring phenotype among the genotypes examined here. We were surprised, however, to find that among the 76 maternal families in which we sought density-dependent responses of seed mass, only nine exhibited significant changes in seed mass across the range of densities and sample sizes we examined. It is likely that had we used a wider range of planting densities or had planted a much larger number of siblings within each density, our statistical tests would have detected a higher proportion of sensitive genotypes. Nevertheless, it appears that the overall decline in seed size as population density increases is due to the strong density dependence exhibited by a relatively small proportion of genotypes present in a natural population.

Are Responsive and Nonresponsive Maternal Siblingships Genetically Distinct?

In the experimental design used here, the paternal genotypes were equally represented among all R and NR maternal sibships. So, while each R sibship shared its paternal genome with its paired NR sibship, the two sibships differed regarding both their maternal parent's genotype and the microenvironment in which their maternal genotypes were raised in the greenhouse. This means that it is not possible to attribute with absolute certainty any observed differences between the R and NR sibships (including the difference between them in the degree of density-dependent F_2 seed mass) to maternally transmitted genetic causes alone. Moreover, even if R and NR sibships do differ regarding the maternally inherited component of their genome, the breeding design used here does not permit us to distinguish between nuclear genes and cytoplasmically inherited genes as the source of this difference.

Whenever significant differences among maternal sibships in a trait are presented as evidence that there is a genetic variation in this trait, one must consider whether environmentally induced differences among sibships in initial seed size might have generated the phenotypic differences between them. This question is of particular importance in wild radish, in which initial seed size can have strong (and positive) effects on subsequent plant performance (Stanton 1984; Mazer 1987b, 1989). In the present study, the differences observed between N and NR sibships regarding F_2 seed mass (figures 19.3 and 19.4) were not the result of differences between them regarding the size of the seeds from which they developed. No significant differences between the two groups were detected regarding either the mass of the F_1 seeds that were sown in the experimental garden or the rates of survivorship within densities (results not shown). Overall, the mean mass of the seeds representing the R sibships was 10.94 mg (sd = 2.38, n = 333); the mean for the NR sibships was 11.03 mg (sd = 3.00, n =338).

Variation in Competitive Ability among Maternal Families of Wild Radish: Maternally Transmitted Genetic Variation or a Response to Variation in Genetic Load?

Variation between R and NR families in their sensitivity to planting density may have two non-mutually exclusive causes. First, the two kinds of families may differ regarding regulatory or structural genes that determine how strongly seed mass is canalized, or at least buffered against the effects of variation in resource availability. In turn, genotypes that produce a constant seed mass regardless of the degree of intraspecific competition may compensate for the reduction in resource availability at high density by producing smaller flowers, shorter internodes, or less nectar, or by allocating fewer resources to any number of other traits not examined here.

Second, the two types of families may differ in the degree of inbreeding, resulting in different levels of genetic load and, consequently, of canalization (Lerner 1954). For example, one explanation for the difference between R and NR families in their sensitivity to density could be that R families are more highly inbred than NR families. If this is the case, and if higher levels of homozygosity within genotypes generate lower levels of canalization, then we would expect to see higher levels of environmentally induced phenotypic variation (including effects on offspring size) in the R relative to the NR families. Wild radish is a possible candidate for this second explanation; the combination of its self-incompatibility and its reliance on insect pollinators presents a strong opportunity for biparental inbreeding. If in natural populations of wild radish the probability of matings between closely related parents varies with local population density due to its potential effects on pollinator flight distances, then it would be inevitable that maternal plants would differ in the degree of inbreeding among their current seed crops.

Given that the 76 maternal families sampled for this study came from a natural population, it seems likely that there was variation among them in the degree of inbreeding. Even though the breeding design used here resulted in at least one generation of outcrossing with the assigned pollen donors, it would not have been sufficient to purge the maternal families of a high genetic load should some of them bear it. On the other hand, if R families are more highly inbred, there is no evidence that they suffer from inbreeding depression: R families have higher phenotypic values than NR families for most traits (except for seed mass at high density), which would not be expected if the former were more highly inbred. Also, while R families showed lower levels of canalization for seed mass than NR families, this was not the case for any of the other traits reported here, for which both family types showed strong responses to density (Figure 19.4; Lifetime flower production, estimated lifetime pollen production, estimated lifetime seed production, and estimated lifetime seed yield; $F > 55$ and $P < 0.0001$ for each). Consequently, we doubt that the differences observed between R and NR families are due primarily to possible differences between them in the level of inbreeding.

Natural populations of wild radish would appear to harbor genetic variation in competitive ability that is open to evolution by natural selection, at least when plants are growing in environments similar to the low-density treatment used here. Under conditions of low intraspecific competition, seeds from R families should be favored over those representing NR families due to the larger seeds they produce with no apparent "cost" in lifetime seed or pollen production. As the mass of sown seeds can have important fitness consequences in wild radish, particularly under competitive conditions (Stanton 1984; Mazer 1987b), this would seem to be an important fitness benefit. A nearly significant density $\times$ sibship type interaction effect on F_2 seed mass ($p < 0.09$) suggests that this advantage would be reversed at high density, where seeds produced by NR sibships are actually larger than those produced by the R sibships (figure 19.4).

This last pair of observations raises the question of why R sibships are so rare. Assume for a moment that the differences between R and NR families observed in figure 19.4 are true differences, despite the lack of statistical significance detected by the ANOVAs reported here. Discounting the possibility that our sampling of wild radish was not sufficient to detect a more widespread degree of sensitivity to density, we suggest that in nature, most of the time, wild radish grows under highly competitive conditions (due to both intra- and interspecific competition) where R sibships enjoy no fitness benefits and may even suffer with respect to mean seed mass (figure 19.4). If wild radish plants in natural populations occupied microhabitats that were mostly low density, we would expect R families to increase in frequency. We would tentatively predict that R families should appear in higher frequencies in wild radish populations in the early stages of recruitment and establishment in new sites. As conditions deteriorated, R families would lose their advantage, particularly if the trend we observed here—whereby NR families produce larger seeds than R families as intraspecific competition increases—is magnified under field conditions. While it is clear that the sample sizes required to detect and to estimate the frequency of R versus NR families in natural populations are quite high, it would be feasible to test this prediction by sampling maternal seed families from a range of wild populations that differed in their resource status or in their age since establishment.

Adaptive Significance of Variation in the Expression of Density-Mediated Maternal Effects on Seed Size

The argument that the expression of maternal environmental effects on offspring size is adaptive amounts to demonstrating that genotypes that produce offspring whose size is environment dependent (the R maternal families) enjoy higher fitness than genotypes for which offspring size is highly canalized (the NR maternal families). This result was not generally observed in this study, although it is important to emphasize that, qualitatively, the relative fitnesses of R versus NR families were environment dependent, and sample sizes available for comparison were low due to the rarity of R families.

In the low-density conditions examined here, the R families exhibited two kinds of consistent (but not statistically significant fitness) advantages over the NR families. First, they produced relatively large seeds, which are likely to perform better than the relatively small seeds produced by their NR counterparts regardless of the environmental conditions encountered by these seeds. Second, these R families exhibited no apparent fitness "cost" for having produced relatively large offspring; they displayed higher levels of current reproduction as well as producing offspring of higher quality than those of the NR families. As the R families produced relatively large seeds under low density, they also grew to become more fecund, to produce more flowers, and to produce more pollen per flower than the NR families, generating an advantage in both female and male reproductive function.

In the high-density conditions examined here, the fitness differential between the R and NR families was reversed, but not regarding both components of fitness (current reproductive yield and offspring size). The NR families did produce offspring that were on average

~10% larger than those of the R families at high density, but there was no clear difference between the two groups regarding lifetime fecundity (male or female).

To interpret these observations in light of the question of whether the maternal environmental effects on offspring size observed in wild radish are "adaptive," it would be trivial (but not sufficient) to state that, given that large seeds survive and perform better than small seeds, the production of relatively large seeds by the R genotypes in the low-density treatment is obviously "adaptive." This reasoning cannot provide a general answer to the question of adaptivity because it does not consider the relative fitnesses of R and NR genotypes across the range of environments encountered by wild radish (i.e., the potential for genotype × environment interactions), nor does it consider their lifetime reproductive success.

For example, while the R genotypes' ability to produce relatively large seeds in low-density plots confers no fitness cost regarding current fecundity (as might be expected if there is an intrinsic trade-off between offspring size and offspring number among genotypes), it does seem to be associated with an inability to produce relatively large seeds at high density. In other words, the fitness "cost" of producing an "adaptive" offspring phenotype under low-density conditions is the production of relatively low-fitness offspring phenotypes at high density. Conversely, the NR genotype produces the "adaptive" phenotype (i.e., relatively large offspring) under high-density conditions, although it is unable to maintain this offspring phenotype at low density.

In sum, whether it is "adaptive" to produce offspring with environment-dependent phenotypes depends not only on whether the particular offspring phenotype produced is one of relatively high fitness for the environment in which it is expressed, but also on whether there is either an immediate cost regarding other fitness components (e.g., current fecundity) or a fitness disadvantage expressed by such a genotype when it encounters other environmental conditions. In this experiment, although we detected no immediate fitness costs for producing relatively large seeds for either the R or NR families, neither of these family types was able to produce offspring of relatively high fitness (here estimated as individual seed mass) across all environmental conditions.

Finally, our combined observations in this and in other studies of wild radish (Mazer and Schick 1991a,b; Mazer 1992; Mazer and Wolfe 1992) underscore the importance of genotype × environment interactions in determining the outcome of selection on fitness-related traits. As for most traits that have been closely observed in wild radish, the answer to the question of whether maternal effects are adaptive appears to depend on the environment in which such effects are expressed.

Acknowledgments This study was supported by a University of California General Research grant to S.J.M. and by a postdoctoral fellowship from the National Science and Engineering Research Council of Canada to L.M.W. In addition, funds from National Science Foundation Award DEB-9157270 to S.J.M. provided crucial support during this work.

References

Aarssen, L. W., and S. M. Burton. 1990. Maternal effects at four levels in *Senecio vulgaris* (Asteraceae) grown on a soil nutrient gradient. Am. J. Bot. 77:1231–1240.

Alexander, H. M., and R. D. Wulff. 1985. Experimental ecological genetics in *Plantago*. X. The effects of maternal temperature on seed and seedling characters in *P. lanceolata* J. Ecol. 73:271–282.

Antonovics, J., and J. Schmitt. 1986. Paternal and maternal effects on propagule size in *Anthoxanthum odoratum*. Oecologia 69:277–282.

Austin, R. B. 1966a. The growth of watercress (*Rorippa nasturtium aquaticum* [L.] Hayek) from seed as affected by the phosphorus nutrition of the parent plant. Plant Soil 24:113–120.

Austin, R. B. 1966b. The influence of phosphorus and nitrogen nutrition of pea plants on the growth of their progeny. Plant Soil 24:359–368.

Bernardo, J. 1996a. Maternal effects in animal ecology. Am. Zool. 36:83–106.

Bernardo, J. 1996b. The particular maternal effect of propagule size, especially egg size: Patterns, models, quality of evidence and interpretations. Am. Zool. 36:216–236.

Black, J. N. 1956. The influence of seed size and depth of sowing on pre-emergence and early vegetative growth of subterranean clover (*Trifolium subterraneum* L.) Austral. J. Agricult. Res. 7:98–109.

Black, J. N. 1957. The early vegetative growth of three strains of subterranean clover (*Trifolium subterraneum* L.) in relation to size of seed. Austral. J. Agricult. Res. 8:1–14.

Bradshaw, A. D. 1972. Evolutionary significance of phenotypic plasticity in plants. Adv. Genet. 13:115–155.

Case, A. L., E. P. Lacey, and R. G. Hopkins. 1996. Parental effects in *Plantago lanceolata* L.:II. Manipulation of grandparental temperature and parental flowering time. Heredity 76:287–295.

Cideciyan, M. A., and A. J. C. Malloch. 1982. Effects of seed size on the germination, growth, and competitive ability of *Rumex crispus* and *Rumex obtusifolius*. J. Ecol. 70:227–232.

Cook, R. E. 1975. The photoinductive control of seed weight in *Chenopodium rubrum* L. Am. J. Bot. 62:427–431.

Corey, L. A., D. F. Matzinger, and C. C. Cockerham. 1976. Maternal and reciprocal effects on seedling characters in *Arabidopsis thaliana* (L.). Heynh. Genet. 82:677–683.

Cresswell, E. G., and J. P. Grime. 1981. Induction of a light requirement during seed development and its ecological consequences. Nature 291:583–585.

Curtis, P. S., A. A. Snow, and A. S. Miller. 1994. Genotype-specific effects of elevated CO_2 on fecundity in wild radish (*Rapanus raphanistrum*). Oecologia 97:100–105.

Delesalle, V. A., and S. Blum. 1994. Variation in germination and survival among families of *Sagittaria latifolia* in response to salinity and temperature. Int. J. Plant Sci. 155:187–195.

Deschamp, P. A., and T. J. Cooke. 1985. Leaf dimorphism in the aquatic Angiosperm *Callitriche heterophylla*. Am. J. Bot. 72:1377–1387.

Devlin, B. 1988. The effects of stress on reproductive characters of *Lobelia cardinalis*. Ecology 69:1716–1720.

Dolan, R. W. 1984. The effect of seed size and maternal source on individual size in a population of *Ludwigia leptocarpa* (Onagraceae). Am. J. Bot. 71:1302–1307.

Drew, R. L. K., and P. A. Brocklehurst. 1990. Effects of temperature of mother-plant environment on yield and germination of seeds of lettuce (*Lactuca sativa*). Ann. Bot. 66:63–71.

Durrant, A. 1962. The environmental induction of heritable changes in *Linum*. Heredity 17:27–61.

Erickson, L. C. 1946. The effect of alfalfa seed size and depth of seeding upon subsequent procurement of stand. J. Am. Soc. Agron. 38:964–973.

Erskine, W., and T. N. Khan. 1977. Genotype, genotype × environmental and environmental effects on grain yield and related characters of cowpea (*Vigna unguiculata* [L.] Walp.). Austral. J. Agricult. Res. 28:609–617.

Falconer, D. S. 1989. Introduction to Quantitative Genetics. 3rd ed. Wiley, New York.

Fenner, M. 1986. The allocation of minerals to seeds in *Senecio vulgaris* plants subjected to nutrient shortage. J. Ecol. 74:385–392.

Garbutt, K., and F. A. Bazzaz. 1987. Population niche structure. Differential response of *Abutilon theophrasti* progeny to a resource gradient. Oecologia 72:291–296.

Gavrilets, S., and S. M. Scheiner. 1993. The genetics of phenotypic plasticity. V. Evolution of reaction norm shape. J. Evol. Biol. 6:31–48.

Gutterman, Y. 1980. Influences on seed germinability: Phenotypic maternal effects during seed maturation. Israel J. Bot. 29:105–117.

Gutterman, Y. 1992. Maternal effects on seeds during development. Pp. 27–57 in Michael Fenner (ed.), Seeds: The Ecology of Regeneration in Plant Communities. CAB International, Wallinford Oxon, UK.

Gutterman, Y., and M. Evenari. 1972. The influence of day length on seed coat colour, an index of water permeability of the desert annual *Ononis sicula* Guss. J. Ecol. 60:713–719.

Gutterman, Y., T. H. Thomas, and W. Hedecker. 1975. Effect on the progeny of applying different daylengths and hormone treatments to parent plants of *Lactuca scariola*. Physiol. Plant. 34:30–38.

Harper, J. L., and M. Obeid. 1967. Influence of seed size and depth of sowing on the establishment and growth of varieties of fiber and oil seed flax. Crop Sci. 7:527–532.

Hendrix, S. D. 1984. Variation in seed weight and its effect on germination in *Pastinaca sativa* L. (Umbelliferae). Am. J. Bot. 71:795–802.

Hill, J. 1967. The environmental induction of heritable changes in *Nicotiana rustica* parental and selection lines. Genetics 55:735–754.

Hume, L. 1994. Maternal environment effects on plant growth and germination of two strains of *Thlaspi arvense* L. Int. J. Plant Sci. 155:180–186.

Kirkpatrick, M., and R. Lande. 1989. The evolution of maternal characters. Evolution 43:485–503.

Lacey, E. P. 1996. Parental effects in *Plantago lanceolata* L. I: A growth chamber experiment to examine pre- and postzygotic temperature effects. Evolution 50:865–878.

Lerner, I. M. 1954. Genetic Homeostasis. Wiley, New York.

Marshall, D. 1986. Effect of seed size on seedling success in three species of *Sesbania* (Fabaceae). Am. J. Bot. 73:457–464.

Marshall, D. L., D. A. Levin, and N. L. Fowler. 1986. Plasticity of yield components in response to stress in *Sesbania macrocarpa* and *Sesbania vesicaria* (Leguminosae). Am. Nat. 127:508–521.

Mazer, S. J. 1987a. Parental effects on seed development and seed yield in *Raphanus raphanistrum*: Implications for natural and sexual selection. Evolution 41:355–371.

Mazer, S. J. 1987b. The quantitative genetics of life history and fitness components in *Raphanus raphanistrum* L. (Brassicaceae): Ecological and evolutionary consequences of seed-weight variation. Am. Nat. 130:891–914.

Mazer, S. J. 1989. Family mean correlations among fitness components in wild radish: Controlling for maternal effects on seed weight. Can. J. Bot. 67:1890–1897.

Mazer, S. J. 1992. Environmental modification of gender allocation in wild radish: Consequences for sexual and natural selection. Pp. 181–225 in Robert Wyatt (ed.), Ecology and Evolution of Plant Reproduction: New Approaches. Chapman and Hall, New York.

Mazer, S. J., and D. L. Gorchov. 1996. Parental effects on progeny phenotype in plants: Distinguishing genetic and environmental causes. Evolution 50:44–53.

Mazer, S. J., and C. T. Schick. 1991a. Constancy of population and genetic parameters for life-history and floral traits in *Raphanus sativus* I. Norms of reaction and the nature of genotype by environment interactions. Heredity 67:143–156.

Mazer, S. J., and C. T. Schick. 1991b. Constancy of population and genetic parameters for life-history and floral traits in *Raphanus sativus* II. Effects of planting density on phenotype and heritability estimates. Evolution 45:1888–1907.

Mazer, S. J., and L. M. Wolfe. 1992. Planting density influences the expression of genetic variation in seed mass in wild radish (*Raphanus sativus* L.: Brassicaceae). Am. J. Bot. 79:1185–1193.

Miao, S. L., F. A. Bazzaz, and R. B. Primack. 1991a. Effects of maternal nutrient pulse on reproduction of two colonizing *Plantago* species. Ecology 72:586–596.

Miao, S. L., F. A. Bazzaz, and R. B. Primack. 1991b. Persistence of maternal nutrient effects in *Plantago major*: The third generation. Ecology 72:1634–1642.

Mousseau, T. A., and H. Dingle. 1991. Maternal effects in insect life histories. Annu. Rev. Entomol. 36:511–534.

Nakamura, R. R., and M. L. Stanton. 1989. Embryo growth and seed size in *Raphanus sativus*: Maternal and paternal effects *in vivo* and *in vitro*. Evolution 43:1435–1443.

Parrish, J. A. D., and F. A. Bazzaz. 1985. Nutrient content of *Abutilon theophrasti* seeds and the competitive ability of the resulting plants. Oecology 65:247–251.

Pet, G., and F. Garretsen. 1983. Genetic and environmental factors influencing seed size of tomato (*Lycopersicon esculentum* Mill.) and the effect of seed size on growth and development of tomato plants. Euphytica 32:711–718.

Plattenkamp, G. A. J., and R. G. Shaw. 1993. Environmental and genetic maternal effects on seed characters in *Nemophila menziesii*. Evolution 47:540–555.

Reznick, D. N. 1991. Maternal effects in fish life histories. Pp. 780–793 in E. C. Dudley (ed.), The Unity of Evolutionary Biology. Dioscorides Press, Portland, Ore.

Roach, D. A., and R. D. Wulff. 1987. Maternal effects in plants. Annu. Rev. Ecol. Syst. 18:209–235.

Sawhney, R., and W. A. Quick. 1985. The effect of temperature during parental vegetative growth on seed germination of wild oats (*Avena fatua* L.). Ann. Bot. 55:25–28.

Schaal, B. A. 1984. Life-history variation, natural selection, and maternal effects in plant populations. Pp. 188–206 in R. Dirzo and J. Sarukhán (eds.), Perspectives in Plant Population Ecology. Sinauer, Sunderland, Mass.

Scheiner, S. M. 1993. The genetics and evolution of phenotypic plasticity. Annu. Rev. Ecol. Syst. 24:35–68.

Schlichting, C. D. 1986. The evolution of phenotypic plasticity in plants. Annu. Rev. Ecol. Syst. 17:667–693.

Schlichting, C. D. 1989. Phenotypic integration and environmental change. Bioscience. 39:460–464.

Schlichting, C. D., and D. A. Levin. 1986. Phenotypic plasticity: An evolving plant character. Biol. J. Linn. Soc. 29:37–47.

Schmitt, J., J. Niles, and R. D. Wulff. 1992. Norms of reaction of seed traits to maternal environments in *Plantago lanceolata*. Amer. Natur. 139:451–466.

Sokal, R. R., and F. J. Rohlf. 1995. Biometry. 3rd ed. Freeman, New York.

Stanton, M. L. 1984. Seed variation in wild radish: Effect of seed size on components of seedling and adult fitness. Ecology 65:1105–1112.

Stanton, M. L. 1985. Seed size and emergence time within a stand of wild radish (*Raphanus raphanistrum* L.): The establishment of a fitness hierarchy. Oecologia 67:524–531.

Stanton, M. L., J. K. Bereczky, and H. D. Hasbrouck. 1987. Pollination thoroughness and maternal yield regulation in wild radish, *Raphanus raphanistrum* (Brassicaceae). Oecologia 74:68–76.

Stearns, S. C. 1989. The evolutionary significance of phenotypic plasticity. Bioscience 39:436–445.

Stratton, D. A. 1989. Competition prolongs expression of maternal effects in seedlings of *Erigeron annuus* (Asteraceae). Am. J. Bot. 76:1646–1653.

Sultan, S. E. 1987. Evolutionary implications of phenotypic plasticity in plants. Evol. Biol. 21:127–178.

Twamley, B. E. 1967. Seed size and seedling vigor in birdsfoot trefoil. Can. J. Plant Sci. 7:603–609.

Van Andel, J., and F. Vera. 1977. Reproductive allocation in *Senecio sylvaticus* and *Chamaenerion angustifolium* in relation to mineral nutrition. J. Ecol. 65:747–758.

Weis, I. M. 1982. The effects of propagule size in germination and seedling growth in *Mirabilis hirsuta*. Can. J. Bot. 60:1868–1874.

Wolfe, L. M. 1992. Why does the size of reproductive structures decline through time in *Hydrophyllum appendiculatum* (Hydrophyllaceae)? Developmental constraints vs. resource limitation. Am. J. Bot. 79:1286–1290.

Wolfe, L. M. 1993. Inbreeding depression in *Hydrophyllum appendiculatum*: Role of maternal effects, crowding, and parental mating history. Evolution 47:374–386.

Wolfe, L.M. 1995. The genetics and ecology of seed size variation in *Hydrophyllum appendiculatum*, a biennial plant. Oecologia 101:343–352.

Wulff, R. D. 1986a. Seed size variation in *Desmodium paniculatum*. I. Factors affecting seed size. J. Ecol. 74:87–97.

Wulff, R. D. 1986b. Seed size variation in *Desmodium paniculatum*. II. Effects on seedling growth and physiological performance. J. Ecol. 74:99–114.

Wulff, R. D., and F. A. Bazzaz. 1992. Effect of the parental nutrient regime on growth of the progeny in *Abutilon theophrasti* (Malvaceae). Am. J. Bot. 79:1102–1107.

Wulff, R. D., A. Caceres, and J. Schmitt. 1994. Seed and seedling responses to maternal and offspring environments in *Plantago lanceolata*. Funct. Ecol. 8:763–769.

Young, K A, and J. Schmitt. 1995. Genetic variation and phenotypic plasticity of pollen release and capture height in *Plantago lanceolata*. Funct. Ecol. 9:725–733.

Concluding Remarks

Generalizations, Implications, and Future Directions

Maternal effects are ubiquitous in nature. They affect the development of organisms, their life histories, behavior, organismal interactions, and population dynamics. Yet—because evolutionary biologists can partition phenotypic variation into genetic and environmental components (and their interactions), with evolutionary responses to natural selection dependent only on the magnitude of natural selection and the amount of genetic variation present within populations—understanding the sources and inheritance of environmental variation is often considered to be of little importance for understanding the evolution of adaptations. Thus, maternal effects are generally treated as little more than common environmental effects that can obscure, or be confused with, genetic variation in characters, obscuring the mechanisms producing the patterns we observe in nature. However, as the contributions to this volume demonstrate, this is a very simplistic and quite incorrect view of maternal effects. Maternal effects can be much more than noise in our data; they are frequently important in generating the phenotypic patterns (and population dynamics) that we observe in nature, and are often important components of population responses to natural selection. In fact, evolutionary responses to selection may be manifest through maternal effects, such that maternal effects may provide mechanisms for adapting to variable environments.

Thus, understanding maternal effects is essential for understanding the evolution of patterns observed in nature. To this end, the contributors to this volume, and the many researchers whose work they discuss, have made tremendous advances in recent years. However, while these advances demonstrate the necessity of understanding maternal effects, they also indicate how much we have yet to learn. Thus, we conclude this book with a brief synopsis of the topics that we believe most deserve the attention of future research. These suggestions are compiled from comments made by many participants at a roundtable discussion following the symposium that spawned this book (Maternal Effects as Adaptations, a symposium organized by T.A. Mousseau and C.W. Fox for the annual meeting of the Society for the Study of Evolution held in St. Louis in June 1996), as well as suggestions scattered throughout the many chapters in this book.

The Evolution of Maternal Effects

The Evolutionary Consequences of Maternal Effects

Other than a few theoretical explorations by Wade, Riska, Kirkpatrick, Lande, and Cheverud (see Wade, Moore et al., and Roitberg, this volume), the evolutionary dynamics of maternal effects (as characters themselves), and characters influenced by maternal effects, have been little explored. Yet, from these few studies it is clear that maternal effects can have dramatic consequences for the evolution of characters (including effects on the rate of evolution and even the direction of a response to selection), and that maternal effects can themselves evolve. For example, early theoretical analyses indicated that the evolutionary dynamics of maternal effects are unusual relative to standard theory; they can result in large time lags in evolutionary responses to selection, and characters subject to maternal effects may even respond to selection in a maladaptive direction. More recent models (e.g., Wade, chapter 1) explore how selection at two levels (within and between families), and epistatic interactions between mothers and their progeny, influence evolutionary dynamics in complex ways not considered by more traditional theoretical explorations. Other recent theoretical analyses have highlighted similarities among genetic models of sexual selection, kin selection, and maternal effects (e.g., Moore et al., chapter 2), stimulating alternative approaches to addressing many evolutionary questions, such as the evolution of parent-offspring interactions.

Genetics of Maternal Effects

For characters to respond to selection, there must be heritable genetic variation. However, to understand the evolutionary dynamics of a character, we need to examine not only genetic variation influencing the phenotype, but also genetic variation in the maternal effects that influence the character in natural populations across a range of natural environments. Although maternal effects have been defined as environmental effects (i.e., independent of progeny genotype), their expression is frequently mediated by an individual progeny's genotype or environment, resulting in many complicated interactions that can be difficult to dissect using standard quantitative genetic designs (see Roff, Shaw and Byers, and Mazer and Wolfe, this volume). Nevertheless, these interactions have significant consequences for the evolution of phenotypic characters and the maternal effects influencing them (Wade, chapter 1). Thus, we must design genetic experiments to adequately dissect genotypic effects from genetically and environmentally based maternal effects (see Rossiter, chapter 8) and their interactions. This can be difficult without well-developed (and easily employed) statistical or experimental techniques for distinguishing maternal effects that persist over multiple generations from genetic variation due to either maternal transmission of cytoplasmic genes (Shaw and Byers, chapter 7) or sex-linked genes affecting characters expressed differentially among the sexes or by only the heterogametic sex (such as many behavioral and life history traits), and can be especially difficult in studies of organisms for which complicated breeding designs are not feasible (e.g., most animals). Development of improved techniques for dissecting phenotypic variance into its components, and the application of these techniques to ecologically and behaviorally interesting characters, may be our most important first step to understanding how truly important maternal effects are in ecology and evolution.

Measuring Selection on Maternal Effects

A key property of maternal effects is that they can be affected by natural selection at both the parental and offspring stages. Selection at these two levels may frequently be different in magnitude and even direction, leading to evolutionary conflicts of interest between parents and their offspring. For example, selection may favor large eggs at the individual level (if progeny developing from large eggs have higher fitness) but small eggs at the parental level (because females that lay small eggs can lay more eggs; Fox and Mousseau, chapter 10). Thus, multilevel studies of selection, in the range of environments experienced by parents and their progeny, are necessary to understand the evolution of characters that exhibit plasticity in response to maternal environments (see Donohue and Schmitt, chapter 9). For maternal effects that may have evolved as mechanisms for transgenerational phenotypic plasticity (see Donohue and Schmitt, and chapter 9, Fox and Mousseau, chapter 10), we must examine how well maternal environments predict progeny environments, and how well parents can make use of this information when producing progeny.

Maternal Effects as Adaptations

Maternal effects are abundant in nature, and selection can be demonstrated to favor or oppose the evolution of maternal effects. However, demonstrating selection favoring the evolution of a maternal effect is not sufficient evidence to demonstrate that a maternal effect has evolved as an adaptation. Throughout this book we have made the bold assertion that many maternal effects have evolved as adaptations, but few of the examples discussed have been adequately demonstrated to represent adaptations. The same rigorous techniques applied to studies of behavior, morphology, and life history need now be applied to the study of maternal effects to determine how pervasive adaptive maternal effects are in nature and how often maternal effects evolve simply as consequences of the evolution of parental and offspring characters (see Lacey, chapter 4, and Fox and Mousseau, chapter 10). Only through rigorous experimental design can we address questions such as how often maternal effects are the mechanisms by which organisms cope with environmental heterogeneity, rather than simply unavoidable consequences of environmental heterogeneity.

Maternal Effects as Traits

Throughout this book we have made reference to maternal effects as characters themselves, rather than simply emergent properties of independently evolving maternal and progeny traits. Although we believe that this approach is illuminating, particularly in that it raises the awareness of researchers to the potential for maternal effects to evolve as adaptations, an acceptable conceptual framework for discussing the evolution of maternal effects has yet to be developed. Recent discussions concerning the genetics and evolution of phenotypic plasticity (references in Fox and Mousseau, chapter 10) highlight the theoretical issues that arise when studying the evolution of characters that may be emergent properties, and thus can provide insights into the difficulties that will be encountered when studying the evolution of maternal effects as characters.

The Ecological Consequences of Maternal Effects

As with examinations of the evolutionary implications of maternal effects, studies of the ecological implications of maternal effects are in their infancy. Recent theoretical studies by Rossiter (chapter 8) and Ginzburg (chapter 3) have demonstrated that maternal effects can have dramatic consequences for population dynamics, and may explain the unusual population dynamics of outbreak species. Similarly, maternal effects can mediate behavioral interactions between individuals, both within and between species. For example, mothers may respond to perceived density or environmental quality by modifying the type of progeny that they produce (e.g., Fox and Mousseau, chapter 10), influencing interactions between her progeny and other individuals (e.g. Messina, chapter 13) or other species, and thus potentially affecting population dynamics. As maternal effects are more widely recognized as ubiquitous in nature, and as many commonly studied traits (such as oviposition preference; Roitberg, chapter 5) are acknowledged to result in maternal effects, the influence of parental phenotype, environment, or population dynamics and other ecological processes should become more widely understood.

The Physiology, Development, and Transmission of Maternal Effects

Other than a large literature on the ecology and evolution of propagule size (see chapters by Heath and Blouw, Kaplan, Sinervo, and Fox and Mousseau for egg size; Mazer, and Donohue and Schmitt, for seed size), the mechanisms by which parents affect the phenotypes of their progeny are understood for very few maternal effects. Several well-understood examples are discussed in this volume (e.g., Denlinger, chapter 16, for diapause in a flesh fly; Clark and Galef, chapter 15, for reproductive behavior of rodents). While the environmental cues that stimulate maternal effects, and the phenotypes they result in, are often well studied and well understood, the physiological and developmental mechanisms by which maternal effects are transmitted from parents to their progeny are poorly understood. In fact, this lack of understanding about how maternal effects are transmitted across generations may be responsible for the perception among many scientists that maternal effects primarily result from environmental or phenotypic constraints on reproduction, preventing females from producing healthy, normal, or "optimal" progeny. The physiology, development, and transmission of maternal effects may thus be one of the most fruitful, illuminating, and as yet relatively unexplored areas of research in evolutionary biology.

As a final thought, we would like to reiterate the principle theme raised throughout this volume. Maternal effects are common in nature, and their ecological and evolutionary implications are often substantial. It is no longer sufficient to design ecological and evolutionary experiments to control for, or eliminate, maternal effects. Instead, we need to design experiments that allow us to understand maternal effects because in these effects we will find a new and rich source of explanation for the puzzles that nature has provided for us to decipher.

Taxonomic Index

Author Index

Subject Index

Printed in the United Kingdom
by Lightning Source UK Ltd.
126324UK00001B/157/A